Jean-François Lamonier (Ed.)

Catalytic Removal of Volatile Organic Compounds

MDPI

This book is a reprint of the Special Issue that appeared in the online, open access journal, *Catalysts* (ISSN 2073-4344) from 2015–2016 (available at: http://www.mdpi.com/journal/catalysts/special_issues/volatile-organic-compounds).

Guest Editor
Jean-François Lamonier
Université Lille, Sciences et Technologies
Unité de Catalyse et Chimie du Solide
France

Editorial Office
MDPI AG
Klybeckstrasse 64
Basel, Switzerland

Publisher
Shu-Kun Lin

Managing Editor
Zu Qiu

1. Edition 2016

MDPI • Basel • Beijing • Wuhan • Barcelona

ISBN 978-3-03842-213-6 (Hbk)
ISBN 978-3-03842-214-3 (PDF)

Table of Contents

List of Contributors

Ouafae Achak Laboratory LGCVR, UAE/L01FST, Faculty of Sciences and Techniques, University Abdelmalek Essaadi, B.P. 416 Tangier, Morocco.

Kaisu Ainassaari Chemical and Environmental Engineering, University of Oulu, POB 4300, 90014 Oulu, Finland.

Mhamed Assebban Laboratory LGCVR, UAE/L01FST, Faculty of Sciences and Techniques, University Abdelmalek Essaadi, B.P. 416 Tangier, Morocco.

Jean-Pierre Bellat Laboratoire Interdisciplinaire Carnot de Bourgogne UMR 6303 CNRS, Université Bourgogne, 21078 Dijon, France.

Igor Bezverkhyy Laboratoire Interdisciplinaire Carnot de Bourgogne UMR 6303 CNRS, Université Bourgogne, 21078 Dijon, France.

Rachid Brahmi Laboratory of Catalysis and Corrosion of the Materials, Department of Chemistry, University of Chouaïb Doukkali, 20 Route de Ben Maachou, 24000 El Jadida, Morocco.

Julien Brunet Unité de Chimie Environnementale et Interactions sur le Vivant (UCEIV, EA 4492), Université du Littoral Côte d'Opale (ULCO), 145 avenue Maurice Schumann, 59140 Dunkerque, France.

Sylvie Capelle Unité de Chimie Environnementale et Interactions sur le Vivant (UCEIV, EA 4492), Université du Littoral Côte d'Opale (ULCO), 145 avenue Maurice Schumann, 59140 Dunkerque, France.

Francisco Carrasco-Marín Department of Inorganic Chemistry, Faculty of Sciences. University of Granada, Avda. Fuentenueva s/n. 18071, Spain.

Sandra Casale Laboratoire de Réactivité de Surface (LRS), CNRS UMR 7197, Sorbonne Université, Université Pierre et Marie CURIE (UPMC), 4 place Jussieu, 75005 Paris, France.

María Haidy Castaño Estado Sólido y Catálisis Ambiental (ESCA), Departamento de Química, Facultad de Ciencias, Universidad Nacional de Colombia, Kra 30 N° 45-03, Bogotá, Colombia.

Tarik Chafik Laboratory LGCVR, UAE/L01FST, Faculty of Sciences and Techniques, University Abdelmalek Essaadi, B.P. 416 Tangier, Morocco.

Dayan Chlala Unité de Catalyse et Chimie du Solide UMR 8181 CNRS, Université Lille1 Sciences et Technologies, 59650 Villeneuve d'Ascq, France; Laboratory of Physical Chemistry of Materials/PR2N, Faculty of Sciences, Lebanese University, Fanar, Lebanon.

Renaud Cousin Unité de Chimie Environnementale et Interactions sur le Vivant (UCEIV, EA 4492), Université du Littoral Côte d'Opale (ULCO), 145 avenue Maurice Schumann, 59140 Dunkerque, France.

Nathalie De Geyter Department of Applied Physics, Research Unit Plasma Technology, Faculty of Engineering and Architecture, Ghent University, Sint-Pietersnieuwstraat 41, B-9000 Ghent, Belgium.

Akira Fujishima Kanagawa Academy of Science and Technology, KSP building East 407, 3-2-1 Sakado, Takatsu-ku, Kawasaki, Kanagawa 213-0012, Japan; Photocatalysis International Research Center, Tokyo University of Science, 2641 Yamazaki, Noda, Chiba 278-8510, Japan.

Jesus Manuel Garcia-Vargas Chimie-Biochimie, Université de Lyon, Lyon, F-69003, France, Université Lyon 1, Villeurbanne, F-69622, CNRS, UMR 5256, IRCELYON, 2 avenue Albert Einstein, Villeurbanne, F-69622, France.

Eric Genty Unité de Chimie Environnementale et Interactions sur le Vivant (UCEIV, EA 4492), Université du Littoral Côte d'Opale (ULCO), 145 avenue Maurice Schumann, 59140 Dunkerque, France.

Sonia Gil Chimie-Biochimie, Université de Lyon, Lyon, F-69003, France, Université Lyon 1, Villeurbanne, F-69622, CNRS, UMR 5256, IRCELYON, 2 avenue Albert Einstein, Villeurbanne, F-69622, France.

Jean-Marc Giraudon Unité de Catalyse et Chimie du Solide UMR 8181 CNRS, Université Lille1 Sciences et Technologies, 59650 Villeneuve d'Ascq, France.

Anne Giroir-Fendler Chimie-Biochimie, Université de Lyon, Lyon, F-69003, France, Université Lyon 1, Villeurbanne, F-69622, CNRS, UMR 5256, IRCELYON, 2 avenue Albert Einstein, Villeurbanne, F-69622, France.

Soukaina Haffane Laboratory LGCVR, UAE/L01FST, Faculty of Sciences and Techniques, University Abdelmalek Essaadi, B.P. 416 Tangier, Morocco.

Mio Hayashi Kanagawa Academy of Science and Technology, KSP building East 407, 3-2-1 Sakado, Takatsu-ku, Kawasaki, Kanagawa 213-0012, Japan.

Toshifumi Hosoya General Planning Division, Sumitomo Electric Industries, Ltd., 1-1-3, Shimaya, Konohana-ku, Osaka 554-0024, Japan.

Achraf El Kasmi Laboratory LGCVR, UAE/L01FST, Faculty of Sciences and Techniques, University Abdelmalek Essaadi, B.P. 416 Tangier, Morocco.

Riitta L. Keiski Chemical and Environmental Engineering, University of Oulu, POB 4300, 90014 Oulu, Finland.

Hyun-Ha Kim National Institute of Advanced Industrial Science and Technology (AIST), 16-1 Onogawa, Tsukuba 305-8569, Japan.

Niina Koivikko Chemical and Environmental Engineering, University of Oulu, POB 4300, 90014 Oulu, Finland.

Abdelhamid Korrir Laboratory LGCVR, UAE/L01FST, Faculty of Sciences and Techniques, University Abdelmalek Essaadi, B.P. 416 Tangier, Morocco.

Said Laassiri WestCHEM, School of Chemistry, Joseph Black Building, University of Glasgow, G12 8QQ Glasgow, UK.

Madona Labaki Laboratory of Physical Chemistry of Materials/PR2N, Faculty of Sciences, Lebanese University, Fanar, Lebanon.

Tiina Laitinen Chemical and Environmental Engineering, University of Oulu, POB 4300, 90014 Oulu, Finland.

Jean-François Lamonier Unité de Catalyse et Chimie du Solide UMR 8181 CNRS, Université Lille1 Sciences et Technologies, 59650 Villeneuve d'Ascq, France; Université de Lille, CNRS, UMR 8181 - UCCS - Unité de Catalyse et Chimie du Solide, F-59000 Lille, France.

Christophe Leys Department of Applied Physics, Research Unit Plasma Technology, Faculty of Engineering and Architecture, Ghent University, Sint-Pietersnieuwstraat 41, B-9000 Ghent, Belgium.

Leonarda Francesca Liotta Istituto per Lo Studio dei Materiali Nanostrutturati (ISMN)-CNR, via Ugo La Malfa, 153, 90146 Palermo, Italy.

Francisco José Maldonado-Hódar Department of Inorganic Chemistry, Faculty of Sciences. University of Granada, Avda. Fuentenueva s/n. 18071, Spain.

Pascale Massiani Laboratoire de Réactivité de Surface (LRS), CNRS UMR 7197, Sorbonne Université, Université Pierre et Marie CURIE (UPMC), 4 place Jussieu, 75005 Paris, France.

Young Sun Mok Department of Chemical and Biological Engineering, Jeju National University, Jeju 690-756, Korea.

Rafael Molina Estado Sólido y Catálisis Ambiental (ESCA), Departamento de Química, Facultad de Ciencias, Universidad Nacional de Colombia, Kra 30 № 45-03, Bogotá, Colombia.

Sergio Morales-Torres Department of Inorganic Chemistry, Faculty of Sciences. University of Granada, Avda. Fuentenueva s/n. 18071, Spain.

Sonia Moreno Estado Sólido y Catálisis Ambiental (ESCA), Departamento de Química, Facultad de Ciencias, Universidad Nacional de Colombia, Kra 30 Nº 45-03, Bogotá, Colombia.

Rino Morent Department of Applied Physics, Research Unit Plasma Technology, Faculty of Engineering and Architecture, Ghent University, Sint-Pietersnieuwstraat 41, B-9000 Ghent, Belgium.

David J. Morgan Cardiff Catalysis Institute, School of Chemistry, Cardiff University, Main Building, Park Place, CF10 3AT Cardiff, UK.

Anass Mouammine Chemical and Environmental Engineering, University of Oulu, POB 4300, 90014 Oulu, Finland.

Nobuaki Negishi National Institute of Advanced Industrial Science and Technology (AIST), 16-1 Onogawa, Tsukuba 305-8569, Japan.

Tsuyoshi Ochiai Kanagawa Academy of Science and Technology, KSP building East 407, 3-2-1 Sakado, Takatsu-ku, Kawasaki, Kanagawa 213-0012, Japan; Photocatalysis International Research Center, Tokyo University of Science, 2641 Yamazaki, Noda, Chiba 278-8510, Japan.

Atsushi Ogata National Institute of Advanced Industrial Science and Technology (AIST), 16-1 Onogawa, Tsukuba 305-8569, Japan.

Satu Ojal Chemical and Environmental Engineering, University of Oulu, POB 4300, 90014 Oulu, Finland.

Mohamed Ousmane Chimie-Biochimie, Université de Lyon, Lyon, F-69003, France, Université Lyon 1, Villeurbanne, F-69622, CNRS, UMR 5256, IRCELYON, 2 avenue Albert Einstein, Villeurbanne, F-69622, France.

Giuseppe Pantaleo Istituto per Lo Studio dei Materiali Nanostrutturati (ISMN)-CNR, via Ugo La Malfa, 153, 90146 Palermo, Italy.

Agustín F. Pérez-Cadenas Department of Inorganic Chemistry, Faculty of Sciences. University of Granada, Avda. Fuentenueva s/n. 18071, Spain.

Christophe Poupin Chimie-Biochimie, Université de Lyon, Lyon, F-69003, France, Université Lyon 1, Villeurbanne, F-69622, CNRS, UMR 5256, IRCELYON, 2 avenue Albert Einstein, Villeurbanne, F-69622, France.

Laurence Retailleau Unité de Chimie Environnementale et Interactions sur le Vivant (UCEIV, EA 4492), Université du Littoral Côte d'Opale (ULCO), 145 avenue Maurice Schumann, 59140 Dunkerque, France.

Douglas Romero Unité de Catalyse et Chimie du Solide UMR 8181 CNRS, Université Lille1 Sciences et Technologies, 59650 Villeneuve d'Ascq, France.

Sébastien Royer Institut de Chimie des Milieux et Matériaux de Poitiers UMR 7285 CNRS, Université Poitiers, 86073 Poitiers, France.

Prem K. Seelam Chemical and Environmental Engineering, University of Oulu, POB 4300, 90014 Oulu, Finland.

David R. Sellick Cardiff Catalysis Institute, School of Chemistry, Cardiff University, Main Building, Park Place, CF10 3AT Cardiff, UK.

Stéphane Siffert Unité de Chimie Environnementale et Interactions sur le Vivant (UCEIV, EA 4492), Université du Littoral Côte d'Opale (ULCO), 145 avenue Maurice Schumann, 59140 Dunkerque, France.

Ahmed Souikny Laboratory LGCVR, UAE/L01FST, Faculty of Sciences and Techniques, University Abdelmalek Essaadi, B.P. 416 Tangier, Morocco.

Sharmin Sultana Department of Applied Physics, Research Unit Plasma Technology, Faculty of Engineering and Architecture, Ghent University, Sint-Pietersnieuwstraat 41, B-9000 Ghent, Belgium.

Shoko Tago Kanagawa Academy of Science and Technology, KSP building East 407, 3-2-1 Sakado, Takatsu-ku, Kawasaki, Kanagawa 213-0012, Japan.

Hiromasa Tawarayama Optical Communications R&D Laboratories, Sumitomo Electric Industries, Ltd., 1 Taya-cho, Sakae-ku, Yokohama 244-8588, Japan.

Stuart H. Taylor Cardiff Catalysis Institute, School of Chemistry, Cardiff University, Main Building, Park Place, CF10 3AT Cardiff, UK.

Yoshiyuki Teramoto National Institute of Advanced Industrial Science and Technology (AIST), 16-1 Onogawa, Tsukuba 305-8569, Japan.

Quang Hung Trinh Department of Chemical and Biological Engineering, Jeju National University, Jeju 690-756, Korea.

Arne M. Vandenbroucke Department of Applied Physics, Research Unit Plasma Technology, Faculty of Engineering and Architecture, Ghent University, Sint-Pietersnieuwstraat 41, B-9000 Ghent, Belgium.

About the Guest Editor

Jean-François Lamonier received his Ph.D. degree in Chemistry from the University of Lille 1 Sciences and Technologies, France, in 1993 on the topic "Selective Catalytic Reduction of NOx". He worked as Assistant Professor at the Littoral Côte d'Opale University in Dunkirk, France, from 1996 to 2007. In 2004 he obtained the accreditation to Supervise Research (Habilitation Thesis). In 2007 he was promoted to Full Professor at the University of Lille 1 and joined the "Unité de Catalyse et Chimie de Solide" (UCCS). He has co-authored over 90 peer-reviewed scientific papers on heterogeneous catalysis applied to the elimination of environmental pollutants (NOx, VOC). Presently his research focuses on the development and characterization of heterogeneous catalysts for application such as indoor and outdoor VOC emissions removal. He obtained two innovative Techniques for the Environment Awards from the French Agency for the Sustainable Development (ADEME) in 2001 and in 2010. He is currently the Deputy Director of the Chevreul Institute and responsible for the UCCS of the "Environmental Catalysis" team.

Preface to "Catalytic Removal of Volatile Organic Compounds"

Jean-François Lamonier

Reprinted from *Catalysts*. Cite as: Lamonier, J.-F. Catalytic Removal of Volatile Organic Compounds. *Catalysts* **2016**, *6*, 7.

The degradation of air quality by the release of volatile organic compounds (VOCs) into the air particularly harms human health and our environment. Regulation of outdoor VOC emissions is required to prevent the formation of ground-level ozone, which is principally responsible for photochemical smog. Indoor emissions of VOCs have been the subject of recent consideration for many governments around the world because of the adverse impact of VOCs on the health of people exposed to them.

Because VOCs are numerous and varied, they include alkanes, aromatics, chlorinated hydrocarbons, alcohols, aldehydes, ketones, esters, *etc.*, many technologies have been developed to control their emissions. Depending on their toxicity, concentration, presence or lack thereof in a mixture, and market value, the removal of VOCs from air can be achieved by recovery or destructive processes

Heterogeneous catalytic oxidation is regarded as the most promising technology to control VOC emissions with low energy consumption and with selective conversion into harmless molecules [1].

The high volume and low VOC concentrations in air could require the coupling of the catalytic oxidation process with other technologies such as adsorption and non-thermal plasma (NTP) in order to control the emissions while reduced operating costs [2].

This special issue covers promising recent research and novel trends in the fields of outdoor and indoor VOC abatement using different technological approaches and including recent developments in material chemistry to achieve more efficient processes.

This special issue collects three reviews and nine articles.

The review by Satu Ojala *et al.* [3] summarizes the commercially existing VOC utilization possibilities and presents utilization applications that are in the research phase. The authors introduce some novel ideas related to the catalytic utilization possibilities of the VOC emissions and underline that catalysis offers not only a way of VOC removal and reduction of their atmospheric reactivity and harms in the working environment but also provides excellent options for novel and sustainable products. Applications of carbon-supported catalysts for VOC catalytic oxidation are reviewed by Francisco José Maldonado-Hódar *et al.* [4]. The authors examine the extent to which carbon-based materials in association with noble metals (Pt and Pd) can be used as catalysts for benzene, toluene and xylenes elimination. In particular the authors point out the influence of the support of hydrophobicity and mesoporosity on the catalyst performances. The review by Sharmin Sultana *et al.* [5] identifies potential research applications of the abatement of VOCs as well as of the regeneration of adsorbents using a newly developed innovative technique, *i.e.*, the cyclic operation of VOC adsorption and plasma-assisted regeneration. The authors clearly show the influence of critical process parameters on the adsorption and regeneration steps and highlight the direction of the future work on this topic which must be focused on for the feasibility and optimization of the duration of the sequential intervals.

Two articles focus on the combination of NTP with catalysis as a feasible way to overcome the poor selectivity towards the target compounds and low energy efficiency of the use of NTP alone. Yoshiyuki Teramoto *et al.* [6] study the reaction mechanism of the zeolite-plasma hybrid system for toluene decomposition. Using different configurations of the zeolite hybrid reactor, the authors show that the main factor enhancing the reaction mechanism is ozone species produced by the plasma which are able to decompose toluene molecules adsorbed onto the zeolite. Mok *et al.* [7] also highlight that the location of catalysts should be carefully considered to observe a positive synergy between plasma and Fe-Mn cordierite honeycomb catalysts. One of the key parameters when designing a catalyst is the support nature which plays an important role in improving the activity and durability of supported noble metals. Two examples are given by Stuart Taylor *et al.* and Leonarda Liotta *et al.* In the total oxidation of naphthalene, Stuart Taylor *et al.* [8] suggest that large platinum particles, in combination with platinum in metallic and oxidized states, are needed to maximize the catalytic activity over SiO_2 support. On the contrary, Leonarda Liotta *et al.* [9], studying different oxides as support, show that formation of highly-dispersed Pd^{2+} species over TiO_2 is required for propene oxidation at lower temperatures. SiO_2 and TiO_2 can be successfully used in combination to produce a simple air and water purification unit. Tsuyoshi Ochiai *et al.* [10] show that a one-end sealed porous amorphous-silica tube coated with TiO_2 photocatalyst layers has great potential for compact and in-line VOC removal.

The use of engineered transition metal containing nanomaterials as catalysts is of interest because of the high price and limited resource of noble metals, most commonly used in practice due to their high intrinsic activity [11,12]. This topic is well illustrated by two articles describing cobalt-based mixed oxides catalysts prepared via the hydrotalcite route. Renaud Cousin *et al.* [13] show that the use of microwaves during catalyst preparation leads to a more efficient catalyst for toluene oxidation while Sonia Moreno *et al.* [14] stress that the catalytic behavior in the total oxidation of the binary mixture of toluene and 2-propanol is dependent on the redox properties and oxygen mobility in Co-Mn mixed oxide.

Finally, two works are devoted to the investigation of VOC adsorption under dynamic conditions. Tarik Chafik *et al.* [15] show clearly that clay mineral is a promising material with interesting adsorptive properties allowing valorization of available local resources with significant value-added application in environmental control. Jean-François Lamonier *et al.* [16] demonstrate that copper-exchanged zeolite material can be considered as a potential hybrid system for the treatment of toluene in low concentrations in air, since this material combines a VOC adsorber with an oxidative catalyst to clean air.

I am very pleased to serve as guest editor of this special issue and I would like to first express my gratitude to Professor Keith Hohn, the Editor-in-Chief of *Catalysts*, for entrusting me with this task. I would like to thank all the authors for their insights and all reviewers for their valuable comments to improve the quality of the papers. Finally, I would like to thank all the staff of the *Catalysts* Editorial Office. I hope that this issue will be a valuable resource for upcoming research on volatile organic compounds removal.

References

1. Quiroz Torres, J.; Royer, S.; Bellat, J.-P.; Giraudon, J.-M.; Lamonier, J.-F. Formaldehyde: Catalytic oxidation as a promising soft way of elimination. *ChemSusChem* **2013**, *6*, 578–592.

2. Vandenbroucke, A.M.; Nguyen Dinh, M.T.; Nuns, N.; Giraudon, J.-M.; de Geyter, N.; Leys, C.; Lamonier, J.-F.; Morent, R. Combination of non-thermal plasma and Pd/LaMnO₃ for dilute TCE abatement. *Chem. Eng. J.* **2016**, *283*, 668–675.

3. Ojala, S.; Koivikko, N.; Laitinen, T.; Mouammine, A.; Seelam, P.K.; Laassiri, S.; Ainassaari, K.; Brahmi, R.; Keiski, R.L. Utilization of volatile organic compounds as an alternative for destructive abatement. *Catalysts* **2015**, *5*, 1092–1151.

4. Morales-Torres, S.; Carrasco-Marín, F.; Pérez-Cadenas, A.F.; Maldonado-Hódar, F.J. Coupling noble metals and carbon supports in the development of combustion catalysts for the abatement of BTX compounds in air streams. *Catalysts* **2015**, *5*, 774–799.

5. Sultana, S.; Vandenbroucke, A.M.; Leys, C.; de Geyter, N.; Morent, R. Abatement of VOCs with alternate adsorption and plasma-assisted regeneration: A review. *Catalysts* **2015**, *5*, 718–746.

6. Teramoto, Y.; Kim, H.-H.; Negishi, N.; Ogata, A. The Role of ozone in the reaction mechanism of a bare zeolite-plasma hybrid system. *Catalysts* **2015**, *5*, 838–850.

7. Trinh, Q.H.; Mok, Y.S. Non-thermal plasma combined with cordierite-supported Mn and Fe based catalysts for the decomposition of diethylether. *Catalysts* **2015**, *5*, 800–814.

8. Sellick, D.R.; Morgan, D.J.; Taylor, S.H. Silica supported platinum catalysts for total oxidation of the polyaromatic hydrocarbon naphthalene: An investigation of metal loading and calcination temperature. *Catalysts* **2015**, *5*, 690–702.

9. Gil, S.; Garcia-Vargas, J.M.; Liotta, L.F.; Pantaleo, G.; Ousmane, M.; Retailleau, L.; Giroir-Fendler, A. catalytic oxidation of propene over Pd catalysts supported on CeO_2, TiO_2, Al_2O_3 and M/Al_2O_3 oxides (M = Ce, Ti, Fe, Mn). *Catalysts* **2015**, *5*, 671–689.

10. Ochiai, T.; Tago, S.; Hayashi, M.; Tawarayama, H.; Hosoya, T.; Fujishima, A. TiO_2-impregnated porous silica tube and its application for compact air- and water-purification units. *Catalysts* **2015**, *5*, 1498–1506.

11. Chlala, D.; Giraudon, J.-M.; Nuns, N.; Lancelot, C.; Vannier, R.-N.; Labaki, M.; Lamonier, J.-F. Active Mn species well dispersed on Ca^{2+} enriched apatite for total oxidation of toluene. *Appl. Catal. B Environ.* **2016**, *184*, 87–95.

12. Quiroz, J.; Giraudon, J.-M.; Gervasini, A.; Dujardin, C.; Lancelot, C.; Trentresaux, M.; Lamonier, J.-F. Total oxidation of formaldehyde over MnO_x-CeO_2 catalysts: The effect of acid treatment. *ACS Catal.* **2015**, *5*, 2260–2269.

13. Genty, E.; Brunet, J.; Poupin, C.; Casale, S.; Capelle, S.; Massiani, P.; Siffert, S.; Cousin, R. Co-Al mixed oxides prepared via LDH route using microwaves or ultrasound: Application for catalytic toluene total oxidation. *Catalysts* **2015**, *5*, 851–867.

14. Castaño, M.H.; Molina, R.; Moreno, S. Oxygen storage capacity and oxygen mobility of Co-Mn-Mg-Al mixed oxides and their relation in the VOC oxidation reaction. *Catalysts* **2015**, *5*, 905–925.

15. Korrir, A.; El Kasmi, A.; Assebban, M.; Souikny, A.; Haffane, S.; Achak, O.; Chafik, T. Non-calorimetric determination of the adsorption heat of volatile organic compounds under dynamic conditions. *Catalysts* **2015**, *5*, 653–670.

16. Romero, D.; Chlala, D.; Labaki, M.; Royer, S.; Bellat, J.-P.; Bezverkhyy, I.; Giraudon, J.-M.; Lamonier, J.-F. Removal of toluene over NaX zeolite exchanged with Cu^{2+}. *Catalysts* **2015**, *5*, 1479–1497.

Non-Calorimetric Determination of the Adsorption Heat of Volatile Organic Compounds under Dynamic Conditions

Abdelhamid Korrir, Achraf El Kasmi, Mhamed Assebban, Ahmed Souikny, Soukaina Haffane, Ouafae Achak and Tarik Chafik

Abstract: Avoiding strong chemical bonding, as indicated by lower heat of adsorption value, is among the selection criteria for Volatile Organic Compounds adsorbents. In this work, we highlight a non-calorimetric approach to estimating the energy of adsorption and desorption based on measurement of involved amounts, under dynamic conditions, with gaseous Fourier Transform Infrared spectroscopy. The collected data were used for obtaining adsorption heat values through the application of three different methods, namely, isosteric, temperature programmed desorption (TPD), and temperature-programmed adsorption equilibrium (TPAE). The resulting values were compared and discussed with the scope of turning determination of the heat of adsorption with non-calorimetric methods into a relevant decision making tool for designing cost-effective and safe operating of adsorption facilities.

Reprinted from *Catalysts*. Cite as: Korrir, A.; El Kasmi, A.; Assebban, M.; Souikny, A.; Haffane, S.; Achak, O.; Chafik, T. Non-Calorimetric Determination of the Adsorption Heat of Volatile Organic Compounds under Dynamic Conditions. *Catalysts* **2015**, *5*, 653–670.

1. Introduction

In the context of current interest in developing low cost and efficient adsorbents for Volatile Organic Compounds (VOCs)-contaminated effluent treatment, avoiding strong chemical bonding as indicated by lower heat of adsorption value is of interest with respect to adsorbent and/or adsorbate recycling [1–3]. This is also of importance for separation processes such as thermal swing adsorption (TSA) [4] or pressure swing adsorption (PSA) [5]. The involved physical adsorptions are mainly of a dispersive or electrostatic nature, depending on adsorbate molecules as well as on adsorbent surface functions and porosity [6]. For such adsorption process, the removal of VOCs-contaminated flow is usually achieved with adsorbent packed in a fixed bed. Thus, investigating the heat effects evolved during VOCs' adsorption/desorption processes under dynamic conditions closer to real situations is of great importance. Micro-calorimetry, coupled with complementary techniques such as volumetry, IR spectroscopy, or chromatography, is reported

1

to enable simultaneous determination of adsorbed/desorbed amounts as well as the evolved heat of adsorption [7]. The latter parameter can be determined, also, with non-calorimetric procedures through the application of the isosteric method and the Clausius–Clapeyron equation [8] or temperature-programmed desorption (TPD) [7]. It is worth noting that different heat values might be found according to calculation methods [1] and/or whether the experiments were carried out under static or dynamic conditions [9]. The application of the isosteric method based on the Clausius–Clapeyron equation using adsorption isotherms is known to be convenient for the assessment of the strength of physical bonding during adsorption [10], whereas for adsorption involving a significant irreversible fraction, the TPD method was found to be more appropriate [11]. However, attention has to be paid regarding experimental conditions that ensure minimal influence of diffusion and readsorption [12].

The aim of this work is to highlight an experimental approach based on measuring adsorbed and desorbed VOC amounts onto natural clay under dynamic conditions using Fourier transform infrared spectroscopy (FTIR). The collected data have been used to determine the heat effects during the adsorption process using isosteric and TPD methods. However, it should be pointed out that a lot of data need to be acquired in order to extract adsorption heat values. This drawback can be alleviated using the so-called temperature-programmed adsorption equilibrium (TPAE) method, as reported in [13]. The heat values are extracted by fitting the experimental curves corresponding to the variation of the equilibrium coverage with theoretical curves using an appropriate model. The values obtained with these three methods have been discussed in the case of o-xylene adsorption and desorption onto bentonite clay in comparison with commercial silica chosen as the reference material. The choice of these materials rather than activated carbon is mainly due to the exothermic phenomenon associated with values of VOC adsorption enthalpy, usually ranging between 40 and 90 kJ.mol^{-1} [4,14]. This can yield to high temperature increases depending on levels of treated VOC concentration and may induce significant local warming to start combustion of either the adsorbate or the adsorbent itself, particularly if activated carbon is used [15,16]. Thus additional consideration of safety conditions is required for appropriate engineering and safe operating of adsorption facilities, which in turn needs accurate knowledge of the evolved heat effect values. Therefore, the choice of mineral adsorbent such as clays might be of interest not only with respect to efficiency/cost ratio but also for safety reasons.

2. Results and Discussion

2.1. Materials

The natural material (bentonite) tested in the present work was extracted from deposits located in the north of Morocco (Nador region). It was used with minimal processing and compared to a commercial material (SiO_2 aerosil-200 from Degussa Corp., Düsseldorf, Germany)

2.2. Textural and Chemical Characterization

The textural properties (surface areas and pore volumes) were measured with a Micromeritics ASAP 2020 apparatus using nitrogen adsorption and desorption isotherms at 77 K.

The chemical composition analysis has been carried out with an X-ray fluorescence spectrometer (Brucker S4 Pionner).

2.3. Measurement of Dynamical Adsorption and Desorption Amounts

Dynamic adsorption/desorption experiments have been carried out under atmospheric pressure using the experimental apparatus as reported elsewhere [17,18]. Prior to adsorption, a model gaseous mixture with a given concentration of o-xylene vapor has been prepared by injection, using a syringe pump, into a controlled nitrogen flow. All lines and valves of the experimental devices have been permanently heated. The resulting o-xylene concentration is expressed as molar fraction C_{in}, corresponding to P/P_{atm}, where P is the vapor pressure of o-xylene and P_{atm} the atmospheric pressure with P_{atm} = 760 torr. Note that in order to avoid vapor condensation, o-xylene partial pressure has been always maintained lower than its vapor pressure value at adsorption temperature. The studied sample (1g of bentonite or 0.3 g of silica) contained in quartz micro-reactor (U-type) was first pre-treated under 100 $cm^3.min^{-1}$ of pure nitrogen at atmospheric pressure and 473 K for 2 h, then cooled to adsorption temperature T_a. The same samples have been used for several experiments and pretreated before each testing. Adsorption has been carried out under o-xylene/N_2 mixture until saturation, followed by isothermal desorption under pure N_2 flow until o-xylene concentration in N_2 flow at the reactor outlet reached zero. Subsequent linear heating has been carried out under nitrogen flow in order to achieve complete desorption of the remaining adsorbed species according to the temperature-programmed desorption (TPD) method.

Another type of experiment, the so-called temperature-programmed adsorption equilibrium (TPAE) method, has been carried out following the procedure reported in details in previous work [13]. It should be pointed out that TPAE experiments were performed using the above operating procedure and conditions to reach adsorption equilibrium. Afterwards, the samples were subjected to heating with different rate

3

(0.5 K/min) under an *o*-xylene/N_2 mixture, which is different from TPD protocol. This approach allows us to maintain quasi-constant adsorption pressure during desorption *o*-xylene while yielding to progressive lowering of adsorption equilibrium coverage. Finally, total removal of *o*-xylene adsorbed species is completed at higher temperature by switching *o*-xylene/N_2 to pure N_2 flow.

During all these experiments, the *o*-xylene concentration in the gas stream at the reactor outlet has been monitored by FTIR Jasco 410 spectrometer (resolution of 4 cm^{-1}) using a homemade Pyrex gas cell equipped with CaF_2 windows. This technique conveys multiple advantages offered by FTIR instrumentation, such as rapid spectra acquisition and easy quantitative analysis, especially when there are no IR bands overlapping [19,20]. The molar fraction of *o*-xylene has been obtained by integration of its IR bands located between 2700 and 3200 cm^{-1}, according to the Beer–Lambert law relating IR bands area to concentration [21,22]. Preliminary calibration with *o*-xylene/N_2 mixtures of known composition has been carried out using reactor bypass. The FTIR detector response has been found to produce a linear plot in the studied concentration range, permitting correlation of IR bands area with *o*-xylene concentration. On the other hand, the use of FTIR has an additional advantage associated with eventual detection of new IR bands, corresponding to the formation of eventual new species in the gas stream indicating adsorbate degradation during the adsorption or desorption steps.

3. Results and Discussion

3.1. Textural and Chemical Characterization

3.1.1. BET Surface Area and Porosity

The N_2 adsorption/desorption isotherms obtained at 77 K with bentonite and silica samples (Figure 1) can be considered as type II according to Rouquerol *et al.* [23]. In the case of silica the Type II character seems to be associated with adsorption on both external surface as well as some interparticles capillary condensation [24]. As for bentonite clay, the displayed hysteresis loop is characteristic of the mesopore network, corresponding more likely to pseudo-Type II (Type IIb), according to Gregg and Sing [25], rather than Type IV, as also reported in the literature [26–28]. The specific surface area S_{BET} has been calculated according to the BET equation using the linear part in the range of $0.05 < P/P_{atm} < 0.3$, while the total porous volume V_t has been estimated from the adsorption at $P/P_{atm} = 0.98$. The resulting BET surface area has been found to be 201.7 cm^2/g and 83.5 cm^2/g, respectively, for silica and bentonite. The microporous volume V_{mic} values have been extracted from the interception of t-plot line and the mesopore volume (V_{meso}) value has been obtained by subtracting the microporous volume from the total porous volume (V_t). The obtained values, as summarized in Table 1, are consistent with aforementioned features confirming the presence of negligible microporosity and hysteresis loop associated with some interparticles' capillary condensation. Note that in the case of isotherms obtained with bentonite, the Type IIb is characteristic of platy particles assemblage such as that usually present in clay mineral.

Table 1. Textural parameters of bentonite and silica adsorbents.

Solids	S_{BET} (m²/g)	V_t [a](cm³(STP)/g)	V_{meso} [b](cm³(STP)/g)	V_{mic} [c](cm³(STP)/g)	S_{ext} [d](m²/g)	S_{mic} [e](m²/g)
Bentonite	83.5	0.213	0.2113	0.0017	81.02	2.469
SiO₂	201.7	0.430	0.4298	0.0002	198.53	1.47

[a] Total pore volume; [b] Mesoporous volume; [c] Microporous volume; [d] External specific surface and [e] Specific micropore surface area.

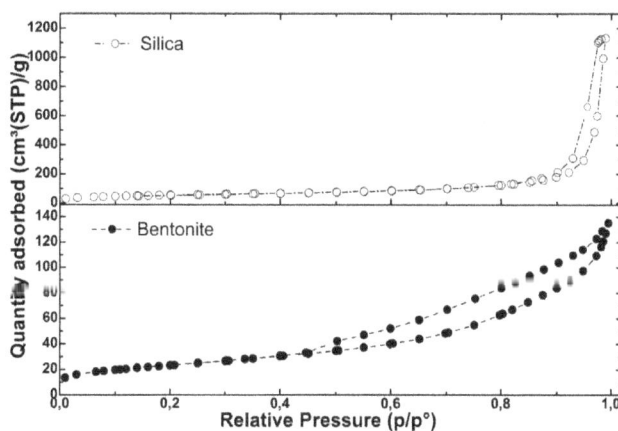

Figure 1. Nitrogen adsorption/desorption isotherms obtained with bentonite and silica at 77 K.

3.1.2. Chemical Composition

The chemical composition of studied bentonite clay as revealed by X-ray fluorescence (Brucker S4 Piooner spectrometer) indicated a higher content of SiO_2 in bentonite (about 74%) in addition to the presence of magnesium oxide with traces of sodium, copper, titanium, and zinc oxides (in Table 2). This justifies the choice of silica as a reference material for comparison with natural clay in the context of investigating a VOC adsorbent associating lower cost and efficiency, which is a challenging issue for Volatile Organic Compounds (VOCs) removal by adsorption.

5

Table 2. Chemical composition (wt. %) of bentonite and commercial silica samples

Solids	SiO_2	Al_2O_3	CaO	MgO	Fe_2O_3	Na_2O	K_2O	SO_3	Cl	CuO	TiO_2	ZnO
Bentonite	74	13.1	5.36	2.9	1.88	1.1	0.52	0.38	0.32	0.13	0.12	0.1
SiO_2	99.4	0.053	0.025	-	0.017	-	0.023	0.2	0.059	-	-	-

3.2. Dynamic Adsorption/Desorption

3.2.1. Determination of Adsorbed and Desorbed Amounts

The evolution of o-xylene concentration at the reactor outlet (C_{out}) obtained from FTIR spectra enabled the monitoring of adsorbent loading as function of time during adsorption (*i.e.*, breakthrough curve) and desorption. Figure 2 shows the variation profile of the of o-xylene concentration in the gas flow at the reactor outlet, represented as relative values (C_{out}/C_{in}) during the aforementioned cycle (adsorption at 300K until saturation (A), followed by isothermal desorption (B) then TPD (C)). The adsorbed amounts corresponding to saturation have been determined by the numerical integration of the breakthrough curve area (part A of Figure 2) according to Equation (1), where F is the gas flow rate ($cm^3.min^{-1}$), C_{in} and C_{out} correspond to the inlet and outlet concentrations of VOC in flow gas, V_m is the molar volume, m is the weight of solid in the reactor (g), and t_a is the time at adsorption saturation:

$$Q_{ads} = \frac{F \times C_{in}}{V_m \times m} \int_0^{t_a} \left(1 - \frac{C_{out}}{C_{in}}\right) dt \tag{1}$$

Accordingly, the studied concentration of o-xylene in N_2 corresponding to 0.395% yields to Q_{ads} values of 395 and 823 µmol/g, respectively, for bentonite and silica. Note that the bentonite loading capacity was lower than that of silica due to its lower BET surface area and total pore volume. These values are therefore higher than those reported in the literature for adsorbents of silicate type; e.g., 111, 61, 80, and 165µmol/g obtained for the adsorption of 3 torr of o-xylene (corresponds to 0.395% o-xylene/N_2) at 298 K respectively for zeolite of MCM-22 type [29], Webster soil (2.6 m^2/g) [30], the kaolinite (13.6 m^2/g) [30] and Webster HP (33 m^2/g) [30]. In the case of higher surface area adsorbents such as activated carbons—AC40 (1300 m^2/g) [31] and CA (990 m^2/g) [32]—and zeolite types—Al-Meso 100 (915 m^2/g) [33], UL-ZSM5-100-2 (840 m^2/g) [33], and AlPO4-5 [34]—, the reported o-xylene loading capacities are 4666, 1000, 2800, 1800, and 1166 µmol/g, respectively, which are higher than those obtained with bentonite and silica. Nevertheless, the expression of these values in µmol/m^2, considering their higher specific surface areas, yields to 3.5, 1, 3, and 2.6 µmol/m^2 for AC40, CA, zeolite types Al-Meso 100, UL-ZSM5-100-2, and AlPO4-5, respectively, which are lower when compared to the adsorption capacities obtained for bentonite (4.7 µmol/m^2)

6

and silica (4.1 $\mu mol/m^2$). Furthermore, taking into account the apparent density values of 550 and 50 kg/m^3, respectively, for bentonite and silica [3], the expression of adsorptive capacity values in $\mu mol/m^3$ becomes even better by a factor of 7 in favor of bentonite as compared to silica, which permits significant reduction in terms of adsorbent bed volume.

On the other hand, the amount involved in desorption were determined first during the isothermal desorption step conducted following the sorbent saturation by switching the o-xylene/N_2 mixture to pure nitrogen flow. This led to a gradual decrease of the o-xylene IR bands as a result of its concentration decrease at the reactor outlet, following the profile showed in part B of Figure 2. The application of Equation (2) determined Q_{des} values of 351 and 777 $\mu mol/g$, respectively, for bentonite and silica, corresponding to the weakly adsorbed fraction released during isothermal desorption:

$$Q_{des} = \frac{F}{V_m \times m} \left(\int_{t_a}^{t_b} C_{Out} dt \right) \qquad (2)$$

Here, t_a and t_b correspond to the starting and ending times of isothermal desorption. It is to be noted that although the o-xylene concentration reached zero, the calculated desorbed fraction represents only 88% and 94% of the total adsorbed amount, respectively, for bentonite and silica. The remaining adsorbed o-xylene is likely more strongly retained and requires subsequent heating under N_2 flow according to TPD method. The integration of the resulting TPD curve area (part C in Figure 2) using Equation (3) permits us to obtain Q_{DTP} values of 50 and 39 $\mu mol/g$, respectively, for bentonite silica:

$$Q_{DTP} = \frac{F}{V_m \times m} \left(\int_{t_b}^{t_c} C_{Out} dt \right) \qquad (3)$$

These adsorbed amounts are apparently associated with slow diffusion; the o-xylene fraction was adsorbed in the less accessible porosity of the inner interparticle space, as discussed in our previous work [9]. Therefore, the data corresponding to the total adsorbed amount (Q_{ads}) have been found to fit the mass balance equation ($Q_{ads} \approx Q_{des} + Q_{DTP}$) with a precision around 2% (Figure 3). On the other hand, it should be pointed out that the repetition of the cycles of adsorption/desorption under N_2 flow, at least during three successive regeneration cycles, yields to reproducible results without significant alteration of the adsorptive properties.

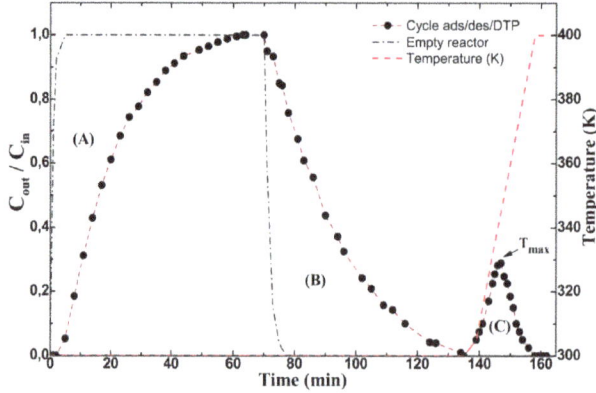

Figure 2. Profile of *o*-xylene concentration variation at reactor outlet obtained with bentonite clay during the different experiment steps: isothermal adsorption then desorption followed by DTP.

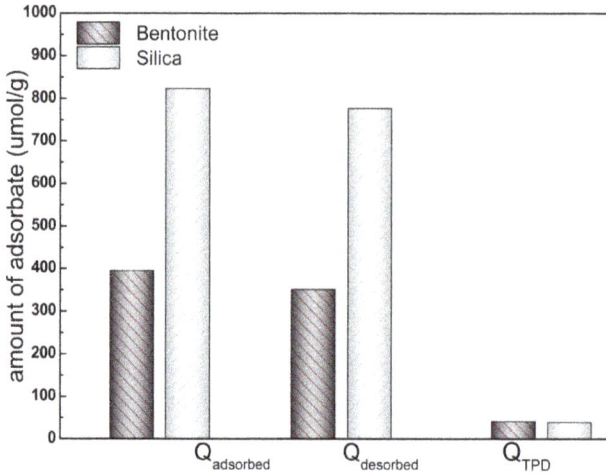

Figure 3. Amount of *o*-xylene adsorbed and desorbed obtained with bentonite and silica adsorbents using a 0.395% *o*-xylene/N_2 mixture.

The same experimental device has been used to determine the adsorbed and desorbed amounts following the TPAE methodology previously reported [12,29]. Note that the design of the TPAE experiments is simpler than TPD, particularly in terms of the required number of experiments [35]. Furthermore, there is no need for perquisite of assumption concerning the validity of experimental conditions with minimal contribution of diffusion and readsorption. The TPAE experiments were conducted with a mixture of 0.395% *o*-xylene/N_2 at 300 K until reaching adsorption equilibrium, then the adsorption temperature (T_a) was slightly increased, with a

8

ramp of 0.5 K/min with bentonite and 1 K/min in the case of silica, in order to allow a quasi-constant partial pressure of o-xylene (P_a). Finally, isothermal desorption is carried out under nitrogen flow to ensure whether o-xylene is fully desorbed. Figure 4 shows the o-xylene evolution profile at the reactor outlet during different steps of the TPAE experiment conducted with bentonite (a similar profile was obtained with silica and not shown here). The desorbed amount as indicated by a positive peak is determined using Equation (4):

$$Q_{TPAE} = \frac{F}{V_m \times m} \int_{t_b}^{t_a} (C_{t_0} - C_{t_{eq}}) dt \qquad (4)$$

Accordingly, the numerical integration of the peak area indicated in part B of Figure 4 yields Q_{TPAE} values of 358.81 µmol/g and 705.75 µmol/g, respectively, for bentonite and silica.

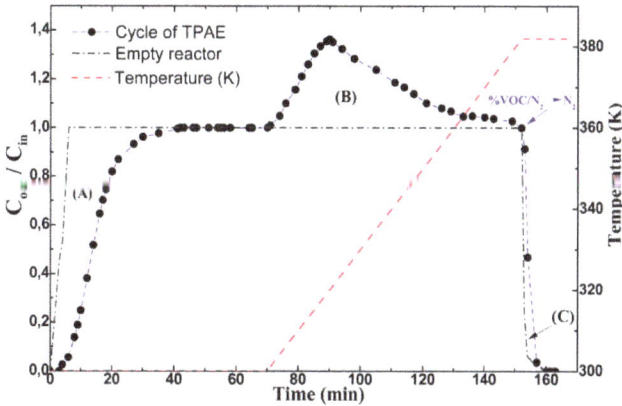

Figure 4. Evolution of o-xylene concentration at the reactor outlet obtained with bentonite clay during TPAE experiments using a 0.395% o-xylene/N_2 mixture.

3.2.2. Adsorption Isotherms

This section is devoted to the investigation of isotherms obtained at temperatures ranging between 300 and 348 K using data corresponding to the adsorption equilibrium loading data obtained as described in Section 3.2.1. The used data has been collected under partial pressures of o-xylene ranging between 0.5 and 9 torr permitted by the available instrumentation used to generate partial pressure of o-xylene vapor in the N_2 flow. Note that considerable experimental effort has been devoted in order to collect the saturation capacity data needed for the adsorption isotherms shown in Figure 5, which constitutes the major drawback of the adopted approach. The resulting isotherms have been fitted by Langmuir,

Freundlich, and Temkin models according to the nonlinear regression method using Mathcad software. In the case of the Langmuir model [36], two parameters have been used as expressed by Equation (5):

$$Q = \frac{Q_m bP}{1 + bP} \tag{5}$$

where Q is the adsorbed loading (µmol/g), Q_m is the adsorbed amount corresponding to monolayer(µmol/g), P is the adsorbate pressure (torr), and b is the empirical isotherm parameter (torr^{-1}). The values of the constant b and Q_m have been obtained, respectively, from the intercepts and the slopes.

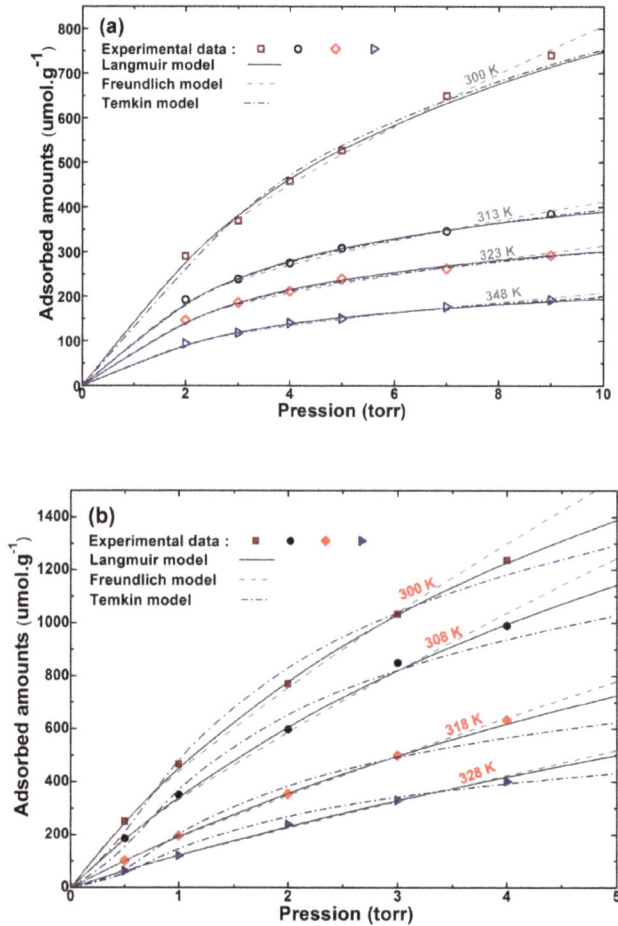

Figure 5. (a) Experimental and modeled adsorption isotherms obtained with bentonite clay at different temperatures. (b) Experimental and Modeled adsorption isotherms obtained with silica at different temperatures.

Concerning the Freundlich model [37], the empirical expression (Equation 6) has been used:

$$Q = kP^{\frac{1}{n}} \tag{6}$$

where Q is the adsorbed amount, k is the equilibrium constant ,and n is an empirical constant (with $n > 1$).Whereas for the Temkin model [38], the following equation has been used:

$$Q = a.LnP + c \tag{7}$$

where a is a constant depending on Temperature T and b is a constant related to the heat of adsorption.

Table 3. Langmuir, Freundlich, and Temkin parameters for adsorption of natural and commercial adsorbents.

Solids	Adsorption Temperatures	Adsorption amounts	Langmuir parameters			Freundlich parameters			Temkin parameters		
	$T(K)$	$Q_{m\,exp}$	Q_m	b	R^2	k	N	R^2	a	c	R^2
Bentonite	300	831	1261	0.29	0.999	256	0.55	0.997	296	167	0.991
	313	385	526	0.285	0.998	143	0.459	0.993	128	100	0.998
	323	292	408	0.281	0.998	110	0.453	0.985	95	80	0.997
	348	192	264	0.276	0.998	70	0.474	0.99	65	47	0.998
SiO$_2$	300	1298	2807	0.31	0.990	851.50	0.713	0.978	857.66	1044	0.996
	308	971	2387	0.29	0.997	500.70	0.694	0.986	493.71	581.1	0.993
	318	633	1132	0.28	0.994	230.90	0.690	0.986	225.80	267.42	0.990
	328	402	631	0.267	0.994	130.45	0.679	0.991	127.49	149	0.986

$Q_{exp}(\mu mol/g)$; $Q_m(\mu mol/g)$; b (torr^{-1}); k (torr.$\mu mol/g$); n (torr^{-1}.$\mu mol.g^{-1}$), a (torr^{-1}. $\mu mol.\,g^{-1}$), and c in $\mu mol/g$.

Table 3 summaries the fitted parameter b at different T as well as Q_m and Q_{mexp} values corresponding to adsorbent loading at monolayer and at saturation measured using an experimental break-through curve. Q_m has been determined using the linearzed form of the Langmuir equation and values of adsorbed o-xylene monolayer has been extracted using the best fitted experimental data. As shown, the resulting values remain lower than those corresponding to the measured maximal adsorption capacity, meaning that the monolayer is not reached at the studied experimental conditions. On the other hand, the fittings parameters reported in Table 3 show relative deviations less than 3%, suggesting that the Langmuir model yields a better fit of o-xylene adsorption on bentonite and silica [36].

3.3. Heat of Adsorption Determination

3.3.1. Application of the Isosteric Heat Method

One of the classical non-calorimetric methods for determining the adsorption heat (ΔH_{ads}) is based on the application of Clausius–Clapeyron equation, seen here as Equation (8):

$$\Delta H_{ads} = -R \left(\frac{\partial \ln P}{\partial (1/T)} \right) \tag{8}$$

where R is the perfect gas constant and P and T correspond to the partial pressure and temperature at equilibrium, respectively.

The method consists of converting isotherm data (Figure 5) into isosteres by plotting $\ln P = f(1/T)$ at constant loading and the value of isosteric heat ΔH_{ads} is extracted from the slope of the obtained straight line. This approach does not require assumptions concerning fitting the experimental data with any model. However, it is necessary to observe very cautiously whether the resulting isosteres match with straight lines, which relies strongly on the experimental conditions' accuracy. As reported, an error of $\pm$ 2K on temperature measurement may result in incertitude of $\pm$ 8 kJ/mol for isosteric heat value [39]. On the other hand, the application of this procedure for different coverage permits us to follow the o-xylene isosteric heat variation adsorption as a function of θ ($\theta = Q_{ads}/Q_m$), where Q_{ads} is the adsorbed amount for a given adsorption pressure and Q_m is the loading capacity at monolayer calculated as previously indicated. Thus, heat of adsorption values ranging from 45.67 to 41.57 kJ/mol and 49.39 to 47.97 kJ/mol have been found for bentonite and silica, respectively, for the corresponding surface coverage indicated in Table 4. These values are slightly higher than the heat of vaporization for o-xylene (43.43 kJ/mol) [9] and characteristic of weaker bonding, which is considered as an important advantage in terms of easier regeneration. Note that the small decrease observed for increased adsorbate coverage, illustrated in Table 4, is usually attributed to the effect of repulsive interactions between the adsorbed molecules, so-called "adsorbat–adsorbate interaction" [3].

Table 4. Isosteric heats of adsorption at different recovery rates of bentonite and silica.

Surface coverage Recovery rate	Bentonite Isosteric heat (kJ/mol)	Silica Isosteric heat (kJ/mol)
$\theta = 0.15$	45.672	49.395
$\theta = 0.17$	43.475	48.572
$\theta = 0.22$	42.73	48.057
$\theta = 0.27$	41.567	47.974

3.3.2. Application of the Temperature-Programmed Desorption (TPD) Method

In the present work, it has been found that about 12% and 6% of o-xylene, on bentonite and silica, respectively, remained adsorbed although the isothermal desorption curve reached zero. This adsorptive behavior suggests the use of the Temperature-Programmed Desorption (TPD) method for determination of desorption energy E_d value, as firstly reported by Cvetanovic, then by several authors [1,9,40–42]. The method is based on measuring the desorption profile with temperature increase, which goes through a maximum and finally back to zero, indicating that the surface is cleared of adsorbate. Thus, the energy of desorption E_d value is derived from the line slope obtained from Cvetanovic Equation (9) by following the shift of temperature at desorption peak maximum T_m as a function of linear heating rate (β):

$$2 Ln T_m - Ln \beta = \frac{E_d}{RT_m} + Const \tag{9}$$

where T_m is the desorption temperature at peak maximum (K), β is the linear rate of temperature rise (K/min), E_d is the desorption energy (kJ/mol), and R is the perfect gas constant (kJ/mol.K^{-1}).

Note that in order to obtain exploitable data, the TPD curves need to show clearly detectable T_m positions. This was the case in our experiments; the profile displayed in part C of Figure 2 was reproducible for all the sets of adsorption/desorption cycles using different linear heating rates β. The plot slope representing $2ln T_m - ln \beta$ as a function of $1/T_m$ yielded E_d values of 76.46 kJ/mol and 74.01 kJ/mol for bentonite and silica, respectively. Nevertheless, diffusion and re-adsorption phenomena need to be taken into account, particularly for a microporous adsorbent. Otherwise, the resulting apparent energy values might not be able to be systematically correlated with the heat of adsorption, which is a thermodynamic parameter relevant to the adsorption equilibrium. Therefore, if correctly reliable, these data are of interest from the viewpoint of adsorbent and adsorbat regeneration.

3.3.3. Application of Temperature-Programmed Adsorption Equilibrium (TPAE) Method

As previously reported [13], the application of the TPAE method is based on determining the surface coverage evolution θ_e with adsorption temperature T_a and pressure P_a at equilibrium according to Equation (10).

$$\theta_e (T_a) = \frac{Q_{ads} - \left(\frac{F}{m}\right) \int_{t_{eq}}^{t_0} (C_{t_0} - C_{teq}) dt}{Q_m} \tag{10}$$

where Q_{ads} is the measured adsorbed loading, Q_m is the amount corresponding to the monolayer, F is the gaseous flow rate (cm^3 min^{-1}), m is the adsorbent weight (g), t_0 is the time at the beginning of increasing adsorption at temperature T_a, t_{eq} is the time at which equilibrium is established (with $t > t_{eq}$), and C_{t0} and C_{teq} correspond to the inlet and outlet o-xylene concentrations in N_2 flow.

The obtained curves showing the surface coverage profile θ_e variation as a function of adsorption temperature T_e, presented in Figure 6a,b, have been modeled in order to extract the adsorption coefficients and heat of adsorption values through best fit with the theoretical curves. As previously pointed out, particular attention must be paid to experiment design in order to select appropriate experimental temperature T_a and pressure P_a of adsorption values that permit us to obtain an amount of adsorbed o-xylene species significantly lower than Q_m at monolayer. This avoids the formation of multi-layers and consideration of related models such as BET. On the other hand, although Langmuir's model was found to reasonably fit the experimental data, the application of the Temkin model seems to be more realistic considering sorbate–sorbate interaction assumed based on isosteric method data. Accordingly, the variation of surface coverage θ as a function of T_a has been modeled using Equation (11) following the Temkin model:

$$\theta_e = \frac{R.T_a}{\Delta E} \ln \left(\frac{1 + K_{\theta_0}.P_a}{1 + K_{\theta_1}.P_a} \right) \tag{11}$$

where R is the gas constant and $K_{\theta 0}$ and $K_{\theta 1}$ are the adsorption coefficients at low ($\theta = 0$) and high ($\theta = 1$) coverage, respectively, calculated based on statistical thermodynamics consideration as detailed elsewhere, [13]:

$$K_{\theta_0} = \frac{h^3}{(2\Pi M)^{\frac{3}{2}}.(k.T_a)^{\frac{5}{2}}} \exp \left(\frac{E_{\theta_0}}{R.T_a} \right), \text{ and } K_{\theta_1} = \frac{h^3}{(2\Pi M)^{\frac{3}{2}}.(k.T_a)^{\frac{5}{2}}} \exp \left(\frac{E_{\theta_1}}{R.T_a} \right) \tag{12}$$

$\Delta E = E_{\theta 0} - E_{\theta 1}$ where $E_{\theta 0}$ and $E_{\theta 1}$ are, respectively, the heats of adsorption at low and high loading ($\theta = 0$ and $\theta = 1$). Thus, the best fitting permitted extraction of the heats of adsorption values are $E_{\theta=0} = 63$ kJ/mol and $E_{\theta=1} = 61$ kJ/mol for bentonite, and quite similar values are obtained for silica.

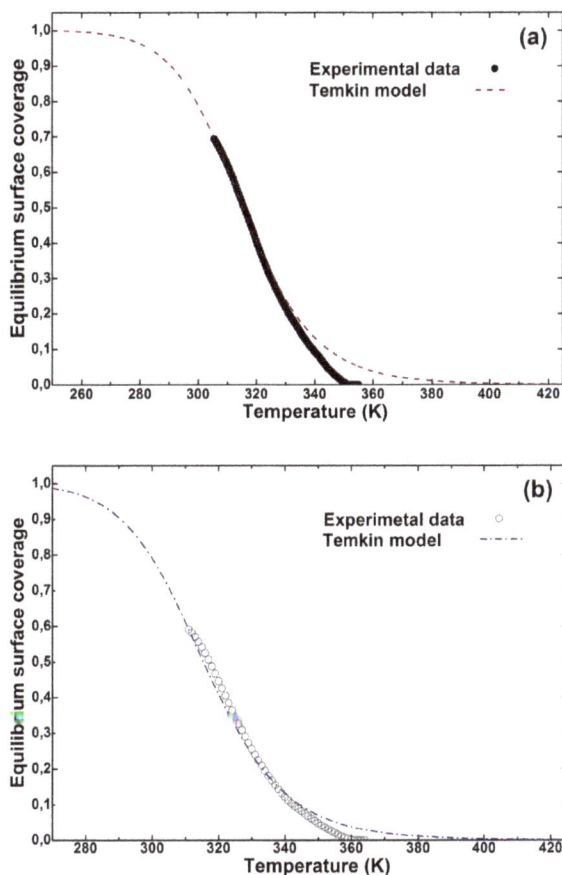

Figure 6. Confrontation of the evolution of *o*-xylene equilibrium coverage with the Temkin model during a TPAE experiment carried out with bentonite (**a**) and (**b**) with silica.

Figure 7 shows a graphical illustration comparing the heat values given by the three methods. As mentioned above, the isosteric method yields values ranging from 45.67 to 41.57 k/mol and from 49.39 to 47.97 kJ/mol, respectively, for bentonite and silica and for surface coverage (θ) ranging between 0.15 and 0.27. However, slightly higher values are obtained with the TPAE method at both lower and higher coverage: $E_{\theta=0} = 63$ kJ/mol and $E_{\theta=1} = 61$ kJ/mol. Note that the application of this method is based on mathematical formalism, taking into account adsorption equilibrium conditions, whereas the TPD method yields even higher values such as 76.46 kJ/mol and 74.01 kJ/mol, respectively, for bentonite and for silica.

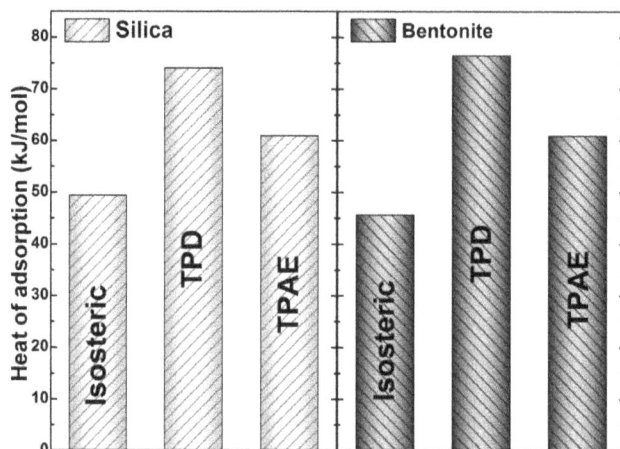

Figure 7. Graphical illustration of the heat of adsorption values obtained using isosteric, TPD, and TPAE methods with bentonite and silica.

4. Conclusions

The present work is devoted to the investigation of the heat of *o*-xylene vapor adsorption under dynamic conditions onto natural clay (bentonite) and commercial silica. The adopted approach is based on the use of FTIR spectroscopy for the measurement of adsorbed and desorbed amounts. The collected data have been used to extract heat of adsorption values following three different methods namely isosteric, TPAE, and TPD. Note that a great number of experiments were required in order to obtain the adsorption heat values, which constitutes a major drawback of these methods. Nevertheless, this can be alleviated using a simpler TPAE method that allows simultaneous determination of adsorbed/desorbed amounts and adsorption heat values.

Therefore, this study highlights the following aspects:

The first one is related to the significant difference that could be found between values of the heat of adsorption depending on the adopted method. This issue requires particular attention because the knowledge of accurate values of the heat effect involved during adsorption is important for the engineering and operating of adsorption facilities, particularly with respect to safety considerations as well as adsorbent/adsorbat regeneration and recycling.

The second aspect is related to the promising potential revealed by clay mineral with interesting adsorptive properties allowing significant reduction in terms of adsorbent bed volume. The obtained result may help local sustainable development through valorization of available local resources such as clays with significant value-added application in environmental control and engineering technologies.

Acknowledgments: Tarik Chafik dedicates this paper in honor to Daniel Bianchi, on the occasion of his nomination as Professor Emeritus and in recognition of his achievement, for the way he inspires his students, co-workers, and colleagues with his ideas and scientific pragmatism. The National Scientific Research Center (CNRST-Maroc) is acknowledged for the financial support through AEK and MA doctorate fellowships.

Author Contributions: A.K. and A.S. carried out experimental work related to measurement of adsorbed and desorbed amounts as well as data analysis; A.E.K. and M.A. took care of textural properties determination and figures preparations; S.H. and O.A. took care of samples characterizations; T.C. designed the experiments and wrote the article.

Conflicts of Interest: The authors declare no conflict of interest.

References

1. Chafik, T.; Zaitan, H.; Harti, S.; Darir, A.; Achak, O. Determination of the Heat of Adsorption and Desorption of a Volatile Organic Compound under Dynamic Conditions Using Fourier-Transform Infrared Spectroscopy. *Spectrosc. Lett.* **2007**, *40*, 763–775.

2. Gupta, V.K.; Suhas. Application of low-cost adsorbents for dye removal—A review. *J. Environ. Manage.* **2009**, *90*, 2313–2342.

3. Zaitan, H.; Korrir, A.; Chafik, T.; Bianchi, D. Evaluation of the potential of volatile organic compound (di-methyl benzene) removal using adsorption on natural minerals compared to commercial oxides. *J. Hazard. Mater.* **2013**, *262*, 365–376.

4. Kane, A.; Giraudet, S.; Vilmain, J.-B.; Le Cloirec, P. Intensification of the temperature-swing adsorption process with a heat pump for the recovery of dichloromethane. *J. Environ. Chem. Eng.* **2015**. in press.

5. Dunne, J.A.; Rao, M.; Sircar, S.; Gorte, R.J.; Myers, A.L. Calorimetric Heats of Adsorption and Adsorption Isotherms. 3. Mixtures of CH4 and C2H6 in Silicalite and Mixtures of CO_2 and C_2H_6 in NaX. *Langmuir* **1997**, *13*, 4333–4341.

6. Rouquerol, J.; Rouquerol, F.; Llewellyn, P.; Maurin, G.; Sing, K.S.W. *Adsorption by Powders and Porous Solids: Principles, Methodology and Applications*; Academic Press: Waltham, MA, USA, 2013.

7. *Calorimetry and Thermal Methods in Catalysis*; Auroux, A., Ed.; Springer: Berlin&Heidelberg, 2013; Volume 154.

8. Otero Areán, C.; Manoilova, O.V.; Turnes Palomino, G.; Rodriguez Delgado, M.; Tsyganenko, A.A.; Bonelli, B.; Garrone, E. Variable-temperature infrared spectroscopy: An access to adsorption thermodynamics of weakly interacting systems. *Phys. Chem. Chem. Phys.* **2002**, *4*, 5713–5715.

9. Chafik, T.; Darir, A.; Achak, O.; Carvalho, A.P.; Pires, J. Determination of the heat effects involved during toluene vapor adsorption and desorption from microporous activated carbon. *Comptes Rendus Chim.* **2012**, *15*, 474–481.

10. Ruthven, D.M. *Principles of Adsorption and Adsorption Processes*; John Wiley & Sons: Weinheim, Germany, 1984.

11. Eley, D.D.; Pines, H.; Weisz, P.B. *Advances in Catalysis*; Academic Press: Waltham, MA, USA, 1967.

12. Niwa, M.; Katada, N.; Sawa, M.; Murakami, Y. Temperature-programmed desorption of ammonia with readsorption based on the derived theoretical equation. *J. Phys. Chem.* **1995**, *99*, 8812–8816.

13. Hachimi, A.; Chafik, T.; Bianchi, D. Adsorption models and heat of adsorption of adsorbed ortho di-methyl benzene species on silica by using Temperature Programmed Adsorption Equilibrium methods. *Appl. Catal. A* **2008**, *335*, 220–229.

14. Chabanel, C.; Nerriere, L.; Pean, A. Composés organiques volatils: réduction des émissions de COV dans l'industrie. Avalable online: http://www.ademe.fr/composes-organiques-volatils-reduction-emissions-cov-lindustrie (accessed on 28 March 2015).

15. Delage, F.; Pré, P.; Le Cloirec, P. Mass Transfer and Warming during Adsorption of High Concentrations of VOCs on an Activated Carbon Bed: Experimental and Theoretical Analysis. *Environ. Sci. Technol.* **2000**, *34*, 4816–4821.

16. Le Cloirec, P.; Pré, P.; Delage, F.; Giraudet, S. Visualization of the exothermal VOC adsorption in a fixed-bed activated carbon adsorber. *Environ. Technol.* **2012**, *33*, 285–290.

17. Zaitan, H.; Chafik, T. FTIR determination of adsorption characteristics for volatile organic compounds removal on diatomite mineral compared to commercial silica. *Comptes Rendus Chim.* **2005**, *8*, 1701–1708.

18. Zaitan, H.; Feronnato, C.; Bianchi, D.; Achak, O.; Chafik, T. Etude des propriétés texturales et adsorbantes d'une diatomite marocaine: Application au traitement d'air charge d'un polluant de type compose organique volatil. *Ann. Chim.* **2006**, *31*, 183–196. (In French)

19. Bianchi, D.; Gass, J.L.; Bouly, C.; Maret, D. *Determination of Efficiency of Exhaust Gas Catalyst by F.T.I.R. Spectroscopy*; SAE International: Warrendale, PA, USA, 1991.

20. Kim, D.J.; Kim, J.M.; Yie, J.E.; Seo, S.G.; Kim, S.-C. Adsorption and conversion of various hydrocarbons on monolithic hydrocarbon adsorber. *J. Colloid Interface Sci.* **2004**, *274*, 538–542.

21. Chafik, T.; Dulaurent, O.; Gass, J.L.; Bianchi, D. Heat of Adsorption of Carbon Monoxide on a Pt/Rh/CeO 2/Al$_2$O$_3$, Three-Way Catalyst Using *in-Situ* Infrared Spectroscopy at High Temperatures. *J. Catal.* **1998**, *179*, 503–514.

22. Zeradine, S.; Bourane, A.; Bianchi, D. Comparison of the Coverage of the Linear CO Species on Cu/Al$_2$O$_3$ Measured under Adsorption Equilibrium Conditions by Using FTIR and Mass Spectroscopy. *J. Phys. Chem. B* **2001**, *105*, 7254–7257.

23. Rouquerol, F.; Rouquerol, J.; Sing, K.S.W. *Adsorption by Powders and Porous Solids Principles, Methodology, and Applications*; Academic Press: Waltham, MA, USA, 1999.

24. Unger, K.K.; Rodríguez-Reinoso, F.; Rouquerol, J.; Sing, K.S.W. *Characterization of Porous Solids II*; Elsevier: Amsterdam, The Netherlands, 1991.

25. Gregg, S.J.; Sing, K.S.W. *Adsorption, Surface Area and Porosity*, 2nd ed.; Academic Press: London, UK, 1982.

26. Groen, J.C.; Peffer, L.A.A.; Pérez-Ramírez, J. Pore size determination in modified micro- and mesoporous materials. Pitfalls and limitations in gas adsorption data analysis. *Microporous Mesoporous Mater.* **2003**, *60*, 1–17.

27. Chevalier, S.; Franck, R.; Lambert, J.F.; Barthomeuf, D.; Suquet, H. Characterization of the porous structure and cracking activity of Al-pillared saponites. *Appl. Catal. A* **1994**, *110*, 153–165.

28. Brunauer, S.; Emmett, P.H.; Teller, E. Adsorption of gases in multimolecular layers. *J. Am. Chem. Soc.* **1938**, *60*, 309–319.

29. Corma, A.; Corell, C.; Pérez-Pariente, J.; Guil, J.M.; Guil-López, R.; Nicolopoulos, S.; Calbet, J.G.; Vallet-Regi, M. Adsorption and catalytic properties of MCM-22: The influence of zeolite structure. *Zeolites* **1996**, *16*, 7–14.

30. Pennell, K.D.; Rhue, R.D.; Rao, P.S.C.; Johnston, C.T. Vapor-phase sorption of p-xylene and water on soils and clay minerals. *Environ. Sci. Technol.* **1992**, *26*, 756–763.

31. Benkhedda, J.; Jaubert, J.-N.; Barth, D.; Perrin, L.; Bailly, M. Adsorption isotherms of m-xylene on activated carbon: measurements and correlation with different models. *J. Chem. Thermodyn.* **2000**, *32*, 401–411.

32. Wang, C.-M.; Chang, K.-S.; Chung, T.-W.; Wu, H. Adsorption Equilibria of Aromatic Compounds on Activated Carbon, Silica Gel, and 13X Zeolite. *J. Chem. Eng. Data* **2004**, *49*, 527–531.

33. Huang, Q.; Vinh-Thang, H.; Malekian, A.; Eić, M.; Trong-On, D.; Kaliaguine, S. Adsorption of *n*-heptane, toluene and o-xylene on mesoporous UL-ZSM5 materials. *Microporous Mesoporous Mater.* **2006**, *87*, 224–234.

34. Chiang, A.S.T.; Lee, C.-K.; Chang, Z.-H. Adsorption and diffusion of aromatics in AlPO₄ 5. *Zeolites* **1991**, *11*, 380–386.

35. Derrouiche, S.; Bianchi, D. Heats of Adsorption Using Temperature Programmed Adsorption Equilibrium: Application to the Adsorption of CO on Cu/Al$_2$O$_3$ and H$_2$ on Pt/Al$_2$O$_3$. *Langmuir* **2004**, *20*, 4489–4497.

36. Langmuir, I. The Constitution and Fundamental Properties of Solids and Liquids. Part I. Solids. *J. Am. Chem. Soc.* **1916**, *38*, 2221–2295.

37. Freundlich, H. *Ueber die Adsorption in Loesungen*; Wilhelm Engelmann: Leipzig, Germany, 1906.

38. Gundry, P.M.; Tompkins, F.C. Chemisorption of gases on metals. *Q Rev. Chem. Soc.* **1960**, *14*, 257–291.

39. Dulaurent, O.; Bianchi, D. Adsorption isobars for CO on a Pt/Al$_2$O$_3$ catalyst at high temperatures using FTIR spectroscopy: isosteric heat of adsorption and adsorption model. *Appl. Catal. A* **2000**, *196*, 271–280.

40. Yoshimoto, R.; Hara, K.; Okumura, K.; Katada, N.; Niwa, M. Analysis of Toluene Adsorption on Na-Form Zeolite with a Temperature-Programmed Desorption Method. *J. Phys. Chem. C* **2007**, *111*, 1474–1479.

41. Joly, J.-P.; Perrard, A. Determination of the Heat of Adsorption of Ammonia on Zeolites from Temperature-Programmed Desorption Experiments. *Langmuir* **2001**, *17*, 1538–1542.

42. Kanervo, J.; Keskitalo, T.; Slioor, R.; Krause, A. Temperature-programmed desorption as a tool to extract quantitative kinetic or energetic information for porous catalysts. *J. Catal.* **2006**, *238*, 382–393.

Catalytic Oxidation of Propene over Pd Catalysts Supported on CeO$_2$, TiO$_2$, Al$_2$O$_3$ and M/Al$_2$O$_3$ Oxides (M = Ce, Ti, Fe, Mn)

Sonia Gil, Jesus Manuel Garcia-Vargas, Leonarda Francesca Liotta,
Giuseppe Pantaleo, Mohamed Ousmane, Laurence Retailleau and
Anne Giroir-Fendler

Abstract: In the following work, the catalytic behavior of Pd catalysts prepared using different oxides as support (Al$_2$O$_3$, CeO$_2$ and TiO$_2$) in the catalytic combustion of propene, in low concentration in excess of oxygen, to mimic the conditions of catalytic decomposition of a volatile organic compound of hydrocarbon-type is reported. In addition, the influence of different promoters (Ce, Ti, Fe and Mn) when added to a Pd/Al$_2$O$_3$ catalyst was analyzed. Catalysts were prepared by the impregnation method and were characterized by ICP-OES, N$_2$ adsorption, temperature-programmed reduction, temperature-programmed oxidation, X-ray diffraction, X-ray photoelectron spectroscopy and transmission electron microscopy. Catalyst prepared using CeO$_2$ as the support was less easily reducible, due to the stabilization effect of CeO$_2$ over the palladium oxides. Small PdO particles and, therefore, high Pd dispersion were observed for all of the catalysts, as confirmed by XRD and TEM. The addition of Ce to the Pd/Al$_2$O$_3$ catalysts increased the metal-support interaction and the formation of highly-dispersed Pd species. The addition of Ce and Fe improved the catalytic behavior of the Pd/Al$_2$O$_3$ catalyst; however, the addition of Mn and Ti decreased the catalytic activity in the propene oxidation. Pd/TiO$_2$ showed the highest catalytic activity, probably due to the high capacity of this catalyst to reoxidize Pd into PdO, as has been found in the temperature-programmed oxidation (TPO) experiments.

Reprinted from *Catalysts*. Cite as: Gil, S.; Garcia-Vargas, J.M.; Liotta, L.F.; Pantaleo, G.; Ousmane, M.; Retailleau, L.; Giroir-Fendler, A. Catalytic Oxidation of Propene over Pd Catalysts Supported on CeO$_2$, TiO$_2$, Al$_2$O$_3$ and M/Al$_2$O$_3$ Oxides (M = Ce, Ti, Fe, Mn). *Catalysts* **2015**, *5*, 671–689.

1. Introduction

One of the major challenges of today is the environmental problem associated with the atmospheric emissions of pollutants after the combustion of fossil fuels. In this way, volatile organic compounds (VOCs) represent a serious environmental problem, because of their direct (carcinogenic or mutagen) and indirect effects (ozone and smog precursors) on the environment and health [1–3]. Several different compounds are usually included in the VOC group, such as aromatic and aliphatic hydrocarbons, alcohols, ketones and aldehydes; compounds that share ease in being oxidized. Therefore, catalytic oxidation of VOCs is an interesting technique in order to reduce the emissions of pollutants during the fossil fuel combustion process, as the low concentration of VOCs in most of the cases does not allow direct combustion [4–6]. Propene is among the most investigated probe molecules in the field of VOC catalytic oxidation, because alkenes are among the major families in automotive exhausts and because of its high photochemical ozone creativity potential (POCP) [7].

Noble metals are often selected as the active phase in most of the VOC catalytic oxidation studies, despite its high price, low thermal stability and tendency of poisoning due to the superior catalytic activity of these species compared to non-noble metals [8]. Most of the catalysts employed in the catalytic oxidation of VOCs are based on the noble metal system, but the development of fuels with a lower quantity of lead, one of the major poisons in palladium catalytic systems [9], has also increased the interest in using Pd as the active phase, due to its lower price compared to other noble metals [10] and its high catalytic activity.

One of the key parameters when developing a catalyst is the nature of the support, as it is well-known that it plays an important role in improving the activity and durability of supported noble metals, with great importance of the surface properties and the metal-support interaction. The catalytic behavior of Pd-supported catalysts in catalytic oxidation strongly depends on the acid-base properties of the support [11,12], as well as on the metal-support interaction [13,14]. The metal-support interaction plays an important role in the oxidation state of Pd during the oxidation of VOCs, as reducible oxides, like TiO_2, CeO_2, Nb_2O_5 and La_2O_3, favor the oxidation of Pd into PdO, while non-reducible oxides do not [15,16]. Pd/Al_2O_3 catalyst has been studied in the combustion of hydrocarbons by several authors [17–20], showing high catalytic activity despite having different catalytic behavior depending on the hydrocarbon [18]. The addition of reducible oxides (CeO_2, TiO_2, etc.) can modify the reduction behavior of the catalyst, so it is interesting to check the influence of these oxides in the oxidation state of Pd and, therefore, in the catalytic activity of the system Pd/Al_2O_3.

In the present study, the catalytic oxidation of propene was investigated over Pd catalysts supported over CeO_2, TiO_2, Al_2O_3 and M/Al_2O_3 oxides (M = Ce, Ti, Fe, Mn). The experimental conditions for the catalytic tests, low concentration of propene (1,000 ppm) in excess of oxygen 9 vol%, were chosen to mimic the conditions of catalytic decomposition of a volatile organic compound of the hydrocarbon-type [2].

The synthesized catalysts were characterized by ICP-OES (inductively-coupled plasma optical emission spectroscopy), specific surface area measurements through the BET method, TPR (temperature-programmed reduction), TPO (temperature-programmed oxidation), XRD (X-ray diffraction), XPS (X-ray photoelectron spectroscopy) and TEM (transmission electron microscopy), in order to evaluate the influence of the different supports and the metal doped on the catalytic properties. Attention was also focused on the catalyst stability during three consecutive cycles.

2. Results and Discussion

2.1. Support and Catalysts Characterization

The physicochemical properties of the commercial supports, Al_2O_3, TiO_2 and CeO_2, and doped-Al_2O_3 supports are listed in Table 1. N_2 adsorption-desorption isotherms, pore size distributions and pore volumes of the commercial supports, Al_2O_3, TiO_2 and CeO_2, are presented in Figure 1. All commercial supports present a N_2 adsorption/desorption profile (Figure 1a) that can be assigned to the Type IV isotherm, showing a hysteresis loop due to capillary condensation representative of a mesoporous material. The three commercial supports, TiO_2, CeO_2 and Al_2O_3, of a mesoporous nature, had BET surface areas ranging from ~80–160 $m^2 \cdot g^{-1}$ and total pore volumes equal to 0.3, 0.4 and 0.2 cm^3/g, respectively. The pore size distribution (Figure 1b) showed a narrow peak centered at 4.6 nm for the Al_2O_3; a broader peak distribution centered at 8.1 nm was observed for TiO_2, while a very low and large distribution curve ranging from around 15 to 50 nm was found for the CeO_2 support. After doping of alumina support (5 mol% of Ce, Ti, Fe and Mn) and calcining at 500 °C, the shape of the N_2 isotherms remained almost the same as that of the commercial alumina (not shown). The values of BET surface areas associated with the doped-alumina (Table 1) were generally lower than the commercial one, except for Ti/Al_2O_3, showing almost the same BET surface area as alumina. It is likely that a partial blockage of the pores of the material occurs upon deposition of the metal precursor and calcination at 500 °C 3 h. Accordingly, the mean pore diameter shifted to slightly higher values, suggesting filling of the small pores. However, the total pore volumes being constant at 0.2 cm^3/g, it could be speculated that the metal doping was largely on the external surface of the alumina.

After the Pd introduction, the shape of the N_2 isotherms remained almost the same as those of the supports (not shown); however, the specific surface area values were lower (Table 2). This finding could suggest a partial blockage of the porous structure of materials upon Pd deposition and further calcination at 400 °C 4 h. The thermal stability of the doped-alumina supports and Pd catalysts was studied after three tests of propene oxidation (APT: after propene tests), see Table 1 and Table 2. The specific surface of the doped-alumina and the Pd catalysts slightly decreased (~10 $m^2 \cdot g^{-1}$), which means that the properties of both doped-alumina and Pd catalysts were not seriously modified after the heat treatment [21].

Table 1. Physicochemical properties of the supports.

Supports	BET surface area $(m^2 \cdot g^{-1})$	BET surface area APT[1] $(m^2 \cdot g^{-1})$	Total pore volume $(cm^3 \cdot g^{-1})$	Mean pore diameter (nm) [2]
TiO_2	115	-	0.3	8.1
CeO_2	79	-	0.4	15.0–50.0 (broad curve)
Al_2O_3	157	-	0.2	4.6
Ce/Al_2O_3	140	128	0.2	5.0
Ti/Al_2O_3	155	150	0.2	4.5
Fe/Al_2O_3	145	143	0.2	5.0
Mn/Al_2O_3	146	145	0.2	5.0

[1] Used after propene tests (APT). [2] Calculated by the BJH method.

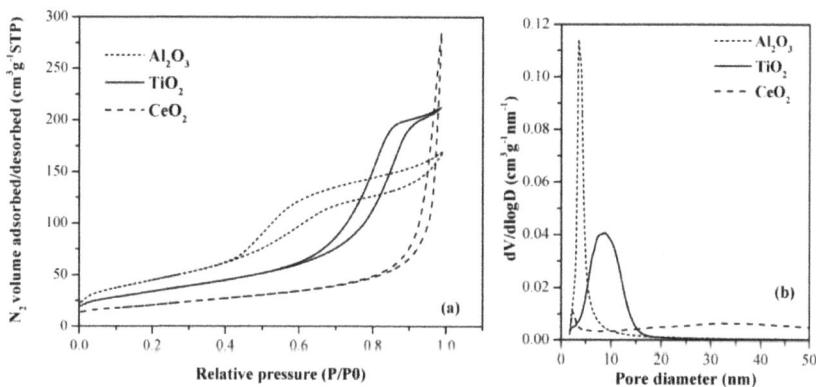

Figure 1. (**a**) N_2 adsorption/desorption isotherms and (**b**) pore size distributions associated with CeO_2, Al_2O_3 and TiO_2 supports.

Table 2. Physicochemical properties of the Pd supported catalysts.

Catalysts	BET surface area $(m^2 \cdot g^{-1})$	BET surface area APT[1] $(m^2 \cdot g^{-1})$	Pd loading (wt%) [2]	$dCeO_2$ (nm) [3]	p-Value [4]
Pd/TiO_2	102	99	0.8	-	0.64
Pd/CeO_2	77	77	0.5	105	0.86
Pd/Al_2O_3	148	144	0.8	-	1.04
$Pd/Ce/Al_2O_3$	137	132	0.8	13	-
$Pd/Ti/Al_2O_3$	154	142	0.8	-	-
$Pd/Fe/Al_2O_3$	138	130	0.8	-	-
$Pd/Mn/Al_2O_3$	135	138	0.8	-	-

[1] Used after propene tests (APT). [2] Pd (wt%) analytical loading determined by ICP-OES.
[3] Average CeO_2 crystal sizes determined by XRD for the half-width of the main CeO_2 peaks corresponding to the (111) reflection. [4] The ratio between the area of the PdO decomposition peaks and the area of the reoxidation peaks, temperature-programmed oxidation (TPO) analysis.

The results obtained in the X-ray diffraction experiments are plotted in Figure 2. The main diffraction peaks corresponding to the support can be observed for the Pd/Al_2O_3, Pd/TiO_2 and Pd/CeO_2 catalysts (Figure 2a), while peaks corresponding to PdO were only obtained for the Pd/Al$_2$O$_3$ catalyst, with a small and broad peak obtained at a 2θ value of 33.9°, which corresponds to the diffraction peak of Miller indices (101). The peaks obtained for the supports correspond to the main diffraction peaks of γ-Al$_2$O$_3$, the main diffraction peaks of TiO$_2$ (brookite) and the typical diffraction peaks related to the cubic lattice of pure CeO$_2$ (CaF$_2$ structural type) [22], for the Pd/Al$_2$O$_3$, Pd/TiO$_2$ and Pd/CeO$_2$ catalysts, respectively. Comparing diffraction profiles obtained for the promoted Al$_2$O$_3$-supported catalysts, the main diffraction peaks are due to the γ-Al$_2$O$_3$ phase. Only Pd/Ce/Al$_2$O$_3$ and Pd/Fe/Al$_2$O$_3$ showed diffraction peaks attributed to CeO$_2$ and Fe$_2$O$_3$ phases, with higher intensity of ceria features suggesting higher crystallinity. The presence of the main diffraction peak corresponding to PdO can be observed in Pd/Ti/Al$_2$O$_3$ and Pd/Mn/Al$_2$O$_3$, while for the other two catalysts, it is not very clear, probably due to the overlap of other diffraction peaks and/or the high dispersion of the PdO particles in these catalysts. It should be remarked that the main diffraction peak of PdO, when present, was very small and broad, which indicates a high dispersion of this metal for all of the catalysts. Moreover, in order to better understand the differences in the catalytic behavior of Pd/CeO$_2$ and Pd/Ce/Al$_2$O$_3$, the crystal size of CeO$_2$ was determined in these two catalysts by using the Debye–Scherrer equation, using the data from the XRD patterns ((111) reflection of CeO$_2$). The data obtained are shown in Table 2, where a much lower value of CeO$_2$ crystal size was found for the catalyst Pd/Ce/Al$_2$O$_3$ (13 nm ± 1.3) compared to that obtained for Pd/CeO$_2$ (105 nm ± 10.5).

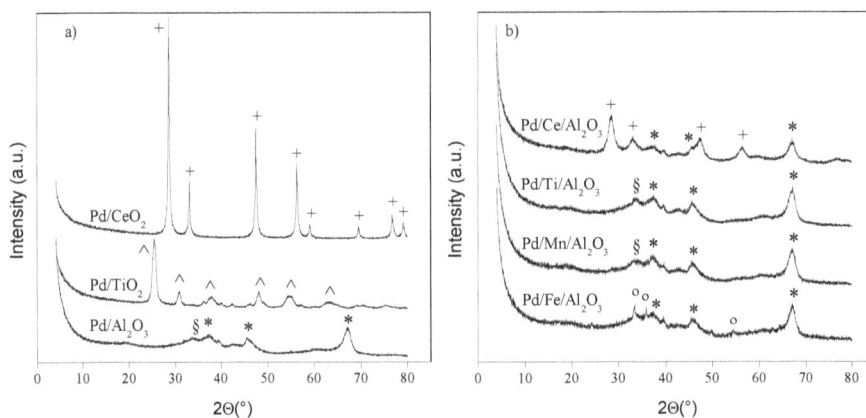

Figure 2. XRD profiles of (**a**) Pd-supported catalysts and (**b**) Pd-doped-supported catalysts, where § denotes reflection of PdO, * denotes reflection of γ-Al$_2$O$_3$, + denotes reflection of CeO$_2$, ˆ denotes reflection of TiO$_2$ and ° denotes reflection of Fe$_2$O$_3$.

In order to investigate the behavior of Pd-supported catalysts under oxidative conditions, TPO heating/cooling ramps were carried out (Figure 3). During the heating ramp, a positive peak at around 450 °C was observed for Pd/TiO_2 and for Pd/CeO_2 that could be attributed to the decomposition of adsorption oxygen species [23], whereas the positive peak in the range of 750–850 °C could be ascribed to the PdO decomposition [23–25]. The obtained results showed that the PdO decomposition occurred over CeO_2 at higher temperatures than over TiO_2 and Al_2O_3, suggesting for Pd/CeO_2 the strongest interaction between the PdO species and the support [24,25]. During the cooling ramp, a single negative peak at 558, 637 and 720 °C, corresponding to the O_2-consumption, was observed for the Pd/Al_2O_3, Pd/TiO_2 and Pd/CeO_2 catalysts, respectively. This negative peak corresponds to the reoxidation of metallic Pd to PdO [23], and the temperature strongly depends on the support. The finding of such a high reoxidation temperature, 720 °C, for Pd/CeO_2 suggests a very high reoxidation tendency of metallic palladium to PdO over ceria. Taking into account the ratio between the area of the PdO decomposition peak and the area of the reoxidation peak, the p-value was calculated [24] and reported in Table 2. The smaller the p-value, the higher is the reoxidation ability of Pd into PdO. A very low p-value was detected for Pd/TiO_2.

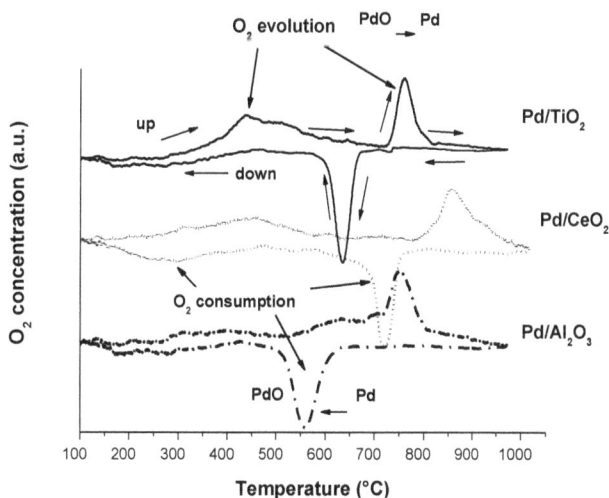

Figure 3. TPO profiles of Pd-supported catalysts.

During the cooling ramp, for the Pd/CeO_2 catalyst, an oxygen consumption peak occurred also at low temperature, at around 350 °C, which could be associated with the reoxidation of the CeO_2 surface partially reduced during the high-temperature heating step.

In Figure 4a, TPR curves of catalysts Pd/CeO_2, Pd/TiO_2 and Pd/Al_2O_3, are displayed. Peaks corresponding to different types of PdO species were observed at different temperatures depending on the support. Very small PdO particles (easily reducible, peak at $-45\,°C$) were detected over TiO_2. A second peak appearing at $8\,°C$, also found for Pd/Al_2O_3, was attributed to the reduction of bulky PdO particles [26]. For both catalysts, Pd/TiO_2 and Pd/Al_2O_3, decomposition of β-hydride species takes place at $60\,°C$ (negative peak). Such species are formed through hydrogen adsorption/diffusion in the Pd° crystallites [27]. Over CeO_2, which is known to stabilize PdO, reduction occurred at a higher temperature ($25\,°C$). Moreover, a second peak, at $50\,°C$, ascribed to the reduction of the ceria surface in contact with palladium, was also found. Above $300\,°C$, CeO_2 and TiO_2 were reduced (surface reduction at around $400–500\,°C$, bulk reduction above $700\,°C$). Figure 4b, shows the reduction profile of the alumina-doped catalysts. In the $Pd/Ti/Al_2O_3$ catalyst, the peak corresponding to the reduction of PdO around $0\,°C$ is much less pronounced than that observed for Pd/TiO_2 or Pd/Al_2O_3, and the decomposition of β-hydride shows two different peaks, probably due to the different interaction of PdH with TiO_2 and Al_2O_3. $Pd/Ce/Al_2O_3$ showed two main reduction peaks for PdO at $11\,°C$ and $40\,°C$, the first one probably due to the interaction of PdO with Al_2O_3 and the other to the interaction with CeO_2. No clear reduction peak of PdO species was found for the $Pd/Fe/Al_2O_3$ catalyst, which is probably due to the reduction of this oxide during the stabilization of the signal due to the easiness of this process in this catalyst. Reduction of PdO takes place at $19\,°C$ for the $Pd/Mn/Al_2O_3$ catalyst, followed by the decomposition of PdH and one big reduction peak at $140\,°C$, assigned to the reduction of manganese oxides [28].

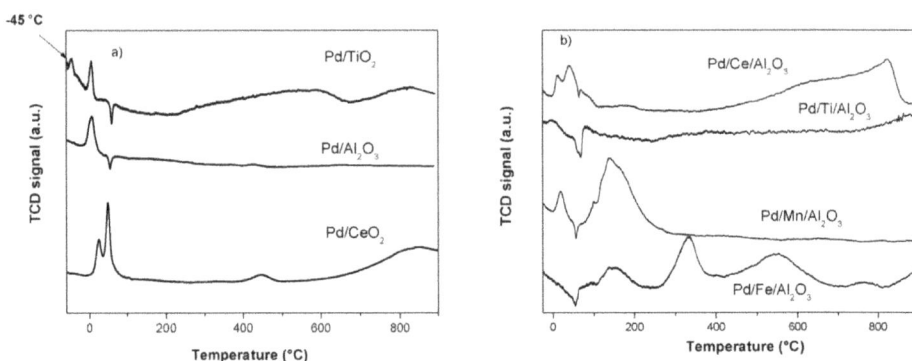

Figure 4. Temperature-programmed reduction (TPR) curves of (**a**) Pd-supported catalysts and (**b**) Pd-doped-supported catalysts.

Figure 5 shows the HR-TEM images of the Pd-supported catalyst and Pd-doped-supported catalysts. All catalysts exhibited small and well-dispersed PdO_x particles with surface area-weighted PdO_x diameters between 4–10 nm ± 1.

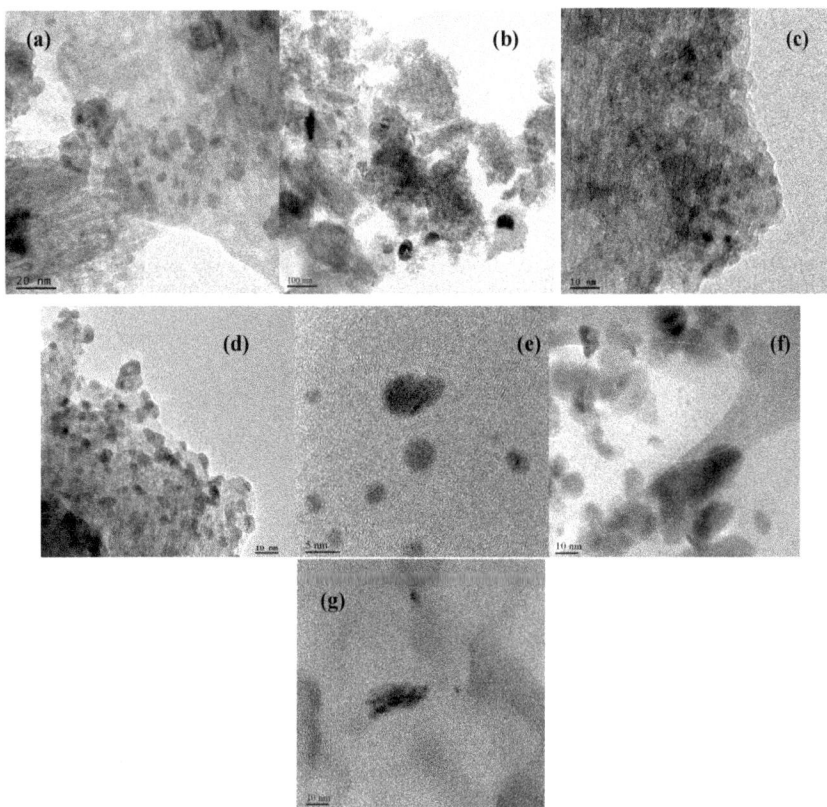

Figure 5. HR-TEM images of the Pd-supported catalysts: (**a**) Pd/Al_2O_3; (**b**) $Pd/Ce/Al_2O_3$; (**c**) $Pd/Fe/Al_2O_3$; (**d**) $Pd/Mn/Al_2O_3$; (**e**) $Pd/Ti/Al_2O_3$; (**f**) Pd/TiO_2; and (**g**) Pd/CeO_2.

The chemical state and the chemical surface composition of the Pd-supported catalysts were evaluated by XPS analyses. The binding energies (BEs) and the atomic ratios are presented in Table 3. XPS profiles of the Al 2p for the Pd/Al_2O_3 and $Pd/M/Al_2O_3$ catalysts are shown in Figure 6. The peaks are centered in the range of 74–74.4 eV, for all Al_2O_3-catalysts, corresponding to the γ-Al_2O_3, in accordance with the XRD results, which mean that the BE of the alumina was not modified after being metal doped. The O/Al atomic ratio showed for the metal-doped catalysts an increased amount of oxygen on the surface. The M/Al atomic ratio changed in the following order: $Pd/Ce/Al_2O_3 < Pd/Mn/Al_2O_3 < Pd/Ti/Al_2O_3 < Pd/Fe/Al_2O_3$;

27

suggesting that among the series of metal-doped catalysts, the highest dispersion on the alumina surface occurred for Fe species.

Table 3. Binding energies (BEs) and atomic ratios obtained from XPS analyses.

| Catalysts | Binding Energies (eV) | | | | | | | | | | Atomic Surface Ratio | |
	Al 2p	[1] Pd 3d$_{5/2}$. Bulk PdO	[1] Pd 3d$_{5/2}$. PdO$_2$	O 1s	Ti 2p$_{3/2}$	Ce 3d$_{5/2}$	[2] Ce^{4+}	[2] Ce^{3+}	Fe 2p$_{3/2}$	Mn 2p$_{3/2}$	O/Al	M/Al
Pd/TiO$_2$	-	336.1 (100)	-	529.7	458.4	-	-	-	-	-	-	
Pd/CeO$_2$	-	-	337.4 (100)	529.1	-	881.9	83	17	-	-	-	
Pd/Al$_2$O$_3$	74.1	336.4 (100)	-	530.8	-	-	-	-	-	-	1.66	
Pd/Ce/Al$_2$O$_3$	74.2	336.1 (83.2)	338.1 (16.4)	530.9	-	882.1	69	31	-	-	1.92	0.025
Pd/Ti/Al$_2$O$_3$	74	336.3 (100)	-	530.8	458.5	-	-	-	-	-	1.79	0.052
Pd/Fe/Al$_2$O$_3$	74	336.1 (100)	-	530.5	-	-	-	-	710	-	1.90	0.09
Pd/Mn/Al$_2$O$_3$	74	336.7 (100)	-	530.9	-	-	-	-	-	641.9	1.74	0.04

[1] The value in parentheses represents the atomic percentages of the palladium chemical components. [2] The percentages of the chemical states of cerium (Ce^{3+} and Ce^{4+}) on the surface.

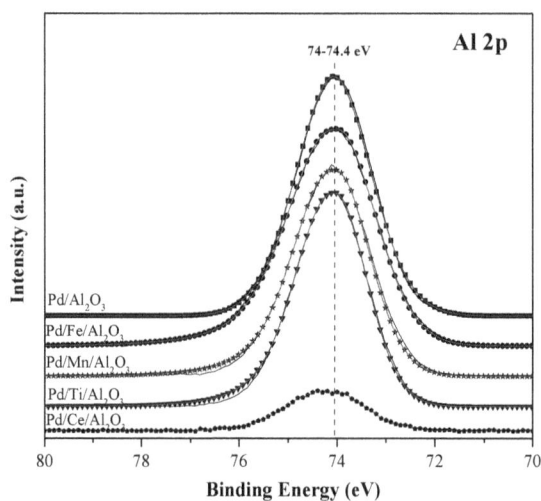

Figure 6. Al 2p XPS spectra of Pd/Al$_2$O$_3$ and Pd-doped-Al$_2$O$_3$ catalysts.

Figure 7 shows the chemical state of palladium for all catalysts. The Pd 3d spectra of the Pd/Al$_2$O$_3$, Pd/M/Al$_2$O$_3$ and Pd/TiO$_2$ catalysts (Figure 7) was characterized by the Pd 3d$_{5/2}$ peak at 336.1 $\pm$ 0.6 eV, corresponding to the Pd^{2+} (bulk PdO) [12,29–31], whereas the Pd/CeO$_2$ catalyst presented the Pd 3d$_{5/2}$ binding energy at 337.4 eV. This peak has been associated by Anna MariaVenezia *et al.* and Yushui Bi *et al.* with Pd^{4+}, as in PdO$_2$, due to the oxygen incorporation into the PdO crystal lattice during calcinations [12,32].

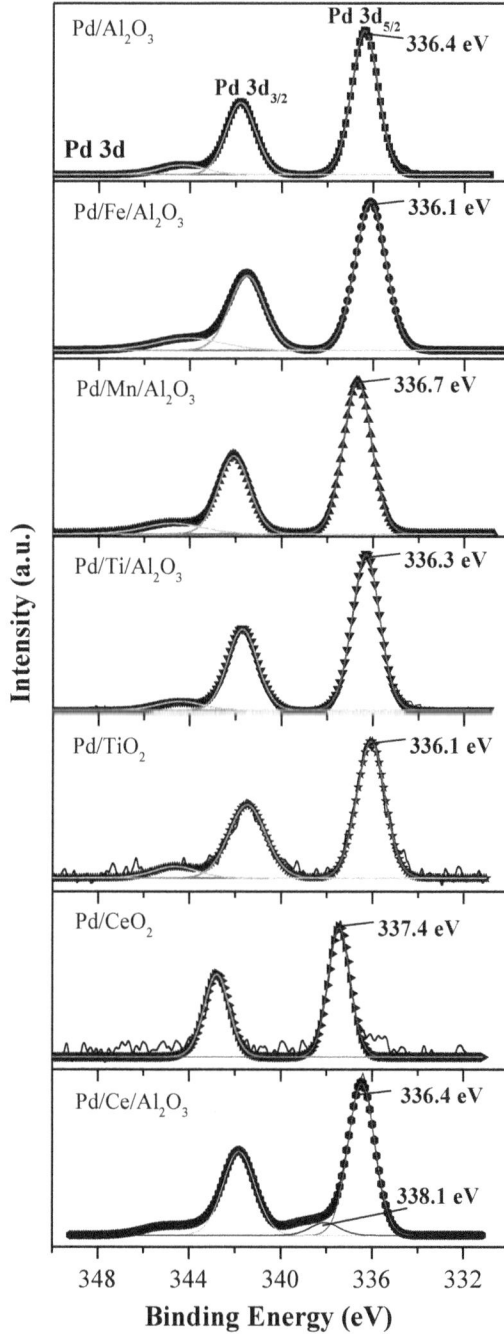

Figure 7. Pd 3d XPS spectra of Pd-supported catalysts.

Moreover, XPS profiles of the $Pd/Ce/Al_2O_3$ catalyst showed that the oxidized palladium was presented in two chemical states, bulk PdO at 336.1 eV and highly-dispersed and deficiently-coordinated Pd^{2+} in contact with the support to form palladium-aluminate structures at 338.1 eV [31] or PdO_2 [12,32], inducing the formation of new interfacial sites for the oxidation reaction. The percentage of the chemical states of palladium is presented in Table 3. The coexistence of these two oxidation states of palladium and/or the formation of palladium-aluminate structures suggests a greater Pd/Ce/support interaction, as has been described by several authors [31,33]. This Pd-Ce interaction could be promoted by ceria species in the reduced state (Ce^{3+}) presented in this catalyst (Table 3) [33]. Moreover, the presence of the Ce^{3+} species results in greater oxygen vacancies due to the formation of a defective ceria structure, as has been reported in previous articles [34]. The increase of the lattice oxygen was related to high metal-particle nucleation sites. Thus, the Ce incorporation on the $Pd/M/Al_2O_3$-supported catalyst structure favors the formation of highly-dispersed and deficient Pd^{2+}, increasing the metal-support interaction and the stabilization of metal particle deposition. Figure 8a shows the Ce 3d spectra of the Pd/CeO_2 and $Pd/Ce/Al_2O_3$ catalysts, where both catalyst presented the same characteristic peaks of pure CeO_2, according to the convention established by Burroughs [35,36]. Thus, peaks 1, 3, 4, 5 and 6, 8, 9, 10 refer to $3d_{3/2}$ and $3d_{5/2}$ binding energies, respectively, and are characteristic of Ce^{4+} states, whereas peaks 2 and 7 refer to $3d_{3/2}$ and $3d_{5/2}$, respectively, and are attributed to the Ce^{3+} states. Moreover, The Ti 2p spectra of the Pd/TiO_2 and $Pd/Ti/Al_2O_3$ are shown in Figure 8b. The Ti $2p_{3/2}$ peak of both catalysts was centered at 458.4–458.5 eV, corresponding to the Ti^{4+} species (TiO_2), in accordance with the XRD results. In addition, according to Figure 8c,d, the binding energies of the Fe $2p_{3/2}$ (710 eV) and Mn-doped $2p_{3/2}$ (641.9 eV) could be ascribed to the Fe^{3+} (Fe_2O_3) and Mn^{4+} (MnO_2) [37,38].

2.2. Catalytic Tests: Oxidation of Propene

The above characterized catalysts based on Pd supported on Al_2O_3, TiO_2 and CeO_2 were tested in the propene oxidation. The catalytic oxidation of propene was investigated in a tubular fixed-bed reactor under a reactive gas mixture containing 1000 ppm of C_3H_6 and 9 vol% O_2 in He at a gas hourly space velocity (GHSV) of 35,000 h^{-1}. Figure 9 shows the catalytic activity in terms of propene conversion as a function of the temperature (light-off curve) during the cooling ramp. The following sequence in activity was detected: $Pd/TiO_2 > Pd/Al_2O_3 > Pd/CeO_2$; suggesting that the nature of the support strongly influences the activity.

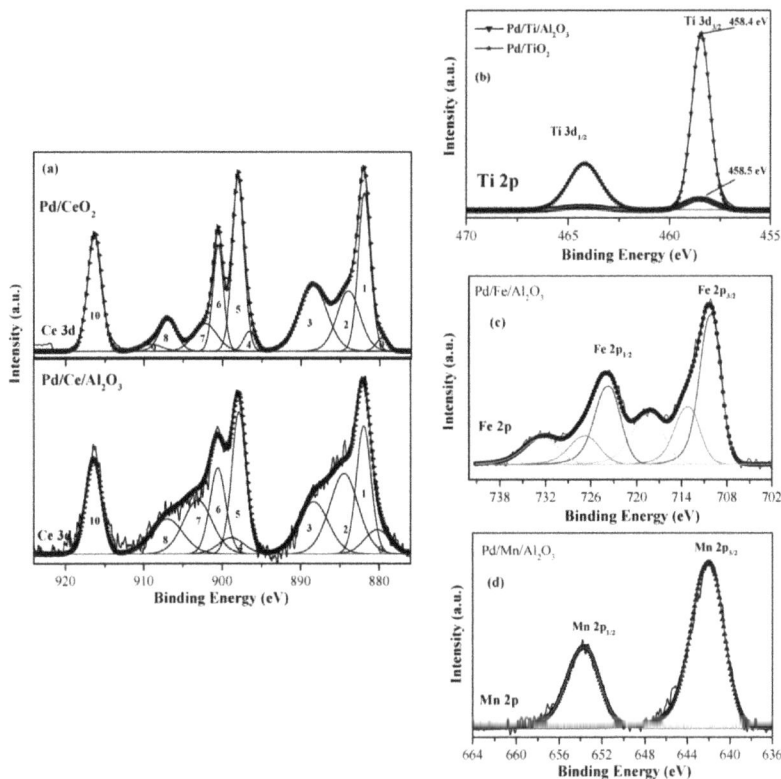

Figure 8. XPS spectra of Pd-supported catalysts: (**a**) Ti 2p; (**b**) Ce 3d; (**c**) Fe 2p; and (**d**) Mn 2p.

In order to evaluate the real effect of the support and alumina doping, the catalytic results were expressed as specific reaction rates calculated at 135 °C ($mmolC_3H_6$ s^{-1} mol$_{Pd}$ $^{-1}$), Figure 10. The same catalytic trend was obtained, where Pd/TiO$_2$ was the most active, Pd/CeO$_2$ was the worst sample and Pd alumina-based catalysts showed an intermediate activity. This suggests that the support also plays an important role in the catalytic activity and stability of the Pd catalysts. Thus, the support used has a strong influence on the oxidation state of the active phase, as has been confirmed by XPS. Therefore, the higher catalytic activity of the Pd/TiO$_2$ and Pd/Al$_2$O$_3$ catalysts could be attributed to the Pd^{2+} species detected in these catalysts, while for the case of Pd/CeO$_2$ catalyst, the presence of Pd^{4+} species could be responsible for its low catalytic activity. Moreover, the higher activity of Pd/TiO$_2$ could be attributed to the higher ability of this catalyst to reoxidize the metallic Pd into PdO, as has been corroborated by TPO from the parameter p-value (Table 2), according to the redox properties of this catalyst and the oxygen mobility [12,39].

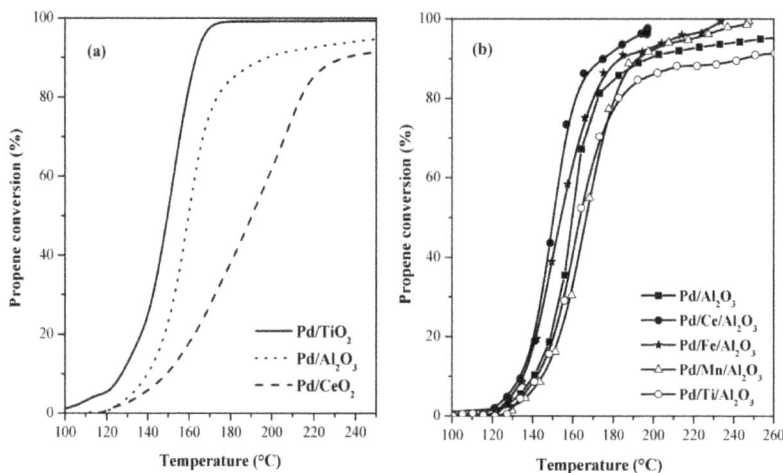

Figure 9. C_3H_6 conversion (%) *versus* temperature (light-off curve) of (**a**) Pd-supported catalysts and (**b**) Pd-doped-supported catalysts, during the cooling ramp.

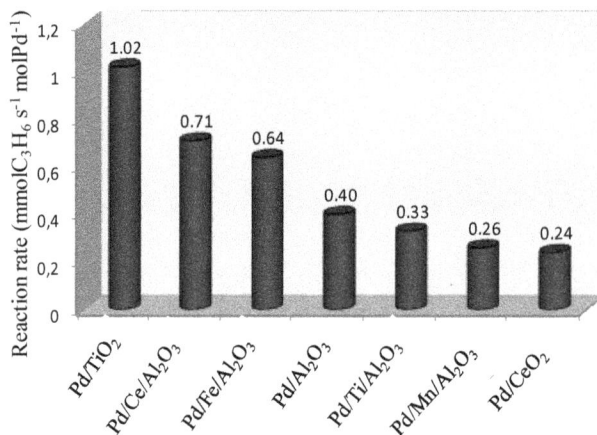

Figure 10. Catalytic activity of Pd-supported catalysts as specific reaction rates calculated at 135 °C.

In addition, the obtained results showed that the addition of Ce and Fe improves the catalytic activity of Pd/doped-alumina catalysts, while no positive effect was played by doping of Ti and Mn; Figure 9b and Figure 10. Several factors may contribute to the observed catalytic trend, such as the dispersion of the metal dopant over the alumina, as well as the amount of available oxygen on the catalyst surface. Accordingly, Pd/Fe/Al$_2$O$_3$, exhibiting quite appreciable activity, was characterized by the highest M/Al surface atomic ratio and quite a high O/Al value.

On the other hand, it is likely that other factors influence the activity; indeed, the Pd/Ce/Al$_2$O$_3$ catalyst with a very low Ce amount on the surface was the most active Pd-doped-Al$_2$O$_3$-supported catalysts. This finding could be attributed to the small CeO$_2$ crystallites formed in this catalyst and the formation of the highly-dispersed and deficiently-coordinated Pd^{2+} in contact with the support, which favors the higher Pd/Ce/support interaction, as has been observed by XPS and XRD analyses, inducing the formation of new interfacial sites for the oxidation reaction [31].

In other words, the high catalytic activity of the Pd/Ce/Al$_2$O$_3$ catalyst could be related to the high Pd/Ce/Al$_2$O$_3$ interaction, generated by the increased oxygen vacancy (defects), which promoted the palladium oxide stabilization and more active sites for the oxidation reaction [31,34]. Moreover, Pd/CeO$_2$ catalyst was less active than the Pd/Ce/Al$_2$O$_3$ according to the higher CeO$_2$ crystallite size and lower surface area [2,31].

In addition, the catalyst stability was evaluated by performing three consecutive catalytic runs for each catalyst. Table 4 lists the temperatures corresponding to 20% (T$_{20}$), 50% (T$_{50}$) and 80% (T$_{80}$) of propene conversion. No appreciable catalytic deactivation was detected after three consecutive catalytic runs, where the different temperatures are close ($\pm 10\ ^\circ$C).

Table 4. Catalytic performances of the Pd-supported catalysts on the propene oxidation during the cooling ramp (third catalytic run).

Catalysts	T$_{20}$ ($^\circ$C) [1]			T$_{50}$ ($^\circ$C) [1]			T$_{80}$ ($^\circ$C) [1]		
	I run	II run	III run	I run	II run	III run	I run	II run	III run
Pd/TiO$_2$	136	139	142	150	154	157	158	163	166
Pd/CeO$_2$	162	162	166	189	190	192	213	214	217
Pd/Al$_2$O$_3$	149	152	155	160	164	170	172	175	182
Pd/Ce/Al$_2$O$_3$	141	144	148	150	153	159	161	164	169
Pd/Fe/Al$_2$O$_3$	142	145	153	154	157	165	170	172	174
Pd/Mn/Al$_2$O$_3$	153	155	159	166	169	175	180	183	189
Pd/Ti/Al$_2$O$_3$	150	152	158	163	166	170	183	186	187

[1] Light-off temperatures at 20%, 50% and 80% of propene conversion, respectively.

Finally, the apparent activation energies were determined for the different Pd-supported catalysts, where the integration equation of first order with respect to propene was used to calculate the reaction rate constants k. The corresponding Arrhenius plots (not shown) were linear for conversion values between 10% and 80%, excluding the occurrence of diffusion limitation. In Table 5, the apparent activation energies E_{act} along with the pre-exponential factor A of the Arrhenius equation, $k = A\ e^{(-E_{act}/RT)}$, are listed. The calculated activation energies and the pre-exponential factors are comparable with values previously reported for propene oxidation [40–42]. Differences in the activation energies and pre-exponential factors, A, reflect different active site ensembles in the catalysts, as a function of the support.

Table 5. Arrhenius parameters [a], activation energy (E_{act}) and pre-exponential factor A (s^{-1}) of the Pd-supported catalysts.

Catalysts	E_{act} (KJ·mol^{-1})	LnA
Pd/CeO$_2$	93.7	26.7
Pd/Al$_2$O$_3$	132	38.4
Pd/TiO$_2$	95.3	28.7
Pd/Ce/Al$_2$O$_3$	162.6	48.1
Pd/Fe/Al$_2$O$_3$	161.1	47.4
Pd/Ti/Al$_2$O$_3$	135.1	39.1
Pd/Mn/Al$_2$O$_3$	130.7	37.6

[a] E_{act} and A calculated from the Arrhenius plot in the temperature range 100–170 °C.

3. Experimental Section

3.1. Catalyst Preparation

Commercial oxides, CeO$_2$ (Aldrich, Milano, Italy), γ-Al$_2$O$_3$ (Aldrich) and TiO$_2$ (Euro Support Manufacturing B.V., AMERSFOORT, The Netherlands) were used as supports. The doped alumina oxides, M (5 mol%)/Al$_2$O$_3$ (95 mol%), were prepared by wet impregnation of commercial Al$_2$O$_3$ with the corresponding metal (Ce, Fe, Mn) nitrate water solution (Ce(NO$_3$)$_3$·6H$_2$O, Fe(NO$_3$)$_3$·H$_2$O, Mn(NO$_3$)$_2$·xH$_2$O, respectively) or by grafting with Ti(iso-OC$_3$H$_7$)$_4$, then filtered, dried at 120 °C overnight and calcined at 500 °C for 3 h. Pd catalysts were prepared over the above supports by the impregnation method with Pd(NO$_3$)$_2$·xH$_2$O. Then, the obtained powders were calcined at 400 °C for 4 h.

3.2. Catalysts Characterization

The real palladium loading in the catalysts was quantified by ICP-OES Activa (Horiba Jobin-Yvon, Palaiseau, France).

N$_2$ adsorption-desorption at −196 °C was performed on a Sorptomatic 1900 (Carlo Erba, Trezzano sul Naviglio, Italy) instrument. Before the measurement, the samples were degassed at 300 °C for 3 h. The specific surface area (SSA) of each sample was obtained by the Brunauer–Emmett–Teller (BET) method. The mean pore size diameter was calculated by the Brunauer, Emmet, and Teller (BJH) method applied to the desorption curve.

Powder X-ray diffraction (XRD) patterns were recorded on a Bruker (Milano, Italy) D5005 diffractometer equipped with a Cu-Kα radiation (λ = 1.5418 Å) and a graphite monochromator on the diffracted beam, and the XRD data were generally collected in the 2θ range of 4°–80° with a scanning step size of 0.02° and 0.5 s. The instrumental broadening was determined by collecting the diffraction pattern of the standard, lanthanum hexaboride, LaB$_6$. The mean crystallite size (d$_s$) of CeO$_2$ was calculated with a precision of $\pm 10\%$ from the line broadening of the most intense reflections using the Scherrer Equation (1):

$$d_s = \frac{k \cdot \lambda}{\beta \cdot \cos(\vartheta)} \tag{1}$$

where d_s is the mean size of the ordered (crystalline) domains, which may be smaller or equal to the grain size, k (0.9) is the shape factor, λ (1.54 Å) is the X-ray wavelength, β is the line broadening at half the maximum intensity (FWHM) in radians and θ is the Bragg angle. A corundum probe (Bruker, Milano, Italy) of known crystallinity has been used as the internal standard.

Temperature-programmed reduction/oxidation (TPR/TPO) experiments were carried out with a Micromeritics Autochem 2910 apparatus equipped with a thermal conductivity detector (TCD). TPR experiments were carried out with a flow rate of 50 mL·min^{-1}. The gas mixture with composition 5 vol% H_2 in Ar was used to reduce the catalysts (100 mg). Firstly, the sample was cooled down to $-60\,^\circ$C in a cold trap, and the TCD signal was registered by gradually increasing the temperature (rate 5 $^\circ$C/min) from $-60\,^\circ$C up to room temperature. After that, the temperature was raised up to 1000 $^\circ$C at the rate of 10 $^\circ$C·min^{-1}. Before the TPR was started, the catalysts were pretreated with a flowing gas mixture of 5% O_2 in He (50 mL·min^{-1}) at 120 $^\circ$C for 1 h in order to purge the surface, then cooling down under Ar. The re-oxidation properties of the catalysts were measured by the TPO experiments. The oxidation gas of 5 vol% O_2 in He (50 mL·min^{-1}) was used to oxidize the catalysts (100 mg), increasing the temperature from room temperature to 1000 $^\circ$C at a heating rate of 10 $^\circ$C/min and then decreasing the temperature from 1000 to 100 $^\circ$C at a cooling rate of 10 $^\circ$C·min^{-1}.

Chemical states of the atoms in the catalyst surface were investigated by X-ray photoelectron spectroscopy (XPS) on an AXIS Ultra DLD spectrometer marketed by Kratos Analytical, operating with Al (Ka) radiation. XPS data were calibrated using the binding energy of C1s (284.6 eV) as the standard.

High-resolution transmission electron microscopy (HRTEM) analyses were carried out using a JEOL 2100 LaB6 equipped with an Oxford Instruments Inca energy dispersive X-ray (EDX) spectrometer. Mean PdO_x particle size, evaluated as the surface-area weighted diameter $(\bar{d}_{PdO_x})$, was computed according to the following Equation (2):

$$\bar{d}_{PdO_x} = \frac{\sum_i n_i d_i^3}{n_i d_i^2} \tag{2}$$

where n_i represents the number of particles with diameter d_i. On average, 50 images for each catalyst have been collected in order to estimate the particle size.

3.3. Catalytic Tests

The catalytic activity measurements of the propene oxidation were carried out in a tubular fixed-bed reactor under the reactive gas mixture containing 1000 ppm of C_3H_6 and 9 vol% O_2 in He. The total gas flow rate was 7.2 Lh^{-1} and the amount of catalyst 0.2 g, equivalent to a GHSV of 35,000 h^{-1}. In a typical experiment, the fresh catalyst (with grain diameters between 50 and 100 µm) was loaded onto a fine-quartz fritted disk, and the reaction temperature was continuously monitored by a thermocouple inserted inside the furnace. Each catalytic run was performed as follows: introduction of the reaction mixture at room temperature, heating at a rate of 1 °C· min^{-1} up to the final temperature of 450 °C, then cooling down under the reaction mixture. Three consecutive catalytic cycles in propene catalytic oxidation were performed for each catalyst. The total conversion of propene (X) was defined as in Equation (3):

$$X_{C_3H_6}(\%) = 100 \times \frac{[C_3H_6]_{in} - [C_3H_6]_{out}}{[C_3H_6]_{in}} \tag{3}$$

where $[C_3H_6]_{in}$ and $[C_3H_6]_{out}$ denoted the inlet and outlet concentrations of propene, respectively.

3.4. Analysis of Products

The reactants and products of the propene oxidation were analyzed by gas chromatography, using a dual CTR1 column from Alltech (Porapak and molecular sieve) and a TCD for CO, O_2 and CO_2. A Porapak column and an FID were employed for C_3H_6 detection. The only reaction products were CO_2 and H_2O, and the carbon balance was close to $\pm 5\%$ in all of the catalytic tests.

4. Conclusions

The catalytic oxidation of propene was investigated over Pd-supported catalysts over CeO_2-, TiO_2-, Al_2O_3- and Pd-doped-supported catalysts (M/Al_2O_3 oxides; M = Ce, Ti, Fe, Mn). All catalysts presented well-dispersed PdO_x nanoparticles, as confirmed by XRD and TEM. The obtained results showed that the support used has a strong influence on the oxidation state of palladium species and, as a consequence, on the catalytic activity. Pd/TiO_2 and $Pd/Ce/Al_2O_3$ catalysts were the best performing catalysts, while Pd/CeO_2 was performed poorly. Among the series of Pd/doped alumina catalysts, the addition of Ce and Fe improves the catalytic activity with respect to Pd/Al_2O_3, while no positive effect was played by doping of Ti and Mn. The highest activity of the Pd/TiO_2 was attributed to the presence of highly-dispersed Pd^{2+} species easily reducible at -45 °C (as detected by TPR), as well as to the high capability of this catalyst to reoxidize Pd into PdO, as found by TPO. Conversely, in the Pd/CeO_2 catalyst, the presence of Pd^{4+} species only, interacting too strongly with the support, seems responsible for its low catalytic activity.

The addition of Ce to the Pd/Al_2O_3 catalyst increased the Pd-CeO_2 interaction, which promoted the palladium oxide stabilization and the formation of more active sites for the oxidation reaction. The presence in the small crystallites of ceria of high oxygen vacancies associated with the presence of Ce^{3+} also contributes to enhancing the oxygen mobility and increasing the catalytic activity of the $Pd/Ce/Al_2O_3$, making such a catalyst more active than Pd over bare ceria.

Acknowledgments: The authors gratefully acknowledge the scientific services of Institut de Recherches sur la Catalyse et l'Environnement de Lyon (IRCELYON) and, in particular, Prakash Swamy, Laurence Massin and Luis Cardenas (XPS measurements) for stimulating discussions.

Author Contributions: The work was coordinated by Anne Giroir Fendler and Leonarda Francesca Liotta, who contributed equally to the data interpretation and discussion. The manuscript was written by Sonia Gil, who also contributed to the XPS characterization. Mohamed Ousmane performed the synthesis of catalysts and textural characterization. Giuseppe Pantaleo carried out the TPR and TPO measurements. Jesus Manuel Garcia-Vargas and Laurance Retailleau performed the TEM analyses and catalytic tests.

Conflicts of Interest: The authors declare no conflict of interest.

References

1. Amann, M.; Lutz, M. The revision of the air quality legislation in the European Union related to ground-level ozone. *J. Hazard. Mater.* **2000**, *78*, 41–62.

2. Liotta, L.F.; Ousmane, M.; Di Carlo, G.; Pantaleo, G.; Deganello, G.; Marcì, G.; Retailleau, L.; Giroir-Fendler, A. Total oxidation of propene at low temperature over Co_3O_4–CeO_2 mixed oxides: Role of surface oxygen vacancies and bulk oxygen mobility in the catalytic activity. *Appl. Catal. A* **2008**, *347*, 81–88.

3. Pérez, A.; Molina, R.; Moreno, S. Enhanced voc oxidation over Ce/CoMgAl mixed oxides using a reconstruction method with edta precursors. *Appl. Catal. A* **2014**, *477*, 109–116.

4. Lahousse, C.; Bernier, A.; Grange, P.; Delmon, B.; Papaefthimiou, P.; Ioannides, T.; Verykios, X. Evaluation of γ-MnO_2 as a VOC removal catalyst: Comparison with a noble metal catalyst. *J. Catal.* **1998**, *178*, 214–225.

5. Cloirec, P.L. *Les Composés Organiques Volatils Dans L'environnement*; Lavoisier, Tec et Doc: Paris, France, 1998.

6. Hosseini, M.; Barakat, T.; Cousin, R.; Aboukaïs, A.; Su, B.L.; De Weireld, G.; Siffert, S. Catalytic performance of core-shell and alloy Pd-Au nanoparticles for total oxidation of VOC: The effect of metal deposition. *Appl. Catal. B* **2012**, *111–112*, 218–224.

7. Derwent, R.G.; Jenkin, M.E.; Saunders, S.M.; Pilling, M.J. Photochemical ozone creation potentials for organic compounds in northwest europe calculated with a master chemical mechanism. *Atmos. Environ.* **1998**, *32*, 2429–2441.

8. Thevenin, P.O.; Ersson, A.G.; Kušar, H.; Menon, P.; Järås, S.G. Deactivation of high temperature combustion catalysts. *Appl. Catal. A* **2001**, *212*, 189–197.

9. Williamson, W.B.; Lewis, D.; Perry, J.; Gandhi, H.S. Durability of palladium automotive catalysts: Effects of trace lead levels, exhaust composition, and misfueling. *Ind. Eng. Chem. Prod. Res. Dev.* **1984**, *23*, 531–536.

10. Cowley, A. *Platinum 2013. Interim Review*; Johnson Matthey: Hertfordshire, UK, 2013.

11. Okumura, K.; Kobayashi, T.; Tanaka, H.; Niwa, M. Toluene combustion over palladium supported on various metal oxide supports. *Appl. Catal. B* **2003**, *44*, 325–331.

12. Venezia, A.M.; Di Carlo, G.; Liotta, L.F.; Pantaleo, G.; Kantcheva, M. Effect of Ti(IV) loading on CH_4 oxidation activity and SO_2 tolerance of Pd catalysts supported on silica SBA-15 and HMS. *Appl. Catal. B* **2011**, *106*, 529–539.

13. Simplício, L.M.; Brandao, S.T.; Domingos, D.; Bozon-Verduraz, F.; Sales, E.A. Catalytic combustion of methane at high temperatures: Cerium effect on PdO/Al_2O_3 catalysts. *Appl. Catal. A* **2009**, *360*, 2–7.

14. Rooke, J.C.; Barakat, T.; Siffert, S.; Su, B.L. Total catalytic oxidation of toluene using Pd impregnated on hierarchically porous Nb_2O_5 and Ta_2O_5 supports. *Catal. Today* **2012**, *192*, 183–188.

15. Sanchez, M.G.; Gazquez, J.L. Oxygen vacancy model in strong metal-support interaction. *J. Catal.* **1987**, *104*, 120–135.

16. Hong, J.; Chu, W.; Chen, M.; Wang, X.; Zhang, T. Preparation of novel titania supported palladium catalysts for selective hydrogenation of acetylene to ethylene. *Catal. Commun.* **2007**, *8*, 593–597.

17. Marécot, P.; Fakche, A.; Kellali, B.; Mabilon, G.; Prigent, P.; Barbier, J. Propane and propene oxidation over platinum and palladium on alumina: Effects of chloride and water. *Appl. Catal. B* **1994**, *3*, 283–294.

18. Maillet, T.; Solleau, C.; Barbier, J.; Duprez, D. Oxidation of carbon monoxide, propene, propane and methane over a Pd/Al_2O_3 catalyst. Effect of the chemical state of pd. *Appl. Catal. B* **1997**, *14*, 85–95.

19. Ivanova, A.; Korneeva, E.; Slavinskaya, E.; Zyuzin, D.; Moroz, E.; Danilova, I.; Gulyaev, R.; Boronin, A.; Stonkus, O.; Zaikovskii, V. Role of the support in the formation of the properties of a Pd/Al_2O_3 catalyst for the low-temperature oxidation of carbon monoxide. *Kinet. Catal.* **2014**, *55*, 748–762.

20. Chen, X.; Cheng, Y.; Seo, C.Y.; Schwank, J.W.; McCabe, R.W. Aging, re-dispersion, and catalytic oxidation characteristics of model Pd/Al_2O_3 automotive three-way catalysts. *Appl. Catal. B* **2015**, *163*, 499–509.

21. Del Angel, G.; Padilla, J.; Cuauhtemoc, I.; Navarrete, J. Toluene combustion on γ-Al_2O_3-CeO_2 catalysts prepared from boehmite and cerium nitrate. *J. Mol. Catal. A* **2008**, *281*, 173–178.

22. Saitzek, S.; Blach, J.F.; Villain, S.; Gavarri, J.R. Nanostructured ceria: A comparative study from X-ray diffraction, raman spectroscopy and bet specific surface measurements. *Phys. Status Solidi (a)* **2008**, *205*, 1534–1539.

23. Yue, B.; Zhou, R.; Wang, Y.; Zheng, X. Study of the methane combustion and TPR/TPO properties of Pd/Ce-Zr-M/Al_2O_3 catalysts with M = Mg, Ca, Sr, Ba. *J. Mol. Catal. A* **2005**, *238*, 241–249.

24. Persson, K.; Thevenin, P.; Jansson, K.; Agrell, J.; Järås, S.G.; Pettersson, L.J. Preparation of alumina-supported palladium catalysts for complete oxidation of methane. *Appl. Catal. A* **2003**, *249*, 165–174.

25. Groppi, G.; Cristiani, C.; Lietti, L.; Forzatti, P. Study of PdO/Pd transformation over alumina supported catalysts for natural gas combustion. *Stud. Surf. Sci. Catal.* **2000**, *130*, 3801–3806.

26. Tidahy, H.; Siffert, S.; Lamonier, J.-F.; Zhilinskaya, E.; Aboukaïs, A.; Yuan, Z.-Y.; Vantomme, A.; Su, B.-L.; Canet, X.; de Weireld, G. New Pd/hierarchical macro-mesoporous ZrO_2, TiO_2 and ZrO_2-TiO_2 catalysts for vocs total oxidation. *Appl. Catal. A* **2006**, *310*, 61–69.

27. De la Peña O'Shea, V.; Alvarez-Galvan, M.; Fierro, J.; Arias, P. Influence of feed composition on the activity of Mn and PdMn/Al_2O_3 catalysts for combustion of formaldehyde/methanol. *Appl. Catal. B* **2005**, *57*, 191–199.

28. Gaya, U.I.; Abdullah, A.H. Heterogeneous photocatalytic degradation of organic contaminants over titanium dioxide: A review of fundamentals, progress and problems. *J. Photochem. Photobiol. C: Photochem. Rev.* **2008**, *9*, 1–12.

29. Huang, H.; Ye, X.; Zhang, L.; Leung, D.Y.C. Mechanistic study on formaldehyde removal over Pd/TiO_2 catalysts: Oxygen transfer and role of water vapor. *Chem. Eng. J.* **2013**, *230*, 73–79.

30. Ihm, S.-K.; Jun, Y.-D.; Kim, D.-C.; Jeong, K.-E. Low-temperature deactivation and oxidation state of Pd/γ-Al_2O_3 catalysts for total oxidation of *n*-hexane. *Catal. Today* **2004**, *93*, 149–154.

31. Aznárez, A.; Korili, S.A.; Gil, A. The promoting effect of cerium on the characteristics and catalytic performance of palladium supported on alumina pillared clays for the combustion of propene. *Appl. Catal. A* **2014**, *474*, 95–99.

32. Bi, Y.; Lu, G. Catalytic CO oxidation over palladium supported NaZSM-5 catalysts. *Appl. Catal. B* **2003**, *41*, 279–286.

33. Monteiro, R.S.; Dieguez, L.C.; Schmal, M. The role of pd precursors in the oxidation of carbon monoxide over Pd/Al_2O_3 and Pd/CeO_2/Al_2O_3 catalysts. *Catal. Today* **2001**, *65*, 77–89.

34. Ousmane, M.; Liotta, L.F.; di Carlo, G.; Pantaleo, G.; Venezia, A.; Deganello, G.; Retailleau, L.; Boreave, A.; Giroir-Fendler, A. Supported au catalysts for low-temperature abatement of propene and toluene, as model VOCs: Support effect. *Appl. Catal. B* **2011**, *101*, 629–637.

35. Pfau, A.; Schierbaum, K. The electronic structure of stoichiometric and reduced CeO_2 surfaces: An XPS, UPS and HREELS study. *Surf. Sci.* **1994**, *321*, 71–80.

36. Burroughs, P.; Hamnett, A.; Orchard, A.F.; Thornton, G. Satellite structure in the X-Ray Photoelectron Spectra of some binary and mixed oxides of lanthanum and cerium. *J. Chem. Soc. Dalton Trans.* **1976**, 1686–1698.

37. Grosvenor, A.P.; Kobe, B.A.; Biesinger, M.C.; McIntyre, N.S. Investigation of multiplet splitting of Fe 2p XPS spectra and bonding in iron compounds. *Surf. Interface Anal.* **2004**, *36*, 1564–1574.

38. Ferrandon, M.; Carnö, J.; Järås, S.; Björnbom, E. Total oxidation catalysts based on manganese or copper oxides and platinum or palladium I: Characterisation. *Appl. Catal. A: Gen.* **1999**, *180*, 141–151.

39. Fujimoto, K.-I.; Ribeiro, F.H.; Avalos-Borja, M.; Iglesia, E. Structure and reactivity of PdO_x/ZrO_2 catalysts for methane oxidation at low temperatures. *J. Catal.* **1998**, *179*, 431–442.

40. Alexopoulos, K.; Reyniers, M.-F.; Marin, G.B. Reaction path analysis of propene selective oxidation over V_2O_5 and V_2O_5/TiO_2. *J. Catal.* **2012**, *295*, 195–206.

41. Zhai, Z.; Bell, A.T. The kinetics of selective oxidation of propene on bismuth vanadium molybdenum oxide catalysts. *J. Catal.* **2013**, *308*, 25–36.

42. Kobayashi, M.; Kanno, T.; Takeda, H.; Fujisaki, S. Reaction pathway design and optimization in heterogeneous catalysis: Ii. Oscillating partial oxidation of propene on Pt/SiO_2 designed by widely varying dispersion. *Appl. Catal. A* **1997**, *151*, 207–221.

Silica Supported Platinum Catalysts for Total Oxidation of the Polyaromatic Hydrocarbon Naphthalene: An Investigation of Metal Loading and Calcination Temperature

David R. Sellick, David J. Morgan and Stuart H. Taylor

Abstract: A range of catalysts comprising of platinum supported on silica, prepared by an impregnation method, have been studied for the total oxidation of naphthalene, which is a representative Polycyclic Aromatic Hydrocarbon. The influence of platinum loading and calcination temperature on oxidation activity was evaluated. Increasing the platinum loading up to 2.5 wt.% increased the catalyst activity, whilst a 5.0 wt.% catalyst was slightly less active. The catalyst containing the optimum 2.5 wt.% loading was most active after calcination in air at 550 °C. Characterisation by carbon monoxide chemisorption and X-ray photoelectron spectroscopy showed that low platinum dispersion to form large platinum particles, in combination with platinum in metallic and oxidised states was important for high catalyst activity. Catalyst performance improved after initial use in repeat cycles, whilst there was slight deactivation after prolonged *time-on-stream*.

Reprinted from *Catalysts*. Cite as: Sellick, D.R.; Morgan, D.J.; Taylor, S.H. Silica Supported Platinum Catalysts for Total Oxidation of the Polyaromatic Hydrocarbon Naphthalene: An Investigation of Metal Loading and Calcination Temperature. *Catalysts* **2015**, *5*, 690–702.

1. Introduction

Polycyclic aromatic hydrocarbons (PAHs) are a class of volatile organic compounds (VOCs), that when released into the atmosphere have been identified as detrimental to the environment and harmful to health [1,2]. Since 1998, PAHs have also been specifically listed as persistent organic pollutants, and identified as compounds that must be eliminated from the atmosphere [3]. As a result, legislation has been put in place to reduce their emissions through mechanisms such as European Union Directives, and similar control measures also exist in many other regions of the world. PAHs are released from many combustion processes, and the broader classification of VOCs are emitted even more widely from a very large number of sources. A number of techniques are available to control the release of organic atmospheric pollutants, amongst these catalytic oxidation offers a number of significant advantages over competing options, as it provides a selective and low energy pathway to produce benign products [4]. Both metal oxide and noble metal-based catalysts have been widely investigated for total oxidation of VOCs. However, more specifically the total oxidation of PAHs have not been studied so extensively, but there are a number of studies in the literature, which have investigated the total oxidation of naphthalene as a model PAH [5]. Naphthalene is recognised as a suitable model PAH as it is the major PAH produced in many combustion processes and it is easy to handle due to its relatively low toxicity.

Previously, we have investigated naphthalene total oxidation over a range of catalysts and a number of simple metal oxides have demonstrated high activity. For example, recently manganese oxide has been identified as a potentially effective catalyst [6]. However, the most active metal oxide catalysts identified were those based around ceria. These can be prepared relatively simply using precipitation with urea, and activity structure relationships have been established that nanocrystalline ceria with a small crystallite size, high surface area and a high concentration of oxygen defects leads to high activity for naphthalene total oxidation [7,8]. Ceria can also form the basis of high activity PAH oxidation catalysts, by preparation using nanocasting to prepare high surface area mesoporous structures with improved accessibility [9]. Whilst modification of ceria by incorporation of another element, such as zirconium [10] or copper [11] into the ceria lattice also improve catalyst activity for naphthalene oxidation.

Early studies on catalytic total oxidation of PAHs showed that metal-supported catalysts were more effective than metal oxide catalysts available at the time. For the metal-based catalysts Pt was identified as being more active than Pd and Ru, with Pt-supported on γ-Al_2O_3 being the most active catalyst [12,13]. Considering the identification of Pt as the most active component for naphthalene total oxidation and the high activity exhibited by nanocrystalline ceria, it may be thought that Pt supported on nanocrystalline ceria would be particularly effective. However, the addition of Pt on to nanocrystalline ceria actually suppresses the total oxidation of naphthalene [14]. These findings demonstrate that the interaction between the support and the Pt is an important factor in controlling the total oxidation of naphthalene. In a previous study of naphthalene oxidation we investigated the role of the support for a 0.5 wt.% Pt-based catalyst [15]. The supports investigated were SiO_2, γ-Al_2O_3, CeO_2, TiO_2 and SnO_2, and the SiO_2 supported catalyst was the most active, and remains one of the most active catalysts reported in the literature for naphthalene total oxidation. The metal-support interaction (MSI) of Pt and silica was relatively weak, and surface Pt existed in the metallic and oxidised states. CO chemisorption studies revealed that the Pt crystallite size was relatively large, with a low dispersion. Considering the other supports investigated, it was found that silica was the only support which formed a significant concentration of surface metallic Pt in combination with oxidised Pt. For example, Pt/CeO_2 displayed predominantly surface oxidised Pt species and was amongst the least active of the Pt catalysts investigated in the study.

Hence, the present work aims to build on our previous study of the preparation and activity of 0.5 wt.% Pt/SiO_2. In the current work both the influence of Pt loading and the variation of calcination temperature have been systematically studied and their influence on catalyst activity established.

2. Results and Discussion

The steady-state oxidation of naphthalene to CO_2 as a function of temperature over the SiO_2 supported catalysts containing varying wt.% Pt are shown in Figure 1. The blank experiment for oxidation of 100 vppm naphthalene using an empty reactor tube with a total gas flow of 50 mL min^{-1} has been measured previously [16]. Naphthalene conversion was not detectable below 350 °C, whilst it was <2% at 350 °C and increased to around 20% at 400 °C. Experiments with the silica-based catalysts in this study have been carried out up to 300 °C, and therefore conversion of naphthalene is attributed to surface initiated reactions, with the contribution from purely gas phase homogeneous reactions negligible.

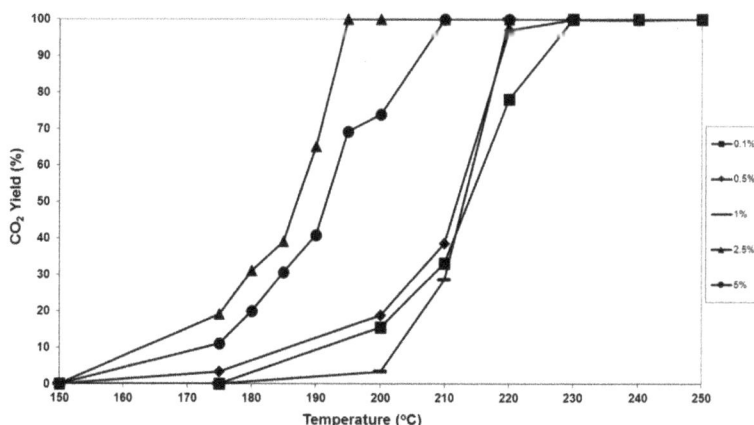

Figure 1. Steady-state conversion of naphthalene to CO_2 as a function of temperature for catalysts with varying platinum loadings on a silica support: ■ 0.1 wt.% Pt; ♦ 0.5 wt.% Pt; ▬ 1.0 wt.% Pt; ▲ 2.5 wt.% Pt; ● 5.0 wt.% Pt.

The data in Figure 1 shows the conversion of naphthalene to CO_2 rather than the naphthalene conversion. The reason for using the CO_2 yield is that naphthalene is a polycyclic aromatic molecule, which can be readily adsorbed onto the surface of the catalyst, potentially leading to over estimation of oxidation activity, especially at lower temperatures [17]. Furthermore, the use of total Naphthalene conversion may also lead to erroneous conclusions due to the possible formation of by-products [13]. Hence, the yield to CO_2 and not the total conversion of naphthalene is a better parameter to establish the total oxidation activity of the catalysts. However, with the Pt/SiO_2 catalysts used in this study the carbon balances were always close to 100%, and the conversion to CO_2 and naphthalene conversion were virtually the same. These observations indicate that selectivity to CO_2 from naphthalene oxidation was around 100%, and there was little or no formation of partially oxidized products under our reaction conditions.

The Pt loading had a significant influence on the catalyst activity for naphthalene total oxidation (Figure 1). The catalysts were not active below 150 °C, but started to show trace conversion to CO_2 at 150 °C. The catalysts containing 0.1, 0.5 and 1.0 wt.% Pt all showed broadly similar naphthalene oxidation activity, and these catalysts were significantly less active than the 2.5 and 5.0 wt.% catalysts. The 0.5 and 1.0 Pt wt.% catalysts demonstrated total (or near total) conversion of naphthalene to CO_2 around 220 °C. The 0.1 Pt wt.% catalyst, with the lowest Pt loading, exhibited some mass transport limited characteristics at higher conversion (as conversion did not rise so steeply to 100%), and total naphthalene conversion was only observed at 230 °C. This observation could be related to the higher surface area of the catalyst, as it most likely has the highest micropore volume. The most active catalyst was 2.5 wt.% Pt, followed closely by 5.0 wt.%. Total oxidation of naphthalene to CO_2 was achieved at 190 °C for 2.5 wt.% Pt, whilst it was around 210 °C for the 5.0 wt.% catalyst, as it exhibited some mass transport limitations at higher naphthalene conversion, the reasons for this are unclear at this stage. The order of catalyst activity can be summarised as:

$$2.5\,Pt/SiO_2 > 5.0\,Pt/SiO_2 > 1.0\,Pt/SiO_2 \approx 0.5\,Pt/SiO_2 > 0.1\,Pt/SiO_2 \qquad (1)$$

Table 1 shows a summary of the characterisation and activity data (T_{50} and T_{100}, representing temperatures for 50% and 100% naphthalene conversion to CO_2 respectively, for the range of silica supported catalysts with varying Pt loading. The BET surface area increased slightly as a low loading of Pt was added to the SiO_2 support, and this is associated with the process of aqueous impregnation and drying. As the loading of Pt was increased there was a decrease of the catalyst surface area, and this would be expected as a consequence of some pore filling by the deposition of the Pt. The Pt dispersion of the catalysts was relatively low, and this is consistent with previous findings for Pt/SiO_2 catalysts prepared by the same method [15]. The greatest Pt dispersion of 28% was observed for the 0.1 Pt/SiO_2 catalyst, and as Pt loading was increased there was a decrease of Pt dispersion. The catalysts containing 2.5 and 5.0 wt.% Pt had very low Pt dispersion of 4 and 3% respectively. CO chemisorption data were also used to calculate average Pt particles sizes, and in accordance with the decrease of dispersion with increasing Pt loading, average particle sizes also increased as Pt loading increased. The average Pt size increased from around 40 Å to 90 Å as Pt loading increased from 0.1 to 1.0 wt.%. When the Pt loading was increased to 2.5 and 5.0 wt.% the average Pt particle size increased significantly to around 280 Å and 740 Å respectively. The number of surface Pt sites also changed as a function of Pt loading, increasing from 0.1 wt.% to a maximum for the 1.0 wt.% Pt/SiO_2 catalyst. When the Pt loading was increased above 1 wt.% there was a decrease of the number of Pt surface sites.

It is interesting to consider the activity of the catalysts in relation to the data for Pt dispersion, average particle size and the number of Pt surface sites. There is no simple relationship between the number of Pt surface sites and total oxidation activity of naphthalene. As Pt loading increased from 0.1 to 1.0 wt.% activity increased, as did the number of surface Pt sites. However the number of surface sites for the 1.0 wt.% catalyst was significantly greater than for the 0.5 wt.% catalyst, but activity was very similar. Furthermore, the number of surface Pt sites decreased as the Pt loading increased from 1.0 to 2.5 and then 5.0 wt.%, but the naphthalene total oxidation activity increased markedly. Significantly the more active 2.5 and 5.0 wt.% catalysts have far larger average Pt particle sizes than the lower loading catalysts, indicating that relatively large Pt particle sizes are required for greater activity.

Table 1. Summary of characterisation data and naphthalene oxidation performance for Pt/SiO_2 catalysts with varying Pt loading prepared by calcination in static air at 550 °C for 6 h (ramp rate 10 °C min^{-1}).

Pt content/wt.%	BET surface area/m^2g^{-1}	Pt dispersion [a]/%	Average Pt particle size [a]/Å	No. Pt sites ($\times10^{18}$) [a]/g	Surface $Pt^0:Pt^{2+}:Pt^{4+}$ ratio [b]	Catalytic performance	
						T_{50} [c]/°C	T_{100} [c]/°C
0	431	-	-	-	-	-	-
0.1	450	28	41	0.88	Not Detected	213	230
0.5	443	19	61	2.63	1.8:1.1:1.0	211	230
1	430	12	91	3.88	1.4:2.3:1.0	212	220
2.5	436	4	276	3.44	9.1:1.6:1.0	187	195
5	417	3	736	2.38	6.2:1.2:1.0	191	210

[a] determined from CO chemisorption; [b] measured by XPS; [c] Temperature for 50 and 100% conversion of naphthalene to CO_2.

The nature of the surface Pt species, and how the relative proportions of different species may vary as a function of Pt loading, can also influence the activity of the catalyst. Accordingly the range of catalysts were characterised using X-ray photoelectron spectroscopy (XPS) and data are summarised in Table 1. For catalysts containing low loadings (<0.5 wt.%) of Pt it was not possible to analyse the Pt species present due to their very low surface concentration. For the catalysts for which it was possible to measure the Pt surface species, it was evident that Pt was present in a range of oxidation states, comprising Pt^0, Pt^{2+} and Pt^{4+}. These general findings are consistent with our previous work [15], as we identified the presence of significant concentrations of metallic Pt along with oxidised Pt on the surface of a Pt/SiO_2 catalyst. Conversely, Pt catalysts prepared by the same method, but supported on TiO_2, CeO_2 and γ-Al_2O_3 only exhibited oxidised surface Pt. The Pt/SiO_2 catalyst was considerably more active than the catalysts prepared using the other supports, suggesting that the presence of metallic Pt, possibly in combination with oxidised Pt is important for naphthalene total oxidation. For the 0.5 and 1.0 wt.% Pt/SiO_2 catalysts the majority of the Pt measured was in the +2 and +4 oxidised states, whilst for the higher loadings of 2.5 and 5.0 wt.% the majority of Pt was in the metallic state, although significant concentrations of Pt^{2+} and Pt^{4+} were also detected. The relative concentration of Pt^0 was greater for the 2.5 wt.% catalyst when compared to the 5.0 wt.% catalyst. These data indicate that the presence of metallic Pt is again important, as the most active catalyst with 2.5 wt.% had the highest concentration of surface Pt^0.

It is difficult to draw too many definitive conclusions between the catalyst structure and activity for the catalysts with different Pt loadings, as Pt dispersion, total number of surface Pt sites, Pt particle size, and the relative and absolute number of surface metallic and oxidised Pt sites all change as a function of loading. However, it is clear that the more active catalysts contain higher loadings of Pt, and these catalysts have very low Pt dispersion, relatively large Pt particles and greater surface concentrations of metallic Pt. Hence, these characteristics are required for more effective naphthalene total oxidation catalysts. In order to try and develop a stronger link between the characteristics of the Pt nanoparticles and naphthalene oxidation, future studies focusing on a range of catalysts containing between 1.0 and 2.5 wt.% Pt would be informative.

In order to try to investigate further the relationship between catalyst structure and activity for naphthalene total oxidation, a range of catalysts were prepared by varying the calcination temperature. As the 2.5 wt.% Pt catalyst was the most active from the study varying the Pt loading, this loading was selected and catalysts were prepared by calcining from 450 to 750 °C.

The steady-state oxidation of naphthalene to CO_2 as a function of temperature over the SiO_2 supported catalysts containing 2.5 wt.% Pt prepared by calcination at different temperatures are shown in Figure 2. Increasing the calcination temperature from 450 to 550 °C resulted in an increase of catalyst activity. Calcination at 550 °C produced a catalyst with optimum activity, as the calcination temperature was increased to 650 and then 750 °C there was successive decreases of catalyst activity. These trends of activity can also be seen by consideration of the T_{50} values (Table 2).

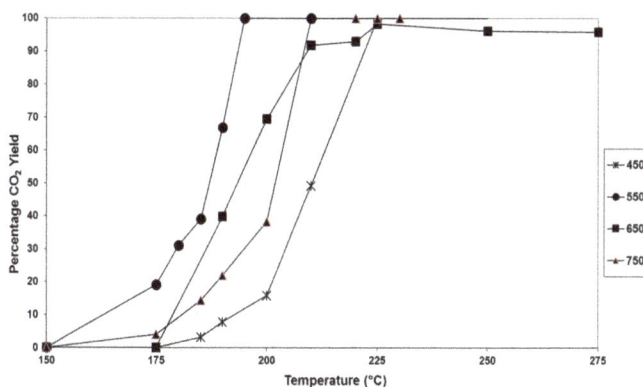

Figure 2. Steady-state conversion of naphthalene to CO_2 as a function of temperature for 2.5 wt.% Pt/SiO_2 catalyst prepared at varying calcination temperatures: $*$ 450 °C; ● 550 °C; ■ 650 °C; ▲ 750 °C.

Table 2. Summary of characterisation data and naphthalene oxidation performance for 2.5 wt.% Pt/SiO_2 catalysts calcined in static air for 6 h at varying temperatures.

Calcination temperature/°C	BET surface area/m^2g^{-1}	Pt dispersion [a]/%	Average Pt particle size [a]/Å	No. Pt sites ($\times 10^{18}$) [a]/g	Surface Pt^0:Pt^{2+}:Pt^{4+} ratio [b]	Catalytic performance	
						T_{50} [c]/°C	T_{100} [c]/°C
450	479	5	203	5.00	6.5:1.3:1.0	210	225
550	436	4	276	3.44	9.1:1.6:1.0	187	195
650	432	4	316	3.23	6.8:2.1:1.0	193	225
750	400	2	417	0.81	8.4:1.5:1.0	202	210

[a] determined from CO chemisorption; [b] measured by XPS; [c] Temperature for 50 and 100% conversion of naphthalene to CO_2.

A summary of the catalysts characteristics are presented in Table 2. The BET catalyst total surface area decreased as the calcination temperature was increased. This was an expected trend as sintering of the silica support will be more extensive at higher temperature. The Pt dispersion remained relatively low, ranging from 5% at 450 °C to 2% at 750 °C. The dispersions for the 2.5 wt.% catalysts, regardless of calcination temperature, were much lower than catalysts calcined at 550 °C with lower Pt content (Table 1). These data indicate that the dispersion of Pt using our preparation method is strongly influenced by the Pt loading and not the calcination temperature. The decrease of Pt dispersion with increasing calcination temperature also results in an increase of the calculated average Pt particle size. This observation is also consistent with the decrease of total surface area, as increasing the size of Pt particles would be expected to increase pore blocking of the silica support. The number of surface Pt sites was greatest for the catalyst calcined at 450 °C, and decreased as the calcination temperature was raised. Hence, as observed for the series of catalysts with varying Pt loading, there was no simple relationship between the number of Pt surface sites and naphthalene total oxidation activity. Rather, it

47

seems there is a more complex relationship with the number of surface sites and Pt particle size, but it is again evident that low dispersion and large Pt particles are required for more active naphthalene total oxidation catalysts.

Analysis results from XPS for catalysts prepared with varying calcination temperature are shown in Table 2 and Figure 3. There was no general simple trend between metallic Pt surface concentration and calcination temperature. In discussion above and in our previous work [15], we consider that the presence of Pt in a combination of oxidised and metallic states is important for enhanced naphthalene total oxidation. The catalyst calcined at 450 °C has the lowest concentration of Pt^0, the concentration increased to a maximum for the catalyst calcined at 550 °C, which was also more active. Therefore it is tempting to conclude that a higher Pt^0 concentration is beneficial for total oxidation, but there is also an effect of Pt dispersion influencing the number of Pt surface sites and Pt particle size. Accordingly, it is interesting to consider the Pt oxidation states for the catalysts calcined at 550 and 650 °C, as these catalysts have similar surface areas, Pt dispersion and Pt average particle sizes. The more active 550 °C calcined catalyst has around 40% more surface Pt in the metallic state, indicating that a higher concentration of Pt^0 in combination with oxidised Pt is an important factor for greater activity.

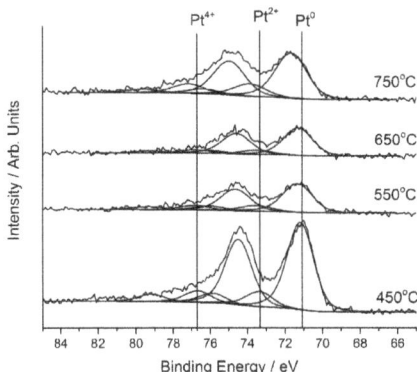

Figure 3. Pt(4f) core-level spectra for a 2.5 wt.% Pt/SiO$_2$ catalyst showing the effect of increasing calcination temperature.

The mechanism of naphthalene total oxidation over Pt-based catalysts remains a subject of debate. A kinetic study of naphthalene oxidation over 1% Pt/γ-Al$_2$O$_3$ by Zhang *et al.* showed that the overall rate of reaction could be expressed using a Langmuir-Hinshelwood kinetic model [13]. The reaction was first order with respect to naphthalene and oxygen. Shie *et al.* also investigated a similar Pt/γ-Al$_2$O$_3$ catalyst to study the total oxidation of naphthalene [18]. However, from their kinetic analysis they concluded that oxidation could best be described by an Eley-Rideal mechanism. In contrast, Radic *et al.* performed kinetic experiments for the total oxidation of the related molecule toluene, over a Pt/Al$_2$O$_3$ catalyst [19], and the kinetics were best described by the Mars van Krevelen mechanism. It was reported that the rate of oxygen chemisorption increased with increasing Pt crystallite size, and that activation energy for oxygen chemisorption also decreased with increasing Pt particle size, which resulted in an increased reaction rate. Relatively weak Pt–O bonds were reported to form on larger Pt particles. No data on Pt oxidation state was reported, but the authors believed that the Pt was metallic since their experiments were performed at temperatures where stable Pt oxide formation was not favoured. Garetto and Apesteguía observed similar behaviour for cyclopentane oxidation [20]. They reported that the turnover frequency values for cyclopentane increased with increasing Pt particle size. Pt existed predominantly as metallic particles and larger Pt crystallite sizes increased the density of active Pt–O species. A redox mechanism was postulated, the first step of which was oxidation of metallic Pt to Pt–O. The dissociative adsorption of oxygen was reported to be the rate determining step. In a time on stream experiment cyclopentane conversion increased with time. This was reported to be due to agglomeration of the Pt particles to larger ones, and hence more active Pt particles. The observations of larger Pt particles being more active, and the importance of metallic Pt are consistent with our observations for the total oxidation of naphthalene. It can be envisaged that for total oxidation adsorption of the naphthalene molecule flat on the surface would be favourable, whereas adsorption end on would more likely result in partial oxidation to products like phthalic anhydride. Hence, larger Pt particles will facilitate such preferential naphthalene adsorption, as well as facilitating O$_2$ chemisorption and increasing the density of more weakly bound Pt–O species [19,20].

The application of catalytic oxidation for VOC control requires catalysts with stable activity. The stability of the most active catalyst containing 2.5 wt.% Pt and calcined at 550 °C was investigated. Firstly the stability of the catalyst activity was studied by cycling through a series of light-off curves, during the cycle the catalyst temperature was reduced to ambient before commencing the next cycle (Figure 4). After the first cycle the catalyst activity increased, with the temperature required to achieve equivalent conversion shifting to approximately 15 °C lower. The second and third cycles showed very similar activity, at each temperature the measured activity was greater for the third cycle, but with the exception of one data point the conversion for each cycle was within experimental error. The reason for the increase of activity after the first cycle is not clear, as it was not possible to recover sufficient catalyst to perform meaningful characterisation. However, speculatively an increase in the Pt particle size, aided by the reaction temperature and the exothermic nature of naphthalene oxidation, may be important as the same effect has been shown to increase the activity of a supported Pt catalyst for cyclopentane oxidation [20].

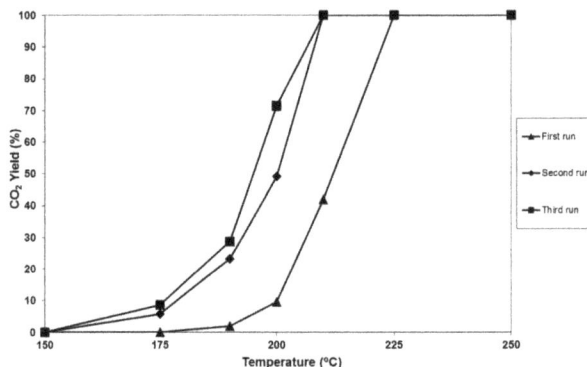

Figure 4. Conversion of naphthalene to CO_2 as a function of temperature for a 2.5 wt.% Pt/SiO_2 catalyst calcined at 550 °C used in repeat cycles: ▲ First cycle; ♦ Second cycle; ■ Third cycle.

Catalyst stability was also probed by measuring activity as a function of time-on-stream (Figure 5). Total naphthalene conversion was observed for approximately the first 1000 minutes on-stream. After this initial period there was a decrease of activity with conversion to CO_2 remaining above 90%. This result is in contrast to the increase of activity after cycling. Again it was not possible to recover sufficient catalyst after use for characterisation, and these deactivation processes are areas for further potential study.

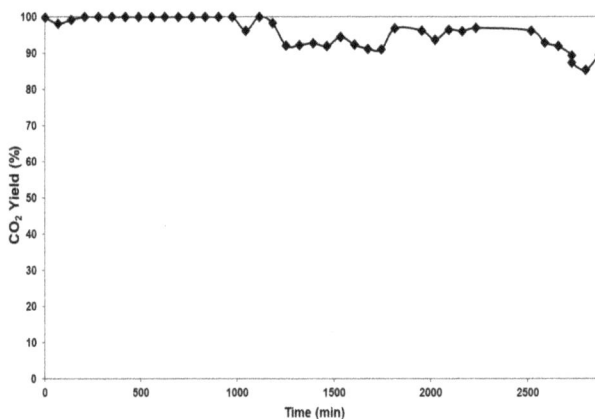

Figure 5. Conversion of naphthalene to CO_2 at 300 °C as a function of time for a 2.5 wt.% Pt/SiO_2 catalyst calcined at 550 °C.

3. Materials and Methods

3.1. Catalyst Preparation

All Pt/SiO$_2$ catalysts were prepared by an impregnation method. An appropriate amount of hexachloroplatinate(IV) hydrate (Aldrich, UK, $\geqslant$99.9%) was dissolved in distilled water (100 mL) then heated to 80 °C with stirring. The quantity of Pt was varied to produce loadings in the final catalyst ranging from 0.1 to 5.0 wt.%. SiO$_2$ (3 g, Silica 60 Å, Fisher, UK) was then added to the solution and the mixture was again stirred at 80 °C until a paste was formed, which was subsequently dried in an oven at 110 °C for 16 h, before calcination in a subsequent step. The initial calcination was performed in static air at 550 °C for 6 h (ramp rate of 10 °C min^{-1}) to produce the final catalyst. In subsequent studies the calcination temperature was varied from 450 to 750 °C for 6 h in static air with a ramp rate from ambient of 10 °C min^{-1}.

3.2. Catalyst Characterisation

Surface areas were determined by multi-point N$_2$ adsorption at −196 °C using a Micromeritics Gemini 2360 apparatus, in accordance with the BET method. CO chemisorption data were obtained using a Quantachrome ChemBET-3000 analyser. The samples were pre-treated in flowing H$_2$ (20 mL min^{-1}) at 400 °C for 1 h, and subsequently cooled to room temperature under the same gas flow. The samples were then exposed to 50 μL pulses of CO at room temperature. The concentration of analyte gas that was not chemisorbed in each pulse was analysed using a thermal conductivity detector. Once CO chemisorption uptake was determined, Pt dispersion was calculated and average Pt particle size was determined using a Pt stoichiometry of 1 with a cross-sectional area of 8 Å^2. It was assumed that the morphology was hemispherical and that 1.25 × 10^{19} Pt surface atoms were present per m^2 of Pt surface area.

X-ray photoelectron spectra were obtained using a Kratos (Manchester, UK) Axis Ultra DLD instrument fitted with a monochromatic Al source (photon energy = 1486.6 eV). Pass energies of 40 eV (high resolution scans) and 160 eV (survey spectra) were used over an analysis area of 700 μm × 300 μm. All spectra were calibrated to C (1 s) of adventitious carbon, taken to be 284.7 eV. All data was analysed and fitted with CasaXPS v2.3.17, using sensitivity factors supplied by the instrument manufacturer.

3.3. Catalyst Testing

The catalytic activity was determined using a laboratory micro-scale reactor coupled with an on-line gas chromatograph (Varian 3400) fitted with thermal conductivity and flame ionisation detectors. Reactants and products were separated using an OV-17 (naphthalene) and Carbosphere 80/100 mesh (O$_2$ and CO$_2$) packed columns. The reactor was composed of a 6.3 mm O.D. 316 stainless steel tube, which was heated using a clam-shell electric furnace. Helium and oxygen flows totalling 50 mL min^{-1} (20% O_2 content) and 100 vppm of naphthalene were used for all experiments. The catalyst was packed to a constant volume in the reactor tube between plugs of silica wool to provide a GHSV of 45,000 h^{-1}. All the lines from the naphthalene source onwards were heated by external heating tapes to prevent condensation. Activity tests were performed in the range of 100–300 °C and the catalyst temperature was controlled by a thermocouple fitted in the centre of the catalyst bed. The low concentration of naphthalene used means that the adiabatic temperature rise during reaction was relatively small and it was possible to control the catalyst bed temperature. Activity of the catalysts was measured at each temperature once steady-state operation was attained; this was established as analyses were performed until a set of 3 consistent data points were obtained. Typically the catalyst was stabilized at each temperature for 30 min to attain steady state operation. The catalyst temperature was increased to the next value and the procedure repeated.

4. Conclusions

Silica supported platinum catalysts prepared by an impregnation method are highly active for the total oxidation of naphthalene to CO_2. Over a range of platinum loadings from 0.1 to 5.0 wt.% the most active catalyst contained 2.5 wt.%. Using a 2.5 wt.% platinum loading the calcination temperature for optimum activity was 550 °C. CO chemisorption and XPS showed that low platinum dispersion to form large platinum particles, in combination with platinum in metallic and oxidised states was important for high catalyst activity. Catalyst performance improved after initial use in repeat cycles, whilst there was slight deactivation after prolonged use on-stream.

Acknowledgments: The authors would like to thank Cardiff University and Cardiff Catalysis Institute for funding.

Author Contributions: D.R.S. and S.H.T. conceived and designed the experiments; D.R.S. and D.J.M. performed the experiments; D.R.S. and D.J.M. analyzed the data; S.H.T. wrote the paper.

Conflicts of Interest: The authors declare no conflict of interest.

References and Notes

1. Levy, J.I.; Houseman, E.A.; Spengler, J.D.; Loh, P.; Ryan, L. Fine particulate matter and polycyclic aromatic hydrocarbon concentration patterns in Roxbury, Massachusetts: A community-based GIS analysis. *Environ. Health Perspect.* **2009**, *109*, 341–347.

2. *European Union PAH Position Paper ISBN92-894-2057-X*; Office for Official Publications of the European Communities: Luxembourg, Luxembourg, 2001.

3. UNO 1998 Protocol to the 1979 convention on long-range transboundary air pollution on Persistent Organic Pollutants.

4. Garcia, T.; Solsona, B.; Taylor, S.H. *The catalytic oxidation of hydrocarbon Volatile Organic Compounds, Handbook of Advanced Methods and Processes in Oxidation Catalysis: From Laboratory to Industry*; Duprez, D., Cavani, F., Eds.; Imperial College Press: London, UK, 2014; Chapter 3; pp. 51–90.

5. Ntainjua, N.E.; Taylor, S.H. The catalytic total oxidation of polycyclic aromatic hydrocarbons. *Topics Catal.* **2009**, *52*, 528–541.

6. Garcia, T.; Sellick, D.R.; Varela, E.; Vázquez, I.; Dejoz, A.; Agouram, S.; Taylor, S.H.; Solsona, B. Total oxidation of naphthalene using bulk manganese oxide catalysts. *Appl. Catal. A* **2013**, *450*, 169–177.

7. Ntainjua, E.; Solsona, B.; Garcia, T.; Taylor, S.H. Influence of preparation conditions of nano-crystalline ceria catalysts on the total oxidation of naphthalene a model Polycyclic Aromatic Hydrocarbon. *Appl. Catal. B* **2007**, *76*, 248–256.

8. Ntainjua, N.E.; Garcia, T.; Solsona, B.; Taylor, S.H. The influence of cerium to urea preparation ratio of nanocrystalline ceria catalysts for the total oxidation of naphthalene. *Catal. Today* **2008**, *137*, 373–378.

9. Puertolas, B.; Solsona, B.; Agouram, S.; Murillo, B.; Mastral, A.M.; Aranda, A.; Taylor, S.H.; Garcia, T. The catalytic performance of mesoporous cerium oxides prepared through a nanocasting route for the total oxidation of naphthalene. *Appl. Catal. B* **2010**, *93*, 395–405.

10. Sellick, D.R.; Aranda, A.; García, T.; López, J.M.; Solsona, B.; Mastral, A.M.; Morgan, D.J.; Carley, A.F.; Taylor, S.H. Influence of the preparation method on the activity of ceria zirconia mixed oxides for naphthalene total oxidation. *Appl. Catal. B* **2013**, *132–133*, 98–106.

11. Aranda, A.; Agouram, S.; López, J.M.; Mastral, A.M.; Sellick, D.R.; Solsona, B.; Taylor, S.H.; García, T. Oxygen defects: The key parameter controlling the activity and selectivity of mesoporous copper-doped ceria for the total oxidation of naphthalene. *Appl. Catal. B* **2012**, *127*, 77–88.

12. Carno, J.; Berg, M.; Sven, J. Catalytic abatement of emissions from small-scale combustion of wood: A comparison of the catalytic effect in model and real flue gases. *Fuel* **1996**, *75*, 959–965.

13. Zhang, X.W.; Shen, S.C.; Yu, L.E.; Kawi, S.; Hidajat, K.; Simon Ng, K.Y. Oxidative decomposition of naphthalene by supported metal catalysts. *Appl. Catal. A* **2003**, *250*, 341–352.

14. Ntainjua, N.E.; Davies, T.E.; Garcia, T.; Solsona, B.; Taylor, S.H. The influence of platinum addition on nano-crystalline ceria catalysts for the total oxidation of naphthalene a model polycyclic Aromatic Hydrocarbon. *Catal. Lett.* **2011**, *141*, 1732–1738.

15. Ntainjua, N.E.; Carley, A.F.; Taylor, S.H. The role of support on the performance of platinum-based catalysts for the total oxidation of Polycyclic Aromatic Hydrocarbons. *Catal. Today* **2008**, *137*, 362–366.

16. Garcia, T.; Solsona, B.; Taylor, S.H. Naphthalene total oxidation over metal oxide catalysts. *Appl. Catal. B* **2006**, *66*, 92–99.

17. Mastral, A.M.; García, T.; Callén, M.S.; Navarro, M.V.; Galbán, J. Removal of naphthalene, phenanthrene, and pyrene by sorbents from hot gas. *Environ. Sci. Technol.* **2001**, *35*, 2395–2400.

18. Shie, J.L.; Chang, C.Y.; Chen, J.H.; Tsai, W.T.; Chen, Y.H.; Chiou, C.S.; Chang, C.F. Catalytic oxidation of naphthalene using a Pt/Al$_2$O$_3$ catalyst. *Appl. Catal. B* **2005**, *58*, 289–297.

19. Radic, N.; Gribic, B.; Terlecki-Baricevic, A. Kinetics of deep oxidation of n-hexane and toluene over Pt/Al$_2$O$_3$ catalysts: Platinum crystallite size effect. *Appl. Catal. B* **2004**, *50*, 153–159.

20. Garetto, T.F.; Apesteguía, C.R. Oxidative catalytic removal of hydrocarbons over Pt/Al$_2$O$_3$ catalysts. *Catal. Today* **2000**, *62*, 189–199.

Abatement of VOCs with Alternate Adsorption and Plasma-Assisted Regeneration: A Review

Sharmin Sultana, Arne M. Vandenbroucke, Christophe Leys, Nathalie De Geyter and Rino Morent

Abstract: Energy consumption is an important concern for the removal of volatile organic compounds (VOCs) from waste air with non-thermal plasma (NTP). Although the combination of NTP with heterogeneous catalysis has shown to reduce the formation of unwanted by-products and improve the energy efficiency of the process, further optimization of these hybrid systems is still necessary to evolve to a competitive air purification technology. A newly developed innovative technique, *i.e.*, the cyclic operation of VOC adsorption and NTP-assisted regeneration has attracted growing interest of researchers due to the optimized energy consumption and cost-effectiveness. This paper reviews this new technique for the abatement of VOCs as well as for regeneration of adsorbents. In the first part, a comparison of the energy consumption between sequential and continuous treatment is given. Next, studies dealing with adsorption followed by NTP oxidation are reviewed. Particular attention is paid to the adsorption mechanisms and the regeneration of catalysts with in-plasma and post-plasma processes. Finally, the influence of critical process parameters on the adsorption and regeneration steps is summarized.

Reprinted from *Catalysts*. Cite as: Sultana, S.; Vandenbroucke, A.M.; Leys, C.; De Geyter, N.; Morent, R. Abatement of VOCs with Alternate Adsorption and Plasma-Assisted Regeneration: A Review. *Catalysts* **2015**, *5*, 718–746.

1. Introduction

Air quality issues have become a huge concern of environmental legislation as a consequence of growing awareness in our global world. Exhausts, form outdoor sources (various chemical industries, painting and printing industries, cars) [1] as well as from indoor sources [2], pollute the air with a variety of harmful substances like volatile organic compounds (VOCs) which pose a threat to human health and the environment [3]. The European legislation related to the reduction of volatile organic compounds (VOCs) emission is getting stringent due to their potential toxicity, carcinogenicity and mutagenicity [4,5]. Furthermore, they are also responsible for odor nuisance, the creation of tropospheric ozone leading to photochemical smog, the intensification of global warming and the depletion of stratospheric ozone layer [3]. Therefore, in order to minimize these adverse effects, indoor air cleaning and end-of-pipe (EOP) treatment using various techniques for the abatement of VOCs is becoming most attractive.

Non-thermal plasma (NTP) technology has attracted growing interest of scientists over the last two decades due to its distinctive characteristic of providing a highly chemical reactive environment (e^-, O^*, HO_2^*, OH^*, N_2^*, O_3, *etc.*) to decompose VOCs at ambient conditions, which repudiates the use of expensive vacuum systems [6,7]. Although other commercial pollution control techniques (thermal incineration, catalytic oxidation, adsorption, biofiltration) [8] are very efficient for the removal of VOCs, these are energetically expensive and difficult to operate in case of moderate flow rates with low VOC concentration in contrast to NTP [9–12]. In a NTP, electrons with high kinetic energies (1–10 eV) are selectively produced consuming almost all the electric energy supplied to the system instead of heating the entire gas unlike thermal and catalytic oxidation. Collisions of these energetic electrons with neutral background molecules close to room temperature, generate active species such as free radicals, metastables, ions and secondary electrons through different chemical processes such as dissociation, excitation and ionization. These active species are able to decompose pollutant molecules to less harmful products (CO_2, H_2O, HX and X_2 with X being a halogen) Additionally, the abatement of low concentrated VOCs (up to 1000 ppm), feasible in indoor air treatment application, is challenging for conventional methods because when the VOC concentrations decrease, the cost per unit pollutant treatment becomes higher in comparison to NTP. Furthermore, NTP systems have several desirable features resulting from their operating conditions, such as a quick start-up, compact system, rapid response to changes in the composition of the waste gas and non-selectivity for the treatment of waste gases with different pollutants such as particulate matters, bacteria and VOCs [13–15].

Unfortunately, industrial implementation of NTP for VOC abatement is impaired by three main bottlenecks such as poor product selectivity, formation of undesired by-products (O_3, NO_x, other VOCs, aerosols) that often increase the overall toxicity of the treated gas stream and low energy efficiency. In order to overcome these limitations, many attempts have been made and have engendered the development of a hybrid system using multiple techniques such as packed bed NTP reactors [16,17], NTP/electrostatic precipitation [18], NTP/catalysis [19,20], photocatalysis [21] and adsorption/NTP [22]. Among these VOC removal techniques, the combination of non-thermal plasma with catalysts/sorbents, *i.e.*, plasma-catalysis, is remarkably investigated during the last decade because of their improved performance such as increased energy efficiency and suppressed unwanted by-products distribution for VOC decomposition. In such a hybrid system, the catalyst can be integrated either inside (IPC-Inside Plasma Catalysis) [23] or downstream (PPC-Post Plasma

Catalysis) [24,25] of the discharge region. In both cases a synergetic effect has been reported in many studies [26–28].

Activation of catalyst by the NTP is expected to play an important role in the synergetic effect. There are different possible activation mechanisms ranging from UV, ozone, work function changes, lattice oxygen activation, heating at local spot, adsorption/desorption, electron-hole pair creation and gas-phase radicals which directly interact with adsorbed pollutants [29]. The abatement of VOCs by combined use of NTP and catalysis is not only governed by gas phase reactions but also by surface reactions on the catalyst. Indeed, by introducing a catalyst in the discharge zone, the discharge gap shortens causing an intensification of the electric field strength. This increases the electron density and mean energy, which increases the concentration of active plasma species, hence promoting high VOC removal efficiency. Furthermore, the introduction of a catalyst inside the plasma can change the nature of the discharge itself. Malik *et al.* reported that volume streamers could change into surface streamers near catalyst surfaces in case of plasma-driven catalysis IPC [30]. These surface streamers show enhanced ionization, which promotes the decomposition of VOCs. However, higher VOC mineralization is directly related with the surface reactivity of the inserted material [31]. If the catalyst has high adsorption capacity, the residence time and concentration of pollutant will increase. This will result in a higher collision probability between pollutant and active species, which will simultaneously stimulate the removal efficiency and mineralization degree. For instance, Song *et al.* have revealed that with NTP treatment, the removal rate of propane with molecular sieve is significantly higher (85%) than is the case with glass beads (17%) or a γ-Al_2O_3 beads (23%) reactor. This was attributed to the higher surface area and smaller pore size of molecular sieve compared to the latter cases, which have weaker adsorption capabilities for propane [32]. On the other hand, partial/soft oxidation of VOCs by active species generated from NTP can significantly enhance the pollutant affinity for the adsorbents [33,34]. During plasma treatment, modification of the VOCs structure can cause an increase of the molecular size of VOCs. It can also modify their chemical nature through formation of new polar functions on molecules which lead to better retention of VOCs on the adsorbents.

Both adsorption and reaction properties of catalysts are used to increase the residence time of VOCs and their reaction probability with active plasma species leading to high mineralization degree [35]. Considering the importance of adsorption in plasma-catalysis, investigation of NTP with the adsorption process has drawn researchers' interest during the last decade. In 1999, an innovative approach, which increased the ratio between deposited energy and treated molecules, was suggested by Ogata *et al.* [31,36]. According to the authors, by using a cycled system of adsorption and NTP discharge, an energy efficient technique can be achieved when operated at optimized intervals. Besides, in comparison to the continuous treatment, the sequential treatment, *i.e.*, adsorption followed by NTP oxidation, is expected to treat waste streams over a wide range of VOC concentrations. In many chemical facilities there is a large variation of VOC concentrations during the operation period. The adsorption step makes this cycled system effective, since energy is not continuously deposited in the treated gas stream, independently of VOC concentration variations. Thus, optimized energy consumption is achieved with this sequential treatment. Furthermore, suppressed formation of harmful inorganic by-products such as CO, O_3 as well as organic by-products is another benefit of this method. This could also be an effective regeneration technique compared to conventional techniques like thermal treatment, which often leads to catalytic deactivation owing to particle agglomerations on catalyst surfaces and poses high a cost. Recently Thevenet *et al.* have partly reviewed the sequential treatment in their publication [37].

This paper presents an intensive literature review dealing with the abatement of VOCs and regeneration of saturated adsorbents with NTP. In the first part the concept of sequential treatment is discussed. Next, the current status and recent achievements of this new technique for the abatement of VOCs is reviewed. The present understanding of the mechanisms involved in adsorption and oxidation pathways during the regeneration process found in literature are summarized. Finally, the influence of critical process parameters on the adsorption and regeneration step is discussed in detail.

2. Sequential Treatment (Adsorption Followed by NTP Oxidation)

A schematic diagram of a continuous flow process is shown in Figure 1. The IPC process is a single stage process where the catalyst is exposed to the active discharge region. In a dry air ozonolysis (PPC) reactor, the polluted gas stream passes through the plasma discharge followed by the catalytic reactor downstream. In both cases, plasma is ignited permanently and the catalyst material can be incorporated into the hybrid reactor in different ways, for instance, as a coating on reactor walls or electrodes, as a packed bed (powder, pellets, coated fibers or porous solid foam) or a layer of catalyst material (usually powder/pellets) [25].

The conceptual diagram of sequential treatment or cycled storage-discharge (CSD) plasma catalytic process for VOC abatement is shown in Figure 2. Firstly, in the adsorption or storage stage, the polluted air stream passes through the catalyst/sorbent bed so that VOCs are removed from the gas phase by adsorption on the catalyst/sorbent until saturation occurs. Next, the saturated catalyst bed is exposed to plasma to oxidize adsorbed VOCs in the regeneration or discharge stage. During this step, the polluted gas stream can be diverted to a fresh catalyst/sorbent bed to ensure continuous operation. In this cyclic approach, surface reactions between adsorbed VOCs/intermediates and plasma generated species such as O_3 or

O radicals on the sorbent are of great importance for the oxidative decomposition of VOCs.

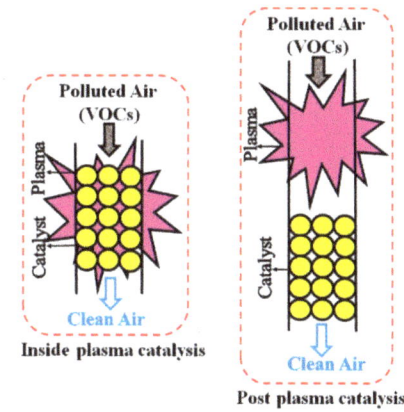

Figure 1. Schematic diagram of continuous plasma-catalysis process.

The energy cost E_c (kWh/m^3) for a sequential treatment is defined as followed:

$$E_c = \frac{P_{discharge} \times t_2}{F_1 \times t_1} \qquad (1)$$

where F_1 is the flow rate (m^3/h) during the storage stage and t_1 and t_2 are the storage and discharge period (h), respectively. In this case, a long storage and a short discharge period are key to achieving low energy cost. The difference in the definition of energy cost between sequential treatment and a continuous flow process is attributed to the energy deposition method. In a continuous flow process, energy density (J/L) *i.e.*, the energy deposited per unit volume of process gas, is required to calculate the energy cost as energy is consumed by the discharge to treat the polluted gas stream. Contrary, in sequential treatment, energy is only alternately deposited during the discharge stage (see Figure 2b). Therefore, the energy consumption for sequential treatment is substantially reduced compared to continuous plasma-catalysis processes. For instance, Sivachandiran *et al.* reported that the energy cost for Isopropanol (IPA-CH$_3$CHOHCH$_3$) removal using a Mn$_x$O$_y$ packed bed NTP reactor, is 14.5 times less with the sequential approach compared to a continuous treatment process [38]. Furthermore, to achieve the same mineralization degree, sequential treatment consumes 10 times less energy.

Since the sequential treatment has gained increased interest over the last decade, an overview of published papers will be given that help to understand the degradation pathway and to optimize the process parameters. Table 1 summarizes the results regarding the adsorption and regeneration stage of these publications found in literature.

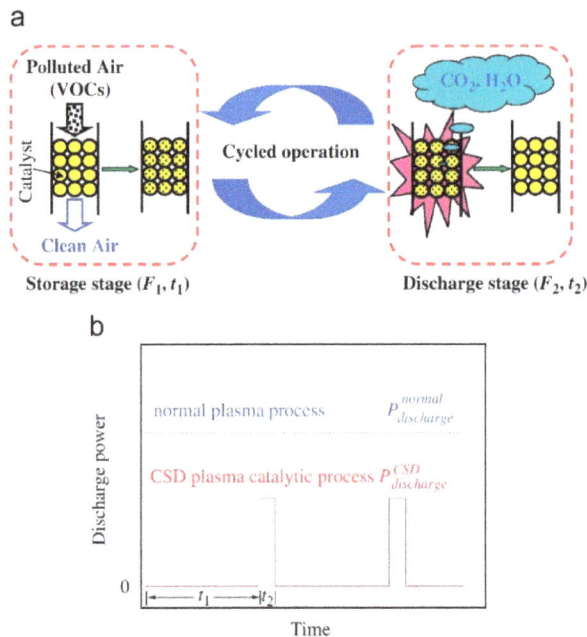

Figure 2. (**a**) Conceptual diagram of sequential treatment, (**b**) difference in discharge power between cycled storage-discharge (CSD) and a continuous plasma catalytic flow process. Reprinted from Ref. [39], with permission from Elsevier.

Zhao *et al.* have investigated the removal of formaldehyde (HCHO) from air using a CSD plasma catalytic process over AgCu/HZSM-5 (HZSM-5, $SiO_2/Al_2O_3 = 360$) catalyst for indoor air purification [39]. Their system combined extremely low energy cost with excellent humidity tolerance while no secondary pollution was detected. In order to achieve 100% conversion of HCHO to CO_2, almost the same discharge period ($t_2 = 10$ min) was required for different storage periods ($t_1 = 100, 300, 600$ min). This proves that long storage periods can be employed to achieve low energy cost. It is reported that the energy cost to remove 6.3 ppm of HCHO from humid air reaches 1.9×10^{-3} kWh/m^3. This cost could be further lowered to 10^{-5}–10^{-4} kWh/m^3 by increasing the storage period to purify diluted HCHO (several hundred ppb) in a typical indoor air environment.

Almost complete oxidation of benzene (C_6H_6) as well as extremely low energy cost were also achieved by using an Ag/HZSM-5 catalyst-packed dielectric barrier discharge (DBD) in sequential treatment [40]. The energy cost for sequential treatment of air containing 4.7 ppm benzene was 3.7×10^{-3} kWh/m^3 while 99.8%

CO$_2$ selectivity was achieved. The catalytic effect of Ag to promote CO oxidation to CO$_2$ and the strong adsorption of Ag with benzene through π-complexation are found to be responsible for this high selectivity [41].

The effect of different catalysts and reactor configurations on plasma catalytic oxidation of stored benzene was also investigated by the same group [42]. The performance of plasma-catalytic oxidation of stored benzene on different metal (Ce, CO, Ag, Mn, Fe, Ni, Cu, Zn) loaded HZSM-5 catalysts was evaluated in terms of carbon balance (B$_c$) and CO$_2$ selectivity (S$_{CO_2}$). Experiments revealed that 0.8 wt.% Ag/HZSM-5 catalyst could significantly improve plasma catalytic oxidation of stored benzene to CO$_2$ ($X_{(C_6H_6)_s} \rightarrow CO_2 \approx 100\%$) in a very short discharge time while the formation of unwanted by-products was almost completely suppressed (Figure 3). However, further increase of Ag loading decreased the carbon balance. Regarding the effect of reactor configuration, both carbon balance and CO$_2$ selectivity reached almost 100% with an IPC reactor (C) within 10 min (Figure 3). In contrast, with a PPC reactor (A and B) the stored benzene was not able to be completely oxidized even after a long discharge time.

Wang *et al.* have tested various metal loaded zeolites (Ag/HZSM-5, Mn/HZSM-5, Ce/HZSM-5, Ag-Mn/HZSM-5) to remove low concentration of toluene (C$_6$H$_5$CH$_3$) by intermittent use of adsorption and non-thermal plasma regeneration [43]. In this investigation, a link tooth wheel-cylinder DC plasma reactor is placed upstream of the adsorption/catalyst reactor, which is supported by a glass sieve plate. It is suggested that owing to the unique characteristics of the orbitals of Ag or Ag$^+$, normal σ bonds to carbon as well as bonds with unsaturated hydrocarbons can be formed in a non-classical manner [44], leading to higher adsorption capacity of Ag loaded zeolites (Ag/HZSM-5, Ag-Mn/HZSM-5). Similar to previous studies, reduced energy consumption (2.2×10^{-3} KWh/m^3) is also reported which is related to high adsorption capacity. Experiments reveal that the catalytic activity for toluene conversion is in the order of Ag-Mn/HZSM-5>Mn/HZSM-5>Ag/HZSM-5>Ce-Mn/HZSM-5>Ce/HZSM-5 which is in accordance with their O$_3$ decomposition ability. Furthermore, it is widely known that oxidation of toluene is significantly affected by lattice oxygen of manganese oxides. Moreover, the ratios of lattice oxygen to surface adsorbed oxygen on the Mn catalysts is enhanced with appropriate Ag loading [45], which explained the high activity and CO$_2$ selectivity (99.9%) of Ag-Mn/HZSM-5 for toluene decomposition.

Mok *et al.* have tested a γ-Al$_2$O$_3$ packed DBD reactor for the treatment of toluene in a sequential approach [46]. The removal efficiency reached 71.4% while ozone was the only by-product of toluene oxidation besides CO and CO$_2$. The energy yield for the treatment of toluene was reported to be 41.2 J/μmol. Experiments revealed that higher discharge power favored formation of CO and CO$_2$ at shorter discharge period.

Table 1. Overview of published papers on sequential treatment.

	Adsorption					Regeneration/NTP Oxidation							
Target pollutant	Catalyst/Adsorbent	Specific surface area (m²/g)	Carrier gas and flow rate (mL/min)	Concentration (ppm)		Plasma reactor type	Carrier Gas and flow rate (mL/min)	P_{dis} (W) SIE (J/L)	Time t (min)	Maximum removal efficiency (%)	Mineralization rate m (%) CO₂ yield c (%)	E_c (kWh/m³)	Ref.
Acetone (C₃H₆O)	TiO₂	38 ± 3	dry air 1000	180		DBD/IPC / DBD/IPC + TPD	dry air 1000	0.33 W	$t_1 = 103$ $t_2 = 30$	27 / 91 [a]	$m = 12$ $c = 11$ / $m = 52$ [a]	-	[47]
Benzene	Ag/HZSM-5	334	80% N₂ + 20% O₂ (50% RH) 600	4.7		DBD/IPC	O₂ 60	4.7 W	$t_1 = 840$ $t_2 = 24$	~100	$c = 99.8$	3.7×10^{-3}	[40]
Toluene	Ag/HZSM-5	-	Air 3000	3		link tooth wheel cylin-der/PPC	synthetic air (40 ± 5% RH) 1000	-	-	62 [a]	$c > 90$	2.2×10^{-3}	[43]
	Mn/HZSM-5									69			
	Ce/HZSM-5									56 [a]			
	Ce-Mn/HZSM-5									93			
	Ag-Mn/HZSM-5									70 [a]	$c = 99.9$		
Formalde-hyde	Ag3.6 wt%-Cu(2.1 wt%)/HZSM-5	229	80% N₂ + 20% O₂ (50% RH) 300	6.8		DBD/IPC	O₂ 60	2.3 W	$t_1 = 690$ $t_2 = 10$	100	$c > 99$	1.9×10^{-3}	[39]
Isopropanol	TiO₂	38 ± 3	Air (50% RH) 1000	163		DBD/IPC	air (50% RH) 1000	-	$t_1 = 1$ 30 $t_2 \approx 60$	91	-	-	[35]
Benzene	HZSM-5	328	80% N₂ + 20% O₂ (50% RH) 600	4.7		DBD/IPC	O₂ 60	4.7 W	$t_1 = 60$ $t_2 = 9$	100	$c = 89$ [a]	-	[42]
	Ag(0.8 wt%)/HZSM-5	334								100	$c = 100$		
	Ag(1.9 wt%)/HZSM-5	329								92 [a]	$c = 100$		
	Ag(4.2 wt%)/HZSM-5	306								63	$c = 100$		

Table 1. *Cont.*

	Adsorption				Regeneration/NTP Oxidation							
Target pollutant	Catalyst/Adsorbent	Specific surface area (m²/g)	Carrier gas and flow rate (mL/min)	Concentration (ppm)	Plasma reactor type	Carrier Gas and flow rate (mL/min)	P_{dis} (W) SIE (J/L)	Time t (min)	Maximum removal efficiency (%)	Mineralization rate m (%) CO₂ yield c (%)	E_c (kWh/m³)	Ref.
Benzene	Ag(1 wt.%)/TiO₂	≤68	air (50% PP) 4000–5000	200	DBD/IPC	O₂ 5000–8000	169 J/L	-	100 [a]	-	-	[29]
	Ag(4 wt.%)/TiO₂		Air (60% PP)				136 J/L		75 [a]	c = 80 [a]		
	0.5%Ag/γ-Al₂O₃	≤210					160 J/L		90 [a]	c = 75 [a]		
	H-Y zeolite		Air (80% PP)10000				140 J/L		100	c ~64 [a]		
	Ag(2 wt.%)/H-Y zeolite	≤520					160 J/L		100	c ~73 [a]		
Formaldehyde	mineral granulate (MP 5)	-	N₂ 176	99	DBD/IPC	N₂ 57	2.2 W	$t_1 = 250$ $t_2 = 4$	94	$m = 38.7$ [b] $c = 26.5$ [b]	-	[48]
Acetaldehyde (CH₃CHO)	α-Al₂O₃	14	95% N₂ + 5% O₂ 100	1000	corona discharge/IPC	95% N₂ + 5% O₂ 100	1 W	-	25	-	-	[49]
Toluene	γ-Al₂O₃	237	N₂ 2500	-	DBD/IPC	O₂ 2500	88 W	$t_1 = 190$ $t_2 = 70$	71.4	-	41.2 J/µmol	[46]
Isopropanol	Mn ₓOᵧ	16 ± 2	dry air 1000	165	DBD/IPC	dry air 1000	0.82 ± 0.02 W	$t_1 = 6$ $t_2 = 30$	66	m = 56 c = 44	-	[50]
					DBD/IPC + TPD				96	m = 84 c = 72		
					DBD/PPC		0.42 ± 0.02 W		41	m = 27 c = 21		
					DBD/PPC + TPD				84	m = 57 c = 51		

63

Table 1. Cont.

	Adsorption					Regeneration/NTP Oxidation							
Target pollutant	Catalyst/Adsorbent	Specific surface area (m^2/g)	Carrier gas and flow rate (mL/min)	Concentration (ppm)		Plasma reactor type	Carrier Gas and flow rate (mL/min)	P_{dis} (W) SIE (J/L)	Time t (min)	Maximum removal efficiency (%)	Mineralization rate m (%) CO$_2$ yield c (%)	E_c (kWh/m³)	Ref.
Isopropanol	TiO$_2$	45	dry air 750	98		DBD/PPC	dry air	68.2 mW	t_1 = 130 t_2 = 70	24	$m = 2$ $c = 1.7$[b]	-	[51]
Acetone				175				72 mW		$m = 6$			
Acetaldehyde	fibrous activated carbon textile	26.32[b]	air 100	200		ac-neon transformer	-	6 J/cm²	-	100	-	-	[22]
Toluene	zeolites	-	air 150000	10-120		DBD/IPC	air 150,000	89 W	-	90-39	-	2.6-13 g/kWh	[52]

TPD—Temperature programmed desorption (296-673 K, 0.4 K/s); PP—Partial pressure of O$_2$, RH- Relative humidity; [a] Approximate value extracted from graphs; [b] Calculated value from equations.

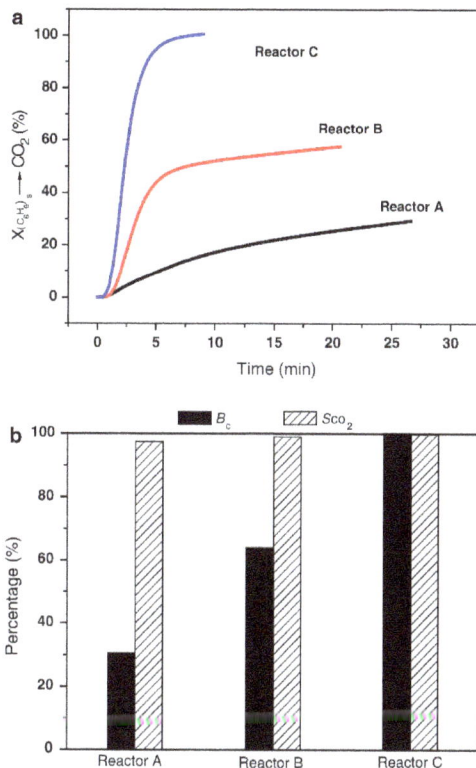

Figure 3. The effect of plasma catalytic reactor configurations on removal of stored benzene over 0.8 wt.% Ag/HZSM-5 catalysts (**a**) the conversion of stored-benzene to CO_2, (**b**) carbon balance and CO_2 selectivity (storage stage with 4.7 ppm benzene, 50% RH (25 °C), 600 mL/min flow rate of simulated air, $t_1 = 1$ h; discharge stage: 60 mL/min O_2, $P = 4.7$ W). Reprinted from Ref. [42], with permission from Elsevier.

Kim *et al.* have compared the decomposition of benzene by using flow-type plasma-driven catalysis (IPC) and a cycled system [9]. In the flow type IPC reactor, the formation of N_xO_y is unavoidable and is correlated with the increase of the conversion of VOCs. In contrast, oxygen plasma completely oxidized the adsorbed benzene on 2.0 wt.% Ag/TiO$_2$ to CO_2 in cycled system, which is impossible with the conventional NTP alone process or the flow-type IPC system.

O_2 plasma treatment as well as O_3 injection were found to be effective for the regeneration of deactivated gold supported TiO$_2$ surface after exposure to 100 ppm toluene or propylene [53]. IPC configuration showed better oxidation performance of adsorbed toluene than PPC configuration. Adsorbed toluene or propylene was preferentially decomposed to CO_2 in both regeneration methods. In case of PPC configuration, direct reaction between gas phase O_3 and adsorbed toluene instead of desorbed toluene is suggested by the authors as a possible pathway for toluene decomposition.

In order to understand the role of adsorption on the removal of acetaldehyde, Klett *et al.* have used sequential treatment [49]. A wire to cylinder configuration packed with α-Al_2O_3 pellets was used to adsorb 1000 ppm CH_3CHO during the adsorption process. After saturation of the adsorbent it was treated with less than 1 W plasma regenerating only 25% of the saturated adsorbent. In order to analyze the role of surface reactivity on the possible decomposition pathway, adsorption of multicomponent mixtures of CH_3CHO, CO, CO_2 and O_3 was carried out and the interactions existing between different compounds and the catalyst surface have been identified. Authors have also used diffuse reflectance infrared Fourier transform spectroscopy (DRIFTS) and have detected several intermediates (acetate, formate) during the adsorption step. During the regeneration of saturated alumina with plasma, the DRIFT study revealed both the presence of acetaldehyde and some intermediates on the surface, which are incorporated in the decomposition reaction. Adsorbed oxygen species, resulting from O_3 decomposition on the surface, lead to CH_3CHO decomposition and CO oxidation to CO_2. Clearly, more acetaldehyde was decomposed to CO and CO_2 due to surface reactions than in the gas phase [49].

The regeneration of IPA and acetone saturated TiO_2 surface under O_3 flow produced by NTP (PPC) was investigated by Barakat *et al.* [51]. To elaborate the oxidation mechanism of IPA and acetone, simultaneous analysis of the gas phase and absorbed phase was carried out with Fourier transform infrared spectroscopy (FTIR) and DRIFTS, respectively. The adsorption of IPA on TiO_2 surface leads to the formation of three surface species. The formation of monodentate isopropoxy groups and surface hydroxyl species indicates dissociative adsorption of IPA through Equation (2) [35], where $S_{(1)}$, $S_{(2)}$ represent two different adsorption sites. Strongly bonded IPA species on surface Lewis acid sites ($Ti^{+\delta}$) and weakly hydrogen bonded IPA species on surface basic sites ($Ti^{+\delta}$) are the result of non-dissociative adsorption of IPA on TiO_2 surface. Regarding IPA oxidation, authors reported that O_3 was simultaneously adsorbed on TiO_2 strong Lewis acid sites and on adsorbed IPA (0.2 O_3 molecule per one IPA molecule). Experiments revealed that independently of the adsorption modes, dissociated and non-dissociated IPA on TiO_2 showed the same reactivity with ozone. Only 2% mineralization was achieved (since seven O_3 molecules are required to release one CO or CO_2 in the gas phase) and 22% of the irreversibly adsorbed IPA was desorbed as intermediate by-product acetone. Agreeing with Arsac *et al.* [54], the authors proposed that acetone, produced from oxidation of adsorbed IPA on $S_{(1)}$ sites, could not remain adsorbed on similar adsorption sites (due to competitive adsorption phenomena between IPA and acetone). Acetone either rapidly desorbed to the gas phase if no surface sites denoted by $S_{(2)}$ were available or diffused toward an available $S_{(2)}$ site. Noting that only a small fraction of TiO_2 surface sites $S_{(2)}$ are specific to the acetone adsorption (there is no competitive chemisorption with IPA). For IPA, literature data [54–56] strongly suggest that acetone is the single route to produce CO_2 and H_2O. Experiments have revealed that when the TiO_2 surface is fully saturated with IPA, small amounts of $S_{(2)}$ sites are accessible for irreversibly adsorbed acetone and its subsequent oxidation into CO_2 [51]. Since, for mineralization, adsorbed IPA needs to go through $S_{(2)}$ adsorbed acetone, oxidation of IPA and acetone into CO_2 have the same limiting step.

$$CH_3CHOHCH_3 + S_{(1)} + S_{(2)} \rightarrow S_{(1)}CH_3CHOCH_3^{ads} + S_{(2)} - H^{ads} \qquad (2)$$

The regeneration of acetone adsorbed TiO_2 surface has also been conducted with an IPC process by Sivachandiran *et al.* [47]. The NTP regeneration is, however, limited to 30%. During the adsorption stage, 30% of adsorbed acetone is converted into adsorbed mesityl oxides by surface aldolization. These mesityl oxides are fragmented into carboxylic acid (e.g., formic acid) by NTP treatment, which was also evidenced by Barakat *et al.* [51]. It was confirmed that formic acid was poorly sensitive to NTP treatment but directly converted into CO_2 under successive thermal treatment. Hence, enhanced mineralization can be achieved using the NTP technique followed by thermal treatment [47].

Recently, three different methods have been used by the same group for regeneration of an IPA saturated Mn_xO_y surface [50]. In order to emphasize the proficiency of NTP treatment compared to other regeneration methods, the total carbon mass balances obtained by each method were reported and are shown in Figure 4. Although 94% regeneration efficiency has been achieved with dry air thermal treatment (DTT), 57% of carbon mass balance is accounted for molecularly by desorbed IPA and acetone, which is considered as a drawback. Dry air ozonolysis (PPC) achieved the lowest regeneration efficiency (41%) with 23% CO_x and 12% acetone contribution. *In-situ* NTP (IPC) treatment is considered as the most efficient method to obtain high mineralization and low VOCs desorption, even though 66% of irreversibly adsorbed IPA has been removed. A possible oxidation/mineralization pathway for adsorbed IPA on Mn_xO_y surface by PPC-DTT and IPC treatment was also proposed in Equations (3–7) and (8–10), respectively. Acetaldehyde and isopropyl esters of formic and acetic acids are the main identified intermediate species in both PPC and DTT methods. This indicates the dependency of the adsorbed IPA decomposition pathway on IPA adsorption modes rather than the regeneration methods. In case of IPC treatment, acetone is the main intermediate species. As reported in [51], IPC treatment prior to thermal treatment significantly improves regeneration efficiency and mineralization degree.

Sivachandiran *et al.* have also studied the influence of air humidity on IPA adsorption on TiO_2 surface and surface plasma regeneration efficiency [35]. During this investigation, only 36% regeneration efficiency has been obtained with dry air NTP treatment. Dry air NTP assisted regeneration of IPA saturated Mn_xO_y showed better performance in terms of total carbon mass balance and CO_x selectivity compared to TiO_2 surface [50]. The superior removal/regeneration efficiency of Mn_xO_y metal oxide is ascribed to the higher ability to decompose ozone. These results revealed the importance of surface reactivity of catalyst placed inside the NTP discharge zone.

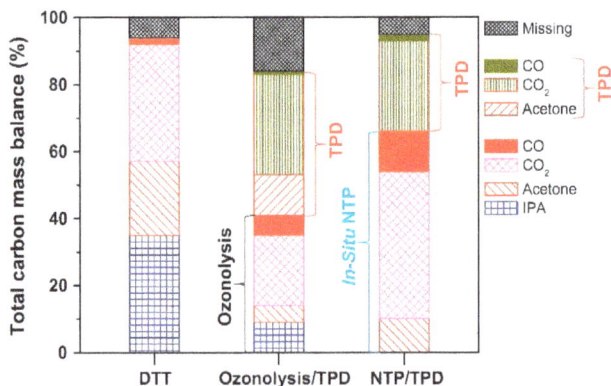

Figure 4. Comparison of adsorbed isopropanol (IPA) mineralization on Mn_xO_y surface using three different regeneration methods. Reprinted from Ref. [50], with permission from Elsevier.

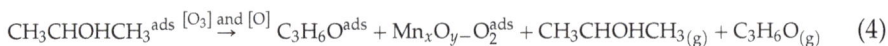

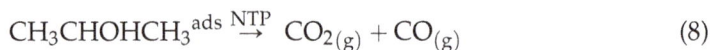

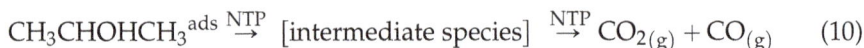

$$O_3 + Mn_xO_y \overset{296\,K}{\rightarrow} Mn_xO_y\text{–}O_2^{ads} + [O] \tag{3}$$

$$CH_3CHOHCH_3{}^{ads} \overset{[O_3]\ and\ [O]}{\rightarrow} C_3H_6O^{ads} + Mn_xO_y\text{–}O_2^{ads} + CH_3CHOHCH_{3(g)} + C_3H_6O_{(g)} \tag{4}$$

$$C_3H_6O^{ads} + Mn_xO_y\text{–}O_2^{ads} \overset{[O]}{\rightarrow} [\text{intermediate species}] + CO_{2(g)} + CO_{(g)} + CO_3^{-ads} + HCOO^{-ads} \tag{5}$$

$$CH_3CHOHCH_3{}^{ads} \overset{[O]}{\rightarrow} [\text{intermediate species}] + CO_{2(g)} + CO_{(g)} + CO_3^{-ads} + HCOO^{-ads} \tag{6}$$

$$[\text{intermediate species}] + CO_3^{-ads} + HCOO^{-ads} \overset{TPD}{\rightarrow} C_3H_6O_{(g)} + CO_{2(g)} + CO_{(g)} \tag{7}$$

$$CH_3CHOHCH_3{}^{ads} \overset{NTP}{\rightarrow} CO_{2(g)} + CO_{(g)} \tag{8}$$

$$CH_3CHOHCH_3{}^{ads} \overset{NTP}{\rightarrow} C_3H_6O^{ads} \overset{NTP}{\rightarrow} CO_{2(g)} + CO_{(g)} \tag{9}$$

$$CH_3CHOHCH_3{}^{ads} \overset{NTP}{\rightarrow} [\text{intermediate species}] \overset{NTP}{\rightarrow} CO_{2(g)} + CO_{(g)} \tag{10}$$

A new hybrid approach consisting of a concentration technique, followed by surface discharge plasma treatment is reported by Yamamoto *et al.* [57]. The process of alternate adsorption and desorption was carried out in the concentration technique where a molecular sieve was used as adsorbent and thermal heat was used to desorb toluene. The purpose of using this concentration technique is to convert low concentrated high flow rate flue gases into a low flow rate gases with high concentrations of VOCs. As a result, the size, energy consumption and operating costs of such cyclic systems are greatly reduced. Experiments revealed that more than 90% toluene decomposition efficiency was achieved by using two surface discharge units at 25 W while the energy efficiency was 34.2 g/kWh. On the other hand, the energy efficiency for continuous plasma treatment was only 1.97 g/kWh. These findings show that the consecutive process of adsorption, desorption and plasma decomposition requires 17 times less energy than the continuous plasma treatment.

The studies mentioned above mainly focus on the adsorbed VOCs oxidation performance by plasma without gas circulation. Few studies have, however, been conducted on plasma-catalytic oxidation performance with a gas circulation. For instance, investigation of plasma-catalytic oxidation of adsorbed toluene using gas circulation with MnO_x and AgO_x catalyst have been reported by Dang et al. [58]. Experiments revealed that when MnO_x/γ-Al_2O_3 is used, continuous cycle mode exhibits better CO_x yield (11% higher) than intermittent cycle mode due to the enhanced utilization of reactive species by gas circulation. It is also observed that CO_2 selectivity of MnO_x/γ-Al_2O_3 and AgO_x/γ-Al_2O_3 catalysts were both close to 100% after a 60 min oxygen plasma treatment.

A pilot-scale test of a toluene oxidation system was developed using an adsorbent (zeolite pellets) and a NTP (surface discharge) with gas circulation by Kuroki et al. [59]. When 1 mL of toluene was oxidized for 90 min, the conversion ratio of toluene to CO_x was 88% wherein 94% CO_2 selectivity was achieved. During 6 mL toluene oxidation for 150 min, these values were decreased to 44% and 89%, respectively. However, the authors reported that the conversion rate of toluene to CO_x and the energy efficiency of the toluene to CO_x conversion were increased with the amount of adsorbed toluene. By the same research group, a xylene decomposition system was also investigated using different NTP plasma sources (60-Hz neon transformer and inverter-type neon transformer) with gas circulation [60]. Inverter-type neon transformer showed better performance (higher conversion ratio of xylene to CO_x and higher energy efficiency) than the 60-Hz neon transformer. However, unstable operation and large amounts of NO_x production makes the inverter-type neon transformer less suitable. A conversion ratio of adsorbed xylene to CO_x of 43% was reached at 60 min and 90% CO_2 selectivity was achieved with the 60-Hz neon transformer.

Another cyclic operation of adsorption and DBD treatment was demonstrated by Yamagata et al. for the abatement of diluted toluene (10–120 ppm) from air [52,61]. Experiments were carried out with and without gas circulation. For the latter case, enhanced decomposition efficiency was obtained with insertion of the honeycomb zeolites in the DBD reactor compared to the plasma alone system. It is suggested that an increase of the residence time due to adsorption of toluene on the zeolites honeycomb sheet is responsible for the enhanced performance. Under gas circulation with 150 L/min flow rate, the decomposition ratio was 70% to 93% while the energy efficiency was 19 to 8.4 g/kWh. Successful regeneration of zeolites honeycomb sheet was also reported with this sequential treatment.

3. Critical Process Parameters for VOC Adsorption

The most important parameters for the selection of an appropriate adsorbent are capacity, selectivity, regenerability and cost. Adsorption capacity is the most prominent characteristic of an adsorbent and is defined as the amount of adsorbate accumulated on the adsorbent surface, per unit mass or volume of the adsorbent. It is evident that better adsorption capacity assists to reduce the overall energy consumption of the process [39]. In most literature, adsorption capacity is expressed by means of an adsorption isotherm, which plots the loading/capacity as a function of concentration at constant temperature. Adsorption isotherms can provide valuable information about the surface chemistry, the fundamentals involved in the adsorption process and estimation of the surface area, pore volume and pore size. Common adsorbents are inorganic materials such as alumina (Al_2O_3), silica gel, zeolites and organic materials such as activated carbon (AC) and polymers. In the following sections, the parameters that affect the adsorption process will be discussed in more detail.

3.1. Physical Properties of Sorbent/Catalyst

The adsorption capacity greatly depends on the surface roughness quantified by the specific surface area (m^2/g). By increasing the specific surface area, the adsorption capacity will be improved. Activated carbon, alumina and silica gel are considered as excellent adsorbents owing to their highly porous structures and large surface areas. Surface area of a powdered adsorbent depends upon its particle size: the particle size is inversely proportional to the specific surface area. Thus, finely divided metals such as nickel and platinum provide a large surface area and are ideal adsorbents. VOC adsorption capacity also depends on material pore volume, pore size, pore size distribution and diffusion rate, cation exchange capacity, pH and surface functional groups [62–64].

The mechanism of adsorption is dependent upon the size of the VOC molecules in comparison with the pore diameter of the adsorbent due to the energetic interactions between the chosen adsorbate and the pores [65]. It has been shown that the adsorption energy depends on the size of pores in adsorbents [66]. The enlargement of pore size reduces the overlapping potential for adsorbate between pore walls leading to lower adsorption energy [67]. Huang *et al.* have demonstrated that the capacity for VOCs adsorption on ACs is controlled by characteristics adsorption energy, which is inversely proportional to the local pore size [68]. VOC molecules tend to adsorb most strongly in areas where the pore diameter of the adsorbent is close to the molecular diameter of the VOCs. For instance, in a silicalite ZSM-5 with narrow pores of 5.5 Å diameter, the methyl ethyl ketone (MEK) molecules with a kinetic diameter of 5.2 Å interact strongly with the channel walls, whereas this interaction is much smaller in large pores of aluminosilicates such as de-aluminated faujasite Y (Fau-Y-7.4 Å) [69]. The effect of the porosity and the surface chemistry of ACs on the adsorption capacity of low concentrated VOCs (benzene and toluene) has been investigated by Lillo-Rodenas *et al.* [70,71]. Experiments revealed that in order to maximize the adsorption capacities of diluted VOCs, an AC surface discharge combined with a large volume of narrow micropores (*i.e.*, pore size below 0.7 nm) and reduced surface oxygen groups are desired.

3.2. Nature of VOCs

The amount of adsorbate depends on the nature of VOCs. In general, the higher its critical temperature or van der Waals' force of attraction, the more readily it will adsorb [64]. This is only valid for physisorption, since in this case the adsorbate physically adsorbs on the surface of adsorbents as a result of van der Waals interactions. Generally three different types of interaction exist among molecules, which contribute to the van der Waal's forces. These are instantaneous dipole-induced dipole interactions, dipole-induced dipole interactions between polar and neutral nonpolar molecules and dipole-dipole interactions between polar molecules. Furthermore, the strength of van der Waals' forces typically depends on three properties, *i.e.*, molecule size, surface area and polarity. For instance, the surface of AC is basically nonpolar and will weakly interact with polar VOC molecules such as acetone [68]. Since chemisorption involves a chemical reaction between the adsorbent and the adsorbate, new types of electronic bonds (ionic or covalent) are created. Due to specificity, the nature of chemisorption depends on the chemical nature of the VOC and the surface morphology.

VOCs vapor pressure or boiling point is generally used to quantify the intermolecular interactions instead of their polarity. Polar compounds have higher boiling points or lower vapor pressures than non-polar compounds. VOCs, which have a high molecular weight and high boiling point can be effectively adsorbed on AC [72]. At higher boiling points, liquefaction and condensation occur more readily leading to increased adsorption capacity [67,68]. The influence of VOC molecular size and shape on adsorption has also been investigated by Yang *et al.* [73]. Experiments revealed that the VOC adsorption capacity on Metal-organic frameworks MIL-101 ($Cr_3F(H_2O)_2OE(O_2C)–C_6H_4–(CO_2)_3 \cdot nH_2O$; n is ~25) decreased with an increase in VOC molecular cross sectional area since large VOC molecules cannot penetrate through the smaller size of MIL-101 cylindrical micropores. Adsorption of VOCs on MIL-101 also showed shape selectivity towards VOCs molecules: The adsorption capacity of p-xylene was higher than that of *m*-xylene and *o*-xylene even though they have almost equal molecule cross-sectional areas (Figure 5).

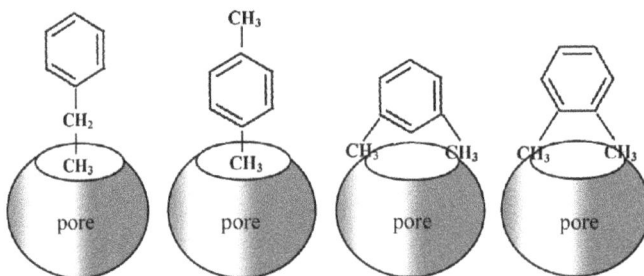

Figure 5. Scheme of ethyl benzene, *p*-xylene, *o*-xylene and *m*-xylene entering into MIL-101 pores. Reprinted from Ref. [73], with permission from Elsevier.

It has also been observed that the VOC concentration can have a positive influence on the adsorption capacity of a material [22,72] and that other chemical properties of VOCs such as oxidation state can also influence adsorption.

3.3. Relative Humidity

The effect of humidity is of great interest for practical applications in industry as well as non-industrial buildings since moisture is always present in to some degree in ambient air. Regarding VOC adsorption, RH might change the chemical state of the sorbent surface or change the adsorption modes and adsorbed amount of VOC on sorbent. It seems that the effect of RH is determined by the VOC-sorbent combination. VOC adsorption can be enhanced, suppressed or remain neutral when air humidity changes, depending on the hydrophobicity of the sorbent and specific VOC properties like e.g., solubility. Kuroki *et al.* have mentioned that the influence of RH is negligible in a toluene adsorption process since hydrophobic zeolite honeycomb is used [74]. Similarly, Zhao *et al.* have investigated HCHO breakthrough capacity over AgCu/HZ catalyst in air with varying RH (dry gas and RH = 20%, 50%, 80% and 93%) [39]. HCHO breakthrough capacity slightly decreases in humid condition compared to dry gas. However, breakthrough capacity remains almost constant when RH changes from 20% to 93% due to the hydrophobic property of high-silica zeolite. Sivachandiran *et al.* have precisely investigated the impact of RH on IPA adsorption over TiO_2 [35]. The influence of air RH on reversibly and irreversibly adsorbed IPA on TiO_2 is shown in Figure 6. The amount of both reversibly as well as irreversibly adsorbed IPA continuously decreases with increasing RH. The result suggests possible competitive adsorption between water vapor and IPA molecules. Even at the lowest moisture (10% RH) content, the partial pressure of H_2O is quite high compared to IPA. Hence, H_2O adsorption is more favorable in this competitive process. Interestingly, the amount of irreversibly adsorbed IPA increases from 84% to 94% when the water content changes from 35% to 65%. It is expected that water forms a multilayer above 35% RH. Due to high water solubility, part of irreversibly adsorbed IPA will dissolve rather than to directly interact with TiO_2 surface at highest RH.

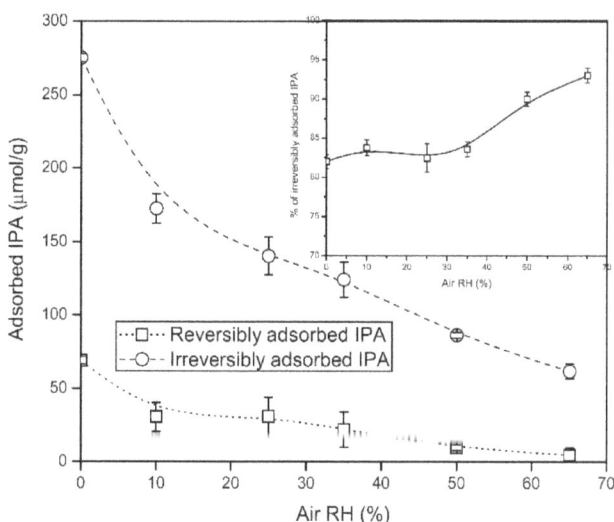

Figure 6. Influence of air relative humidity (RH) on IPA reversible and irreversible adsorbed fractions on TiO_2 at ambient condition. Reprinted from Ref. [35], with permission from Elsevier.

3.4. Temperature

Adsorption process spontaneously occurs when the change in free energy, ΔG, of the system decreases. The entropy change, ΔS, in adsorption process is negative since the adsorbed gas is less mobile than its gaseous state [75]. Therefore, the heat of adsorption, ΔH, must always be negative since $\Delta G = \Delta H - T\Delta S$. Thus, adsorption is always exothermic. The heat of adsorption also defines the extent of the interactions between adsorbate molecules and adsorbent lattice atoms. Physisorption is predominant at lower temperatures due to weak Van Der Waals' forces interactions. Since chemical reaction probability increases with temperature, chemisorption is favored at high temperatures. In most circumstances, the amount of adsorption decreases however with increasing temperature. For instance, Song *et al.* demonstrated that the breakthrough time for the adsorption process was significantly reduced at elevated temperatures compared to room temperature indicating lower adsorption capacity [32]. Brodu *et al.* have reported the influence of temperature on MEK adsorption capacity of two aluminosilicates (Fau-Y, ZSM-5) [69]. The quantity of adsorbed MEK on both Fau-Y and ZSM-5 decreased with increasing temperature. On Fau-Y, the adsorption rate of MEK was 87% at 298 K and 51% at 333 K, respectively. This clearly indicates a higher loss of adsorption capacity with rising temperature. However, some researchers have observed an enhanced adsorption capacity of AC in the high temperature range [76]. In this experiment, adsorption of four VOCs including carbon tetrachloride (CCl_4), chloroform ($CHCl_3$), benzene and methylene chloride (CH_2Cl_2) on AC at different temperatures was carried out. For a given VOC concentration, the adsorbed amount of VOC decreased with increasing temperature except for benzene. This abnormal behavior of benzene is attributed to the activated entry effect, which is responsible for the temperature-dependent nature of equilibrium adsorption [77]. At higher temperature, the diffusion of benzene molecules is facilitated into the narrow micropore channels. This promotes the rate of entry into the micropores leading to enhanced adsorption capacity.

4. Critical Process Parameters for Regeneration

To ensure an economically feasible process with uniform performance of sequential treatment, the regenerability of an adsorbent is considered as the most important parameter. The conventional and most popular regeneration techniques are thermal swing, pressure swing and chemical regeneration (displacement, elution, supercritical extraction) or a combination of those. As this review article is related to plasma regeneration processes, only the influence of process parameters on the regeneration by plasma will be discussed in the following sections.

4.1. Flow Rate

The effect of gas flow rate on the regeneration efficiency has been studied by several groups. When gas flow rate increases the residence time/space time of the gas stream is lowered. This will decrease the plasma exposure time during the regeneration stage leading to a suppression of the decomposition efficiency. Saulich *et al.* have investigated the influence of the N_2 flow rate in a DBD packed with MP 5 on the decomposition of adsorbed HCHO [48]. From Figure 7, it is apparent that with higher N_2 flow rate, the total amount of decomposed HCHO molecules increases and the formation of CO_2 is greatly favored.

However, Kuroki *et al.* have reported opposite phenomena during the adsorption of toluene with a DBD packed with a hydrophobic zeolite honeycomb [74]. According to the authors, at a low gas flow rate, re-adsorption may occur inside the adsorbent during the plasma desorption process, resulting in a lower decomposition efficiency. Therefore, the optimum gas flow rate should be determined.

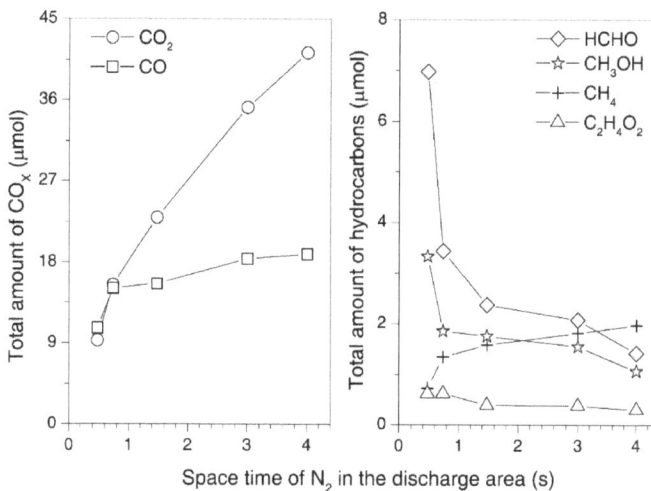

Figure 7. Influence of flow rate on the decomposition of formaldehyde (HCHO) adsorbed on MP 5 (adsorption: 176 mL/min flow rate of polluted N_2 with 99 ppm of HCHO, adsorption time t_1 = 250 min, total amount of adsorbed HCHO: 155 μmol; discharge carrier gas N_2, discharge power P = 2.2 W). Reprinted from Ref. [48], with permission from Elsevier.

4.2. Temperature

Temperature has an influence on the electrical properties of a NTP discharge. With varying temperature, the gas density varies, leading to a change in reduced electric field strength (E/N). Generally, it is known that the kinetic reaction rate of O radicals with VOCs is an increasing function of temperature due to the endothermic reaction behavior. However, the effect of temperature on the production of reactive species such as O radicals still needs detailed studies. Nevertheless, the beneficial influence of temperature on the regeneration process is counteracted by a decrease of the adsorption capacity of the sorbent at higher temperatures. These competitive effects on the plasma-assisted desorption process have been investigated by Song *et al.* who have used a γ-Al_2O_3 packed DBD reactor for toluene removal [32]. As expected, the thermal decomposition of NTP produced ozone increases with elevated temperature. The removal rate of adsorbed toluene on γ-Al_2O_3 beads was also improved at higher temperatures compared to room temperature. In addition, γ-Al_2O_3 was able to reduce some other by-products, such as O_3 and HNO_3.

4.3. Discharge Power

The oxidative ability of NTP treatment for adsorbent regeneration can be improved by increasing input power. Indeed, there is an optimal discharge power at which high oxidation rate and low energy cost can be achieved. The effect of discharge power on the plasma catalytic oxidation of adsorbed benzene over metal supported zeolite (Ag/HZSM-5) catalyst has been studied by Hong-Yu *et al.* [40] (Figure 8a). The optimum DBD operating power is found to be 4.7 W at which almost 100% of adsorbed benzene is oxidized to CO_2. The energy cost at this optimum power was as low as 3.7×10^{-3} kWh/m^3.

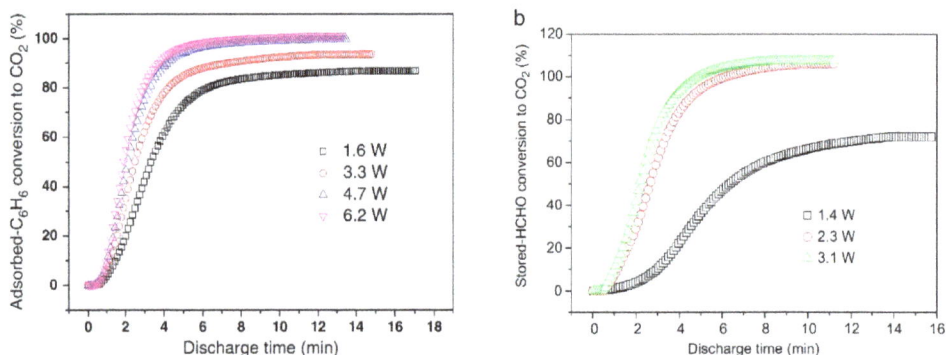

Figure 8. (**a**) Adsorbed-benzene conversion to CO_2 over Ag/HZSM-5 catalysts (**b**) stored-HCHO conversion to CO_2 over AgCu/HZ catalysts as a function of discharge time at various discharge powers. Reprinted from Ref. [39,40], with permission from Elsevier.

Zhao *et al.* have also investigated the effect of discharge power on the plasma catalytic oxidation of stored HCHO on AgCu/HZ catalyst [39]. The conversion of HCHO to CO_2 dramatically increases (Figure 8b) reaching almost 100% after 5 min of operation at 2.3 W. Further increase of the discharge power did not significantly improve the HCHO conversion.

The influence of discharge power on toluene decomposition has also been studied by Mok *et al.* [46] where complete toluene decomposition towards CO and CO_2 was reported instead of quantifying the CO_x selectivity. With increasing discharge power, the concentration of CO and CO_2 further increased indicating faster oxidation of adsorbed toluene.

Sivachandrian *et al.* have precisely investigated the effect of discharge power on the regeneration of adsorbed acetone on TiO_2 surface [47]. They applied a double DBD with TiO_2 coated glass beads. Although the oxidation rate of acetone is improved with increasing input power, the removal efficiency remains constant around 25%. To achieve both sufficient surface regeneration and high mineralization, 0.33 W is considered as an optimum input power. The authors confirmed that even though the regeneration efficiency was limited during NTP treatment, operation at moderate discharge power modified the nature of organic adsorbed species, which facilitated mineralization during successive thermal treatments (Figure 9).

Modification of adsorbed acetone *i.e.*, mesityl oxides into formic acid is only possible above 0.13 W. Hence, acetone desorption is decreased while CO and CO_2 formation are promoted by increasing input power during successive thermal treatment. Any further increase of input power above 0.33 W is useless regarding conversion of mesityl oxide into formic acid.

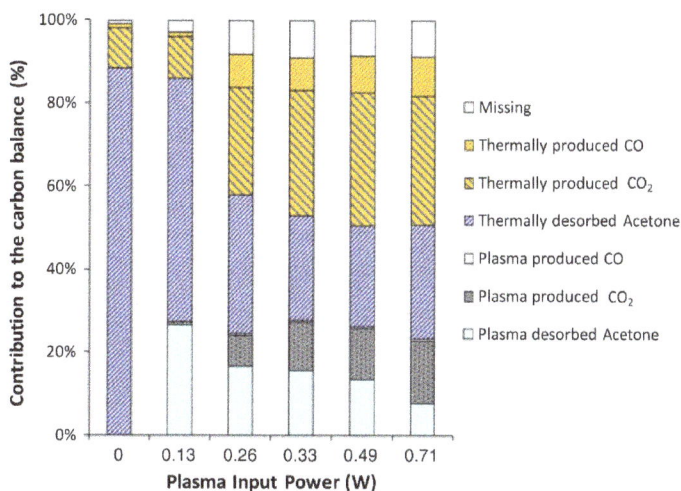

Figure 9. Contributions of CO, CO_2 and acetone to the carbon balance, calculated after successive plasma treatment and thermal treatment. Data are displayed for various plasma input powers. Reprinted from Ref. [47], with permission from Elsevier.

4.4. Relative Humidity

During real-time operation, humidity is permanently present in air, which will impact both the adsorption mechanism and regeneration process unless hydrophobic sorbents are used. Fan *et al.* have studied the removal of low concentrated benzene stored on 0.8% Ag/HZSM-5 catalyst under various humidity conditions (dry air and RH = 20%, 50% and 80%) [42]. Figure 10 shows that the oxidation rate slightly decreases up to 50% RH compared to dry air. However, CO_2 selectivity and carbon balance reached almost 100% in all cases. The plausible explanation of this weak effect of humidity on oxidation rate is the hydrophobic property of high-silica zeolite. Therefore, this result should be limited only to hydrophobic surfaces; moreover, no RH was introduced during the plasma-assisted regeneration process. The presence of humidity can influence the electrical and physical properties of the discharge as well as the plasma chemistry. For instance, the RH can affect the input power and change the nature and amounts of reactive plasma species, which will have an influence on the VOC oxidation pathway.

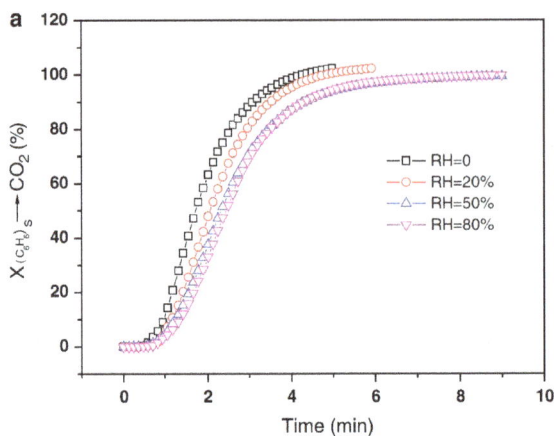

Figure 10. Effect of relative humidity (RH) on plasma catalytic oxidation of stored benzene over 0.8 wt.% Ag/HZ catalysts to CO_2 (storage stage: 4.7 ppm benzene, 0%–80% RH (25 °C), 600 mL/min flow rate of simulated air, t_1 = 1 h; discharge stage with 60 mL/min O_2, P = 4.7 W. Reprinted from Ref. [42], with permission from Elsevier.

To assess the impact of RH on both the adsorption modes and the surface plasma regeneration efficiency, Sivachandiran *et al.* have investigated the removal of IPA with a packed bed reactor coated with TiO_2 under various RH conditions [35]. Increased RH favored IPA mineralization and diminished acetone formation. With increasing RH from 0% to 65%, the carbon balance enhanced from 36% to 91%, clearly proving that moisture has a positive influence on the plasma surface regeneration efficiency. The authors suggest that dissociative electron impact with H_2O produces OH radicals, which are chemisorbed on TiO_2 leading to re-hydroxylation of the pre-treated TiO_2 surface. Eventually, adsorbed IPA will be replaced by hydroxyl groups and desorb. Generally, the oxidation power of OH radicals is much stronger than other oxidants such as oxygen atoms and peroxyl radicals. Increased RH favors production of OH radicals and the subsequent oxidation of VOCs leading to high mineralization rate. In comparison to dry-air experiments, almost 80% acetone production was reduced in the presence of only 10% air RH. Low surface coverage of TiO_2 by H_2O molecules allows the adsorption of the deprotonated CH_3–CHO^*–CH_3 species onto the TiO_2 surface [35]. Indeed, in case of humid air, the most common oxidation pathway for VOC removal with NTP is the H-atom abstraction by OH radicals. [78]. Since large amounts of OH radicals are produced by the discharge in the surface vicinity, it improves the dehydrogenation of IPA hydroxyl groups leading to the formation of acetone [35]. However, above 25% RH, formation of a water multilayer onto the TiO_2 surface impeded the dissociative adsorption of IPA onto the TiO_2 surface obstructing acetone formation.

4.5. O_2 Content/O_2 Partial Pressure

Similar to air RH, the discharge properties and VOC abatement process are sensitive to oxygen content. The presence of oxygen in the discharge generally increases the amount of O radicals leading to a high removal efficiency. However, owing to its electronegative character, oxygen limits the electron density and also reduces the formation of other reactive species e.g., N_2 and N_2^*, which are beneficial for VOC abatement [79–81]. Saulich et al. have used a mineral adsorbent packed DBD reactor to see the effect of O_2 on the plasma decomposition process of adsorbed HCHO [48]. Addition of 10% O_2 to N_2 increased both the formaldehyde decomposition efficiency and the mineralization rate compared to pure N_2. Kim et al. have demonstrated the effect of oxygen content on benzene decomposition using a plasma-catalyst reactor (5% Cu/MOR) [82]. Figure 11 clearly shows that when the plasma is turned on after the adsorption mode, benzene concentration decreases and CO and CO_2 formation simultaneously increases with increasing O_2 percentage. Several catalysts (TiO_2, γ-Al_2O_3, zeolites) loaded with nanoparticles of active metals (Ag, Cu, Zr) or noble metals (Pt, Pd) have also been investigated under various oxygen partial pressures for toluene and benzene abatement [29]. The increase of O_2 content enhanced both the decomposition efficiency (~30%–100%) and the CO_2 selectivity regardless of the catalyst type [29]. Operating the regeneration mode at high O_2 content leads to complete oxidation of VOCs without CO, aerosol and N_xO_y formation and complete decomposition of VOCs to CO_2 is achieved with the CSD system under O_2 plasma in most literature so far.

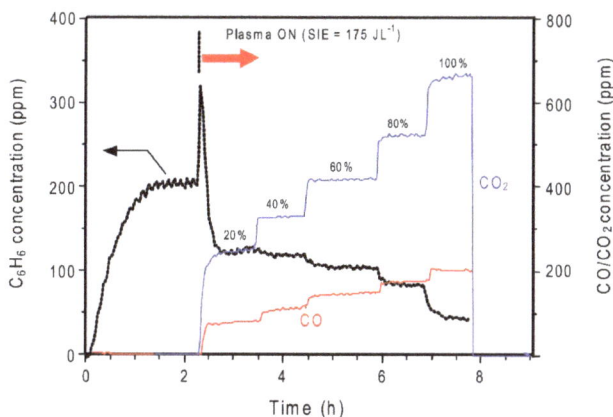

Figure 11. Effect of oxygen content on the volatile organic compounds (VOC) decomposition using a plasma-catalyst reactor (5 wt.% Cu/MOR). Applied voltage and frequency were kept constant at 16 kV and 300 Hz, respectively. Temperature was 100 °C. Space velocity was 33,000/h. Reprinted from Ref. [82], with permission from Elsevier.

4.6. Number of Cycles

Potential irreversible ageing of adsorbent may occur during the regeneration process, which governs the life of an adsorbent. After multiple regeneration cycles, the adsorbent may not fully recover its initial properties, such as the specific surface area or adsorption capacity [83]. Adsorption capacity may decrease due to surface modification by particle agglomerations or polymerization on the catalyst surface, which results in blocking active sites during the regeneration step. Keeping this in mind, researchers should also inspect the effect after the repeatability test of this sequential treatment. For instance, Kuroki *et al.* performed a repeatability test of adsorption and plasma desorption and reported that toluene was almost completely adsorbed over honeycomb zeolites in each adsorption process [84]. Having 75% regeneration efficiency, both desorption and regeneration efficiencies do not deteriorate for 10 repetitions of the adsorption/desorption processes (Figure 12). Similarly, seven repetitive adsorption-discharge cycles were performed for HCHO abatement from mineral adsorbent by Saulich *et al.* [48]. Adsorption capacity and regeneration efficiency were found to be more or less the same in each cycle [48]. The stability of AgCu/HZ catalyst during five consecutive storage-discharge cycles was also examined by D. Zhao *et al.* [39]. HCHO storage capacity was almost constant, and CO_2 selectivity and carbon mass balance were maintained at ~100%.

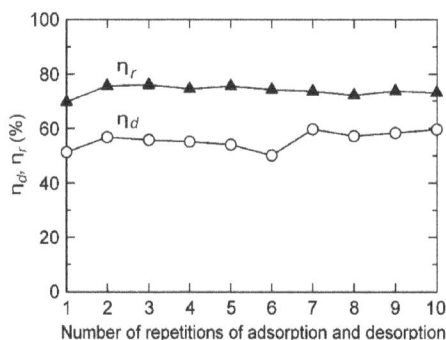

Figure 12. Desorption and regeneration efficiencies denoted by η_d and η_r, respectively, as a function of repetitive operation. Reprinted from Ref. [84], with permission from Elsevier.

5. Conclusions

From the review above, it can be concluded that sequential treatment has attracted growing interest among researchers and has proven to be an effective method for VOC abatement as well as for regeneration of saturated adsorbents. However, choosing an appropriate catalyst to achieve the best performance is still a challenge since it has to combine high adsorption capacity with high catalytic activity for oxidation of adsorbed VOCs. For instance, hard-to-adsorb VOC molecules are not appropriate for adsorption processes [61]. The key factor for the abatement of this kind of compounds is either to choose appropriate catalyst materials with a large adsorption capacity or to remove them through continuous treatment.

Although the review of experimental results of cyclic VOC adsorption and plasma-assisted regeneration shows substantial progress, many questions still remain unanswered regarding the impact of process parameters and VOCs degradation mechanisms. In this regard, direct *in-situ* surface monitoring of the catalyst/adsorbent will be crucial to elucidate these important issues. Furthermore, modeling and simulation studies would also contribute to the understanding of the various underlying plasma-catalytic effects.

The research on sequential plasma-catalysis coupling for VOC abatement has mostly been performed on the laboratory scale. Up to now, only a few studies on pilot scales with gas circulation have been reported [59]. To achieve an economically feasible VOC removal process without gas circulation, cyclic operation of VOC adsorption and catalyst regeneration has been proposed by researchers. In this regard, future work should focus on the feasibility and optimization of the duration of the sequential intervals. Finally, to realize industrial implementation of this energy effective method for VOC abatement scalability studies will also need to be carried out in the near future.

Author Contributions: Sharmin Sultana wrote the first draft of the review which was then refined by the comments and suggestions of all other authors.

Conflicts of Interest: The authors declare no conflict of interest.

References

1. Pal, R.; Kim, K.-H.; Hong, Y.-J.; Jeon, E.-C. The Pollution Status of Atmospheric Carbonyls in a Highly Industrialized Area. *J. Hazard. Mater.* **2008**, *153*, 1122–1135.
2. Sarigiannis, D.A.; Karakitsios, S.P.; Gotti, A.; Liakos, I.L.; Katsoyiannis, A. Exposure to Major Volatile Organic Compounds and Carbonyls in European Indoor Environments and Associated Health Risk. *Environ. Int.* **2011**, *37*, 743–765.
3. Atkinson, R. Atmospheric Chemistry of VOCs and NO$_x$. *Atmos. Environ.* **2000**, *34*, 2063–2101.
4. Vineis, P.; Forastiere, F.; Hoek, G.; Lipsett, M. Outdoor Air Pollution and Lung Cancer: Recent Epidemiologic Evidence. *Int. J. Cancer* **2004**, *111*, 647–652.
5. Tollefsen, P.; Rypdal, K.; Torvanger, A.; Rive, N. Air Pollution Policies in Europe: Efficiency Gains From Integrating Climate Effects with Damage Costs to Health and Crops. *Environ. Sci. Policy* **2009**, *12*, 870–881.
6. Yamamoto, T.; Ramaanathan, K.; Lawless, P.A.; Ensor, D.S.; Newsome, J.R.; Plaks, N.; Ramsey, G.H. Control of Volatile Organic Compounds by an AC Energized Eerroelectric Pellet Reactor and a Pulsed Corona Reactor. *IEEE Trans. Ind. Appl.* **1992**, *28*, 528–534.

7.   Nunez, C.M.; Ramsey, G.H.; Ponder, W.H.; Abbott, J.H.; Hamel, L.E.; Kariher, P.H. Corona Destruction: An Innovative Control Technology for VOCs and Air Toxics. *Air Waste* **1993**, *43*, 242–247.

8.   Sudnick, J.J.; Corwin, D.L. VOC Control Techniques. *Hazard. Waste Hazard.* **1994**, *11*, 129–143.

9.   Kim, H.H.; Ogata, A.; Futamura, S. Complete Oxidation of Volatile Organic Compounds (VOCs) Using Plasma-Diven Catalysis and Oxygen Plasma. *Int. J. Plas. Environ. Sci. Tech.* **2007**, 4–8.

10. Keller, R.A.; Dyer, J.A. Abating Halogenated VOCs. *Chem. Eng.* **1998**, *105*, 100–105.

11. Dyer, J.A.; Mulholland, K. Toxic Air Emissions What is the Full Cost to Your Business? *Chem. Eng.* **1994**, *101*, 4–8.

12. Urashima, K.; Chang, J.S. Removal of Volatile Organic Compounds from Air Streams and Industrial Flue Gases by Non-thermal Plasma Technology. *IEEE Trans. Dielectr. Electr. Insul.* **2000**, *7*, 602–614.

13. Chang, J.S. Next Generation Integrated Electrostatic Gas Cleaning Systems. *J. Electrostat.* **2003**, *57*, 273–291.

14. Dan, Y.; Gao, D.S.; Yu, G.; Shen, X.L.; Gu, F. Investigation of the Treatment of Particulate Matter from Gasoline Engine Exhaust Esing Non-thermal Plasma. *J. Hazard. Mater.* **2005**, *127*, 149–155.

15. Laroussi, M. Low Temperature Plasma-based Ssterilization: Overview and State-of-the-art. *Plasma Process. Polym.* **2005**, *2*, 391–400.

16. Chang, C.L.; Lin, T.S. Decomposition of Toluene and Acetone in Packed Dielectric Barrier Discharge Reactors. *Plasma Chem. Plasma Process.* **2005**, *25*, 227–243.

17. Lee, H.M.; Chang, M.B. Gas-phase Removal of Acetaldehyde Via Packed-bed Dielectric Barrier Discharge Reactor. *Plasma Chem. Plasma Process.* **2001**, *21*, 329–343.

18. Okubo, M.; Yamamoto, T.; Kuroki, T.; Fukumoto, H. Electric Air Cleaner Composed of Nonthermal Plasma Reactor and Electrostatic Precipitator. *IEEE Trans. Ind. Appl.* **2001**, *37*, 1505–1511.

19. Vandenbroucke, A.M.; Mora, M.; Jimenez-Sanchidrian, C.; Romero-Salguero, F.J.; de Geyter, N.; Leys, C.; Morent, R. TCE Abatement with a Plasma-catalytic Combined System Using MnO_2 as Catalyst. *Appl. Catal. B* **2014**, *156*, 94–100.

20. Nguyen Dinh, M.T.; Giraudon, J.M.; Lamonier, J.F.; Vandenbroucke, A.; de Geyter, N.; Leys, C.; Morent, R. Plasma-catalysis of Low TCE Concentration in air Using $LaMnO_3{}^+\delta$ as Catalyst. *Appl. Catal. B* **2014**, *147*, 904–911.

21. Assadi, A.A.; Palau, J.; Bouzaza, A.; Penya-Roja, J.; Martinez-Soriac, V.; Wolbert, D. Abatement of 3-methylbutanal and Trimethylamine with Combined Plasma and Photocatalysis in a Continuous Planar Reactor. *J. Photoch. Photobiol. A* **2014**, *282*, 1–8.

22. Ohshima, T.; Kondo, T.; Kitajima, N.; Sato, M. Adsorption and Plasma Decomposition of Gaseous Acetaldehyde on Fibrous Activated Carbon. *IEEE Trans. Ind. Appl.* **2010**, *46*, 23–28.

23. Kim, H.H.; Ogata, A.; Futamura, S. Effect of Different Catalysts on the Decomposition of VOCs Using Flow-type Plasma-driven Catalysis. *IEEE Trans. Plasma Sci.* **2006**, *34*, 984–995.

24. Vandenbroucke, A.M.; Minh Tuan Nguyen, D.; Giraudon, J.-M.; Morent, R.; De Geyter, N.; Lamonier, J.-F.; Leys, C. Qualitative By-product Identification of Plasma-assisted TCE Abatement by Mass Spectrometry and Fourier-transform Infrared Spectroscopy. *Plasma Chem. Plasma Process.* **2011**, *31*, 707–718.

25. Van Durme, J.; Dewulf, J.; Leys, C.; Van Langenhove, H. Combining Non-thermal Plasma with Heterogeneous Catalysis in Waste Gas Treatment: A Review. *Appl. Catal. B* **2008**, *78*, 324–333.

26. Roland, U.; Holzer, F.; Kopinke, E.D. Combination of Non-thermal Plasma and Heterogeneous Catalysis for Oxidation of Volatile Organic Compounds Part 2. Ozone Decomposition and Deactivation of γ-Al_2O_3. *Appl. Catal. B* **2005**, *58*, 217–226.

27. Subrahmanyam, C. Catalytic Non-thermal Plasma Reactor for Total Oxidation of Volatile Organic Compounds. *Indian J. Chem. A* **2009**, *48*, 1062–1068.

28. Zhu, T.; Li, J.; Liang, W.J.; Jin, Y.Q. Synergistic Effect of Catalyst for Oxidation Removal of Toluene. *J. Hazard. Mater.* **2009**, *165*, 1258–1260.

29. Kim, H.H.; Ogata, A.; Futamura, S. Oxygen Partial Pressure-dependent Behavior of Various Catalysts for the Total Oxidation of VOCs Using Cycled System of Adsorption and Oxygen Plasma. *Appl. Catal. B* **2008**, *79*, 356–367.

30. Malik, M.A.; Minamitani, Y.; Schoenbach, K.H. Comparison of Catalytic Activity of Aluminum Oxide and Silica Gel for Decomposition of Volatile Organic Compounds (VOCs) in a Plasmacatalytic Reactor. *IEEE Trans. Plasma Sci.* **2005**, *33*, 50–56.

31. Ogata, A.; Yamanouchi, K.; Mizuno, K.; Kushiyama, S.; Yamamoto, T. Oxidation of Dilute Benzene in an Alumina Hybrid Plasma Reactor at Atmospheric Pressure. *Plasma Chem. Plasma Process.* **1999**, *19*, 383–394.

32. Song, Y.H.; Kim, S.J.; Choi, K.I.; Yamamoto, T. Effects of Adsorption and Temperature on a Nonthermal Plasma Process for Removing VOCs. *J. Electrostat.* **2002**, *55*, 189–201.

33. Martin, L.; Ognier, S.; Gasthauer, E.; Cavadias, S.; Dresvin, S.; Amouroux, J. Destruction of Highly Diluted Volatile Organic Components (VOCs) in air by Dielectric Barrier Discharge and Mineral Bed Adsorption. *Energy Fuel.* **2008**, *22*, 576–582.

34. Inoue, K.; Okano, H.; Yamagata, Y.; Muraoka, K.; Teraoka, Y. PerformanceTtests of Newly Developed Adsorption/plasma Combined System for Decomposition of Volatile Organic Compounds Under Continuous Flow Condition. *J. Environ. Sci. China* **2011**, *23*, 139–144.

35. Sivachandiran, L.; Thevenet, F.; Gravejat, P.; Rousseau, A. Isopropanol Saturated TiO_2 Surface Regeneration by Non-thermal Plasma: Influence of Air Relative Humidity. *Chem. Eng. J.* **2013**, *214*, 17–26.

36. Ogata, A.; Ito, D.; Mizuno, K.; Kushiyama, S.; Yamamoto, T. Removal of Dilute Benzene Using a Zeolte-hybrid Plasma Reactor. *IEEE Trans. Ind. Appl.* **2001**, *37*, 959–964.

37. Thevenet, F.; Sivachandiran, L.; Guaitella, O.; Barakat, C.; Rousseau, A. Plasma-catalyst Coupling for Volatile Organic Compound Removal and Indoor Air Treatment: A Review. *J. Phys. D* **2014**, *47*, 224011–224024.

38. Sivachandiran, L.; Thevenet, F.; Rousseau, A. Isopropanol Removal Using Mn_xO_y Packed Bed Non-thermal Plasma Reactor: Comparison between Continuous Treatment and Sequential Sorption/regeneration. *Chem. Eng. J.* **2015**, *270*, 327–335.

39. Zhao, D.Z.; Li, X.S.; Shi, C.; Fan, H.Y.; Zhu, A.M. Low-concentration Formaldehyde Removal from Air Using a Cycled Storage–discharge (CSD) Plasma Catalytic Process. *Chem. Eng. Sci.* **2011**, *66*, 3922–3929.

40. Fan, H.Y.; Shi, C.; Li, X.S.; Zhao, D.Z.; Xu, Y. High-efficiency Plasma Catalytic Removal of Dilute Benzene from Air. *J. Phys. D* **2009**, *42*, 225105–225109.

41. Ding, H.X.; Zhu, A.M.; Yang, X.F.; Li, C.H.; Xu, Y. Removal of Formaldehyde from Gas Streams via Packed-bed Dielectric Barrier Discharge Plasmas. *J. Phys. D* **2005**, *38*, 4160–4167.

42. Fan, H.-Y.; Li, X.-S.; Shi, C.; Zhao, D.-Z.; Liu, J.-L.; Liu, Y.-X.; Zhu, A.-M. Plasma Catalytic Oxidation of Stored Benzene in a Cycled Storage-discharge (CSD) Process: Catalysts, Reactors and Operation Conditions. *Plasma Chem. Plasma Process.* **2011**, *31*, 799–810.

43. Wang, W.; Zhu, T.; Fan, X. Removal of Gas Phase Low-concentration Toluene by Intermittent Use of Adsorption and Non-thermal Plasma Regeneration. In Proceedings of 21st International Symposium on Plasma Chemistry (ISPC 21), Queensland, Australia, 4–9 August 2013.

44. Takahashi, A.; Yang, F.H.; Yang, R.T. Aromatics/aliphatics Separation by Adsorption: New Sorbents for Selective Aromatics Adsorption by π-complexation. *Ind. Eng. Chem. Res.* **2000**, *39*, 3856–3867.

45. Qu, Z.; Bu, Y.; Qin, Y.; Wang, Y.; Fu, Q. The Improved Reactivity of Manganese Catalysts by Ag in Catalytic Oxidation of Toluene. *Appl. Catal. B* **2013**, *132*, 353–362.

46. Mok, Y.S.; Kim, D.H. Treatment of Toluene by Using Adsorption and Nonthermal Plasma Oxidation Process. *Curr. Appl. Phys.* **2011**, *11*, S58–S62.

47. Sivachandiran, L.; Thevenet, F.; Rousseau, A. Non-thermal Plasma Assisted Regeneration of Acetone Adsorbed TiO_2 Surface. *Plasma Chem. Plasma Process.* **2013**, *33*, 855–871.

48. Saulich, K.; Mueller, S. Removal of Formaldehyde by Adsorption and Plasma Treatment of Mineral Adsorbent. *J. Phys. D* **2013**, *46*, 045201–045208.

49. Klett, C.; Duten, X.; Tieng, S.; Touchard, S.; Jestin, P.; Hassouni, K.; Vega-González, A. Acetaldehyde Removal Using an Atmospheric Non-thermal Plasma Combined with a Packed Bed: Role of the Adsorption Process. *J. Hazard. Mater.* **2014**, *279*, 356–364.

50. Sivachandiran, L.; Thevenet, F.; Rousseau, A. Regeneration of Isopropyl Alcohol Saturated Mn_xO_y Surface: Comparison of Thermal, Ozonolysis and Non-thermal Plasma Treatments. *Chem. Eng. J.* **2014**, *246*, 184–195.

51. Barakat, C.; Gravejat, P.; Guaitella, O.; Thevenet, F.; Rousseau, A. Oxidation of Isopropanol and Acetone Adsorbed on TIO_2 Under Plasma Generated Ozone Flow: Gas Phase and Adsorbed Species Monitoring. *Appl. Catal. B* **2014**, *147*, 302–313.

52. Yamagata, Y.; Niho, K.; Inoue, K.; Okano, H.; Muraoka, K. Decomposition of Volatile Organic Compounds at Low Concentrations Using Combination of Densification by Zeolite Adsorption and Dielectric Barrier Discharge. *Jpn. J. Appl. Phys.* **2006**, *45*, 8251–8254.

53. Kim, H.H.; Tsubota, S.; Date, M.; Ogata, A.; Futamura, S. Catalyst Regeneration and Activity Enhancement of Au/TiO_2 by Atmospheric Pressure Nonthermal Plasma. *Appl. Catal. A* **2007**, *329*, 93–98.

54. Arsac, F.; Bianchi, D.; Chovelon, J.M.; Ferronato, C.; Herrmann, J.M. Experimental Microkinetic Approach of the Photocatalytic Oxidation of Isopropyl Alcohol on TiO_2. Part 1. Surface Elementary Steps Involving Gaseous and Adsorbed C_3H_xO species. *J. Phys. Chem. A* **2006**, *110*, 4202–4212.

55. Larson, S.A.; Widegren, J.A.; Falconer, J.L. Transient Studies of 2-propanol Photocatalytic Oxidation on Titania. *J. Catal.* **1995**, *157*, 611–625.

56. Xu, W.Z.; Raftery, D.; Francisco, J.S. Effect of Irradiation Sources and Oxygen Concentration on the Photocatalytic Oxidation of 2-propanol and Acetone Studied by in situ FTIR. *J. Phys. Chem. B* **2003**, *107*, 4537–4544.

57. Yamamoto, T.; Asada, S.; Iida, T.; Ehara, Y. Novel NO_x and VOC Treatment Using Concentration and Plasma Decomposition. *IEEE Trans. Ind. Appl.* **2011**, *47*, 2235–2240.

58. Dang, X.; Huang, J.; Cao, L.; Zhou, Y. Plasma-catalytic Oxidation of Adsorbed Toluene with Gas Circulation. *Catal. Commun.* **2013**, *40*, 116–119.

59. Kuroki, T.; Hirai, K.; Matsuoka, S.; Kim, J.Y.; Okubo, M. Oxidation System of Adsorbed VOCs on Adsorbent Using Nonthermal Plasma Flow. *IEEE Trans. Ind. Appl.* **2011**, *47*, 1916–1921.

60. Kuroki, T.; Hirai, K.; Kawabata, R.; Okubo, M.; Yamamoto, T. Decomposition of Adsorbed Xylene on Adsorbents Using Nonthermal Plasma with Gas Circulation. *IEEE Trans. Ind. Appl.* **2010**, *46*, 672–679.

61. Inoue, K.; Furuki, K.; Okano, H.; Yamagata, Y.; Muraoka, K. A New Decomposition System for Volatile Organic Compounds Using Combinations of Dielectric Barrier Discharges with Zeolite Honeycomb Sheets. *Electron. Eng. Jpn.* **2009**, *168*, 1–10.

62. Das, D.; Gaur, V.; Verma, N. Removal of Volatile Organic Compound by Activated Carbon Fiber. *Carbon* **2004**, *42*, 2949–2962.

63. Kim, K.-J.; Ahn, H.-G. The Effect of Pore Structure of Zeolite on the Adsorption of VOCs and their Desorption Properties by Microwave Heating. *Microporous Mesoporous Mater.* **2012**, *152*, 78–83.

64. Chiang, Y.C.; Chiang, P.C.; Chang, E.E. Effects of Surface Characteristics of Activated Carbons on VOC Adsorption. *J. Environ. Eng.* **2001**, *127*, 54–62.

65. Meininghaus, C.K.W.; Prins, R. Sorption of Volatile Organic Compounds on Hydrophobic Zeolites. *Microporous Mesoporous Mater.* **2000**, *35–36*, 349–365.

66. Dubinin, M.M. Fundamentals of the Theory of Adsorption in Micropores of Carbon Adsorbents—Characteristics of their Adsorption Properties and Microporous Structures. *Pure Appl. Chem.* **1989**, *61*, 1841–1843.

67. Mangun, C.L.; Daley, M.A.; Braatz, R.D.; Economy, J. Effect of Pore Size on Adsorption of Hydrocarbons in Phenolic-based Activated Carbon Fibers. *Carbon* **1998**, *36*, 123–129.

68. Huang, M.C.; Chou, C.H.; Teng, H.S. Pore-size Effects on Activated-carbon Capacities for Volatile Organic Compound Adsorption. *AICHE J.* **2002**, *48*, 1804–1810.

69. Brodu, N.; Zaitan, H.; Manero, M.H.; Pic, J.S. Removal of Volatile Organic Compounds by Heterogeneous Ozonation on Microporous Synthetic Alumina Silicate. *Water Sci. Technol.* **2012**, *66*, 2020–2026.

70. Lillo-Rodenas, M.A.; Cazorla-Amoros, D.; Linares-Solano, A. Behaviour of Activated Carbons with Different Pore Size Distributions and Surface Oxygen Groups for Benzene and Toluene Adsorption at Low Concentrations. *Carbon* **2005**, *43*, 1758–1767.

71. Lillo-Rodenas, M.A.; Cazorla-Amoros, D.; Linares-Solano, A. Benzene and Toluene Adsorption at Low Concentration on Activated Carbon Fibres. *Adsorption* **2011**, *17*, 473–481.

72. Li, L.; Sun, Z.; Li, H.; Keener, T.C. Effects of Activated Carbon Surface Properties on the Adsorption of Volatile Organic Compounds. *J. Air Waste Manage. Assoc.* **2012**, *62*, 1196–1202.

73. Yang, K.; Sun, Q.; Xue, F.; Lin, D. Adsorption of Volatile Organic Compounds by Metal-organic Frameworks MILL-101: Influence of Molecular Size and Shape. *J. Hazard. Mater.* **2011**, *195*, 124–131.

74. Kuroki, T.; Fujioka, T.; Okubo, M.; Yamamoto, T. Toluene Concentration Using Honeycomb Nonthermal Plasma Desorption. *Thin Solid Films* **2007**, *515*, 4272–4277.

75. Gundry, P.M.; Tompkins, F.C. Chemisorption of Gases on Metals. *Q. Rev. Chem. Soc.* **1960**, *14*, 257–291.

76. Chiang, Y.C.; Chaing, P.C.; Huang, C.P. Effects of Pore Structure and Temperature on VOC Adsorption on Activated Carbon. *Carbon* **2001**, *39*, 523–534.

77. Gregg, S.J.; Sing, K.S.W. Adsorption, Surface-area and Porosity, 2nd edition. *Text. Res. J.* **1984**, *54*, 792–792.

78. Magne, L.; Pasquiers, S. Lif Spectroscopy Applied to the Study of Non-thermal Plasmas for Atmospheric Pollutant Abatement. *C. R. Phys.* **2005**, *6*, 908–917.

79. Pringle, K.J.; Whitehead, J.C.; Wilman, J.J.; Wu, J.H. The Chemistry of Methane Remediation by a Non-thermal Atmospheric Pressure Plasma. *Plasma Chem. Plasma Process.* **2004**, *24*, 421–434.

80. Ogata, A.; Mizuno, K.; Kushiyama, S.; Yamamoto, T. Methane Decomposition in a Barium Titanate Packed-bed Nonthermal Plasma Reactor. *Plasma Chem. Plasma Process.* **1998**, *18*, 363–373.

81. Mok, Y.S.; Lee, S.B.; Oh, J.H.; Ra, K.S.; Sung, B.H. Abatement of Trichloromethane by Using Nonthermal Plasma Reactors. *Plasma Chem. Plasma Process.* **2008**, *28*, 663–676.

82. Kim, H.H.; Ogata, A. Nonthermal Plasma Activates Catalyst: From Current Understanding and Future Prospects. *Eur. Phys. J.* **2011**, *55*, 13806–13818.

83. Qu, G.; Liang, D.; Qu, D.; Huang, Y.; Li, J. Comparison between Dielectric Barrier Discharge Plasma and Ozone Regenerations of Activated Carbon Exhausted with Pentachlorophenol. *Plasma Sci. Technol.* **2014**, *16*, 608–613.

84. Kuroki, T.; Fujioka, T.; Kawabata, R.; Okubo, M.; Yamamoto, T. Regeneration of Honeycomb Zeolite by Nonthermal Plasma Desorption of Toluene. *IEEE Trans. Ind. Appl.* **2009**, *45*, 10–15.

Coupling Noble Metals and Carbon Supports in the Development of Combustion Catalysts for the Abatement of BTX Compounds in Air Streams

Sergio Morales-Torres, Francisco Carrasco-Marín, Agustín F. Pérez-Cadenas and Francisco José Maldonado-Hódar

Abstract: The catalytic combustion of volatile organic compounds (VOCs) is one of the most important techniques to remove these pollutants from the air stream, but it should be carried out at the lowest possible temperature, saving energy and avoiding the simultaneous formation of nitrogen oxides (NO_x). Under these experimental conditions, the chemisorption of water generated from VOCs combustion may inhibit hydrophilic catalysts. Nowadays, a wide variety of carbon materials is available to be used in catalysis. The behavior of these hydrophobic materials in the development of highly active and selective combustion catalysts is analyzed in this manuscript. The support characteristics (porosity, hydrophobicity, structure, surface chemistry, *etc.*) and the active phase nature (noble metals: Pt, Pd) and dispersion were analyzed by several techniques and the results correlated with the dual adsorptive and/or catalytic performance of the corresponding catalysts. The coupling of highly active phases and carbon materials (activated carbons, honeycomb coated monoliths, carbon aerogels, *etc.*) with tuneable physicochemical properties leads to the complete abatement of benzene, toluene and xylenes (BTX) from dilute air streams, being selectively oxidized to CO_2 at low temperatures.

Reprinted from *Catalysts*. Cite as: Morales-Torres, S.; Carrasco-Marín, F.; Pérez-Cadenas, A.F.; Maldonado-Hódar, F.J. Coupling Noble Metals and Carbon Supports in the Development of Combustion Catalysts for the Abatement of BTX Compounds in Air Streams. *Catalysts* **2015**, *5*, 774–799.

1. Introduction

Volatile organic compounds (VOCs) are compounds with boiling point ranging from (50–100 °C) to (240–260 °C), which corresponds to saturation vapor pressures larger than 102 kPa at 25 °C [1]. Most VOCs are toxic to both humans and the environment. In fact, the tropospheric ozone and the photochemical smog are formed when VOCs are exposed to nitrogen oxides (NO_x) and sunlight. These pollutants are present outdoor and indoor. Some indoor sources include solvents used in the building and maintenance of construction materials, furnishings, office equipment and consumer products.

87

The overall catalytic oxidation (typically referred to as combustion) is one of the most effective VOCs removal technologies, since it operates with dilute concentrations of VOCs and at lower temperature than traditional thermal incinerators. The employ of catalysts allows significantly decreasing the operating temperature and the formation of undesirable by-products (e.g., dioxins and NO_x), as well as saving energy [2].

Combustion catalysts are commonly based on noble metals (Pt, Pd, Rh, Au and Ag), transition metal oxides (Ni, Cu, Co, Cr, Mn, Mo, V, Zn, Zr, Ce, Ti and Fe) and their mixtures (e.g., MnO_x-CuO_x and MnO_x-CeO_2) [3–9]. Catalysts based on transition metal oxides are much cheaper than noble metals and can be quite active in the oxidation of determined VOCs. Manganese oxides (MnO_x) are the most widely studied metal oxides due to the high activity, stability, relative low toxicity and redox properties. In fact, Mn atoms have a strong ability of switching oxidation states (+II, +III, +IV) and forming structural defects, which lead to a high oxygen mobility and storage [2]. In the catalytic oxidation of benzene and toluene over MnO_x–based catalysts (i.e., Mn_3O_4, Mn_2O_3 and MnO_2), the trend of catalytic activity was as follows: $Mn_3O_4 > Mn_2O_3 > MnO_2$, i.e., as the oxygen mobility of the MnO_x catalyst [10]. Cryptomelane, todorokite and pyrolusite are manganese oxides with different square tunnel structures used also in VOCs oxidation. For instance, cryptomelane (KMn_8O_{16}) was very active towards the oxidation of ethanol, ethyl acetate and toluene; although its catalytic performance was influenced by the presence of other phases, in particular Mn_2O_3 and Mn_3O_4 [11]. Toluene was the least VOCs oxidized, due to the decrease of the oxygen mobility and consequently slow incorporation rate of lattice oxygen. On the other hand, ceria (CeO_2) is the ideal support in terms of outstanding oxygen storage capacity and hence, it is commonly employed as a structural and electronic promoter in catalysis [2], because oxygen molecules can be easily transferred to the CeO_2 surface through a reversible redox process between Ce^{4+} and Ce^{3+} [12]. Ordered mesoporous CrO_3 was very active for toluene oxidation [13], but its high toxicity causes serious concerns. Supported vanadia (V_2O_5) catalysts showed excellent activity for the simultaneous removal of VOCs and NO_x in a Cl_2–HCl environment [14], but their corrosive properties avoid their application. In general, catalysts based on metal oxides present often lower activity than those based on noble metals, however they are more resistant to be poisoned by halogenated compounds, As, Pb or P.

In spite of the expensive cost, noble metal-based catalysts are preferred for catalytic oxidation of VOCs, because of their high specific activity, strong resistance to deactivation and ability to be regenerated [15]. However, the industrial application of these catalysts is still limited by costs and sensitivity to poisoning by chlorine/chloride products in the oxidation of chlorinated VOCs [16]. The catalytic performance of supported catalysts depends on: (i) intrinsic characteristics of the active phase (e.g., dispersion, metal loading and oxidation state); (ii) properties of support (e.g., porosity and hydrophobicity) and (iii) the operating parameters (e.g., concentrations of VOCs and oxygen, VOCs nature, overall flow rate, type of reactor) [15]. Pt and Pd are the most widely noble metals studied due to their high stability in severe conditions (high resistance to irreversible oxidation) and high activity in hydrocarbon oxidations [3,17]. Compared to Pt, Pd is generally more active in methane conversion but worse in the transformation of other compounds [18]. Pd also exhibited higher resistance to thermal and hydrothermal sintering and its price is relatively lower. However, Pd-performance is worse than that for Pt in presence of poisons (e.g., sulfur-containing pollutants) [15], although Pd demonstrated to be less inhibited by CO [19]. In literature, other metals such as, Ag, Au [20–22] or bimetallic catalysts (Pt-Au) [23] were active in VOCs oxidation.

Noble metals are generally supported on transition metal oxides such as γ-Al$_2$O$_3$, SiO$_2$, TiO$_2$, ZrO$_2$, Fe$_2$O$_3$, CeO$_2$, MnO$_x$, etc. [22,24–27]. The selection of the support is an important issue, since properties such as acid or basic character, hydrophobicity, precursor and porosity, influence on the activity of the supported active phase. In this context, γ-Al$_2$O$_3$ and CeO$_2$ were widely studied due to the large stability and surface area, which increases the metal dispersion and VOCs adsorption. Supports with acid strength, namely γ-Al$_2$O$_3$, SiO$_2$ and ZrO$_2$, improved the catalytic performance over supported Pt and Pd catalysts [28,29]. Pt, Rh, Pd and Au catalysts supported on TiO$_2$ showed high conversions in VOCs oxidation, like formaldehyde, which was even oxidized at room temperature by using Pt/TiO$_2$. The performance of these noble metals supported on TiO$_2$ was closely related to their capacities on the formation of formate species and the formate decomposition into CO species [30]. The oxidation of formaldehyde has also been investigated with Pt catalysts supported on Fe$_2$O$_3$ [31] and MnO$_2$ [32]. Among noble metals, Au nanoparticles can be exceptional active when they are highly dispersed on metal oxides, such as Fe$_2$O$_3$, Co$_3$O$_4$, CeO$_2$, TiO$_2$ and Mn$_2$O$_3$ [2,22,33]. Although the catalytic performance of supported Au catalysts markedly depends on the size of Au particles and the support nature, which are affected by the synthesis method and pretreatment conditions [34]. For the case of Au catalysts and other noble metals, CeO$_2$ is an excellent support due to the enhancement of the fixation and final dispersion of Au particles, and mainly the strong oxygen storage capacity which increases the mobility of the lattice oxygen and provides adequate oxidation state of Au particles [35]. For instance, Au catalysts supported on CeO_2 showed better activity than those prepared on TiO_2 and γ-Al_2O_3 for the oxidation of toluene [36]. In another study, benzene, toluene and xylenes (BTX) oxidation was studied over Au particles deposited on ZnO, Al_2O_3 and MgO, Au/ZnO presenting higher activity than Au/Al_2O_3 and finally, Au/MgO. This behavior was attributed to the strong metal–oxide interaction, which was favored from the small lattice parameter difference between Au {111} and ZnO {101} lattice planes [37]. Other supports like zeolites (MCM-41, H-ZSM-5) [38,39], pillared interlayered clays [40,41] or perovskites [42–44] were also employed as catalyst supports in the catalytic combustion of VOCs.

Water vapor is often found in industrial air streams and also is generated from VOCs oxidation. In general, water vapor is considered to act as an inhibitor of VOCs oxidation due to the blockage of active surface sites. Metal oxides have certain hydrophilic character and thereby some affinity by water, especially at low temperatures. In this context, the inhibition of supported Pt and Pd catalysts by water was already reported for the oxidation of both propane and propene [45]. This deactivation by water vapor was also observed for oxidations of ethylacetate and benzene over Pt/γ-Al_2O_3 and Pt/TiO_2-WO_3 catalysts [17,46]. To overcome this limitation, the use of hydrophobic supports, such as porous styrene divinylbenzene copolymer (SDB) was investigated for toluene oxidation [47], the catalytic activity being enhanced by expelling water from the catalyst surface, but Pt sintering being also increased.

Carbon materials are also considered an excellent hydrophobic support with high activity in VOCs oxidations (Table 1). Our research group has proposed the use of carbon materials as noble metal catalyst supports for the gas-phase combustion of VOCs. This type of support presents superior properties, such as a large surface area, high porosity, easily tuneable surface chemistry and hydrophobicity. Thus, the active sites will be deposited throughout the support, avoiding metal sintering and inhibition by water vapor at low temperatures [47]. In the present work, the performance of carbon materials as noble metal supports in the gas-phase combustion of BTX and acetone is reviewed. The influence of the porosity, surface chemistry and hydrophobicity of the carbon support on the active phase dispersion and the catalytic activity were addressed and compared to metal oxide-based catalysts.

Table 1. Catalytic oxidation of benzene, toluene and xylenes (BTX) over noble metal catalysts supported on carbon materials.

Catalyst	VOCs type	Gas mixture	VHSV (h^{-1})	$*T_{50}$ (°C)	Ref.
Pt/AC	Toluene	90 ppm/air mixture	22,200	~160	[47]
		442 ppm/air mixture	13,500		
Pt/AC	Benzene	640 ppm/air mixture	21,500	130	[48]
	Benzene	2000 ppm/air mixture		150	
	Toluene	2000 ppm/air mixture		~165	
	Xylenes	2000 ppm/air mixture		~185	
Pt/AC	Benzene	1000 ppm/air mixture	8000	~175	[49]
	Toluene			~195	
	o-Xylene			~240	
Pd/AC	Benzene	1000 ppm/air mixture	19,000	~290	[50]
	Toluene			~290	
	Xylenes			~425	
Pd/AC	o-Xylene	1500 ppm/air mixture	50,000	~175	[51]
Pd/AC-zeolite-CeO$_2$	p-Xylene	1000 ppm/air mixture	7000	175	[52]
		3000 ppm/air mixture		~180	
Pt/ACF	Toluene	1200 ppm/10% O$_2$	15,000	~140	[53]
Pd/ACF				~145	
Pt/MWCNT	Benzene	100 ppm/air mixture	75,000	~115	[54]
	Toluene			110	
	o-Xylene			~105	
Pt/MWCNT	Toluene	400 ppm/6% O$_2$	156,000	~140	[55]
Pt/CA	Toluene	0.1 vol.%/air mixture	-	~145	[56]

* T_{50} = Light-off temperature for a 50% BTX conversion; AC = activated carbon; ACF = activated carbon fiber; CA = carbon aerogel; MWCNT = multi-walled carbon nanotubes.

2. Results and Discussion

2.1. Supported Catalysts versus Doped Catalysts

The employment of advanced carbon materials extends the ways to develop combustion catalysts by using different approaches and formats, such as powders, grains, pellets, films, coatings on different monolith or foam structures and so on (Figure 1). Supported catalysts are traditionally prepared by a deposition method (e.g., impregnation, chemical vapor deposition–CVD) on the corresponding carbon support, previously prepared. Some years ago, our research group has proposed the synthesis of metal-doped carbon aerogels and their direct application as catalysts. These metal-doped materials are prepared by solving the precursor metal-salt into

91

a starting solution of organic precursors, which is decomposed during a thermal treatment performed over the organic polymeric aerogel or xerogel [57]. Thus, the metal is reduced during carbonization by the organic matrix and an additional pretreatment is not needed, in contrast of supported catalysts. The metals used for doping are also active along all synthesis steps of the material, *i.e.*, polymerization, carbonization and/or activation. Interactions between organic–inorganic phases determine the distribution and dispersion of metals, sintering resistance, porosity, graphitization, formation of different active phases, *etc.* Therefore, new challenges are possible in catalysis because resulting metal particles are active in many catalyzed processes.

Figure 1. Carbon materials obtained in different formats: grains, powders, pellets, films and coatings on monolith and foam structures.

The issue of supported catalysts *versus* doped catalysts was studied with two Pt-catalyst series in the gas-phase combustion reaction of toluene [56]. The materials were prepared according the recipes regarded in Table 7. The supported Pt catalysts (1 wt.% Pt loading) were prepared by impregnation of an organic aerogel carbonized at 500 or 1000 °C with an aqueous solution of $[Pt(NH_3)_4]Cl_2$ (*i.e.*, A500-Pt and A1000-Pt, respectively). The same precursor salt was used in the starting solution of Pt-doped catalysts (*i.e.*, B500-Pt and B1000-Pt), a similar Pt content (around 1.1 wt.%) than the impregnated catalysts being determined by thermogravimetric (TG) analysis after carbonization.

The thermal stability in air of carbon-based catalysts is an important parameter to study in reactions or processes requiring high and moderate temperatures. It depends not only on the previous thermal treatments of supports but also on the porous texture, active phase and metal dispersion. In general, the carbon supports are more stable when they are carbonized at higher temperature, while many metals may catalyze the carbon combustion reaction. For the Pt-catalysts series, the burn-off (B.O.) of the support was practically negligible below 300–400 °C (Figure 2), which was much higher than the temperature used to perform the catalytic oxidation of VOCs and thereby, the support gasification did not occur.

The toluene conversion to CO_2 was obtained by using supported Pt-catalysts (A500-Pt) and Pt-doped catalysts (B500-Pt) on carbon aerogels (Figure 3). Only CO_2 and water were found as reaction products, while partial oxidation products including CO and oxygenated organic compounds, were not detected by both gas chromatography (GC) or mass spectrometry (MS) techniques. The total toluene conversion was achieved below 200 °C in both cases, although differences in the activity were observed to depend on several parameters, e.g., Pt-particle size. For catalysts prepared with supports at 500 °C, Pt-particle size was 3.9 nm for A500-Pt after H_2-pretreatment, while particles of 12 nm were obtained for B500-Pt. The higher activity of supported catalysts (A500-Pt) was related with a better Pt-dispersion and an easier accessibility of pollutants to Pt-particle sites. Moreover, in spite of the previously described advantages of metal-doped carbon gels, some metal particles were encapsulated by the organic matrix and thereby inactive in catalysis. In this context, both preparation methods present clear advantages and drawbacks, and the catalyst preparation (supported *vs.* doped) should be determined by the experimental conditions and the final application.

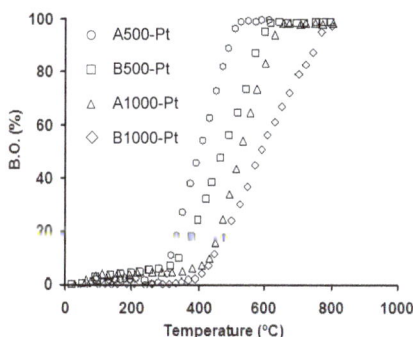

Figure 2. Percentage of burn-off (B.O.) in air against temperature for supported catalysts (A-samples) or doped (B-samples) carbonized at 500 and 1000 °C. Figure reprinted with permission from [56]. Copyright 2004, Elsevier.

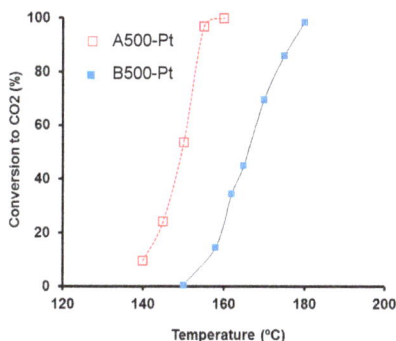

Figure 3. Catalytic performance obtained for supported Pt and Pt-doped catalysts on carbon aerogels in the toluene combustion ([toluene] = 0.45 vol.%, $Q = 3.6 \, L \, h^{-1}$, 0.10 $g_{catalyst}$). Figure adapted with permission from [56]. Copyright 2004, Elsevier.

2.2. Influence of the Support Porosity

Light-off curves for the A1000-Pt catalyst pre-treated in H_2 were obtained under two different toluene concentrations in the feed stream (Figure 4). These results showed that the light-off curve was shifted to a lower temperature when the toluene concentration decreased from 0.45 to 0.10 vol.%, in agreement with VOCs oxidation obtained by other Pt catalysts [48,58]. In addition, the conversion to CO_2 was lower than the total conversion of toluene when the catalyst was tested at low reaction temperatures, showing the participation of VOCs adsorption in the removal of pollutants under these experimental conditions, since partial oxidation products were not detected. Nevertheless, both conversions were matched when the reaction temperature increased due to a less favored adsorption of toluene and the enhancement of the combustion process.

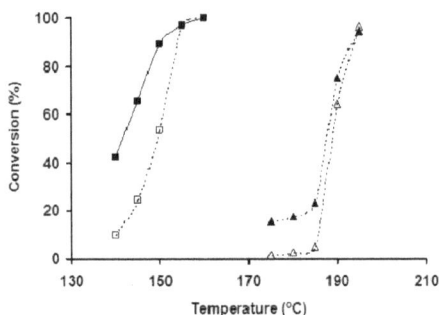

Figure 4. Light-off curves for A1000-Pt at two different toluene concentrations in the feeding gas: 0.10 vol.% (squares) and 0.45 vol.% (triangles). Open symbols: conversion to CO_2. Closed symbols: conversion of toluene (Q = 3.6 L h^{-1}, 0.10 $g_{catalyst}$). Figure reprinted with permission from [56]. Copyright 2004, Elsevier.

VOCs' adsorption on the carbon support is an important key for the respective oxidation because it increases the surface density of adsorbed molecules in the metal-carbon interphase region and enhances the catalytic oxidation at low temperatures [17,48]. VOCs' adsorption is obviously controlled by the physical (porosity) and chemical (interactions) properties of carbon supports.

The macro-mesoporosity of carbon aerogels is commonly associated with the inter-particle voids, and thereby, can be controlled by modifying the synthesis conditions during the polymerization. On the other hand, the microporosity is generated by the release of gases during carbonization [57,59]. In this context, the pore size distribution (PSD) of the supports (carbon aerogels) was modulated by changing the polymerization catalyst (Table 7). Thus, when Li_2CO_3 was used as polymerization catalyst, the resulting carbon aerogel was mesoporous, whereas a macroporous support was produced with Cs_2CO_3. The analysis by mercury

94

porosimetry indicated that both carbon aerogels had a homogenous and unimodal PSD with pore sizes of 25 and 96 nm, respectively. The external and BET surface areas, as well as the meso- and macropore volumes are regarded in Table 2.

Table 2. Textural characterization of the carbon aerogels.

Support	*S_{BET} ($m^2 g^{-1}$)	ρ ($g cm^{-3}$)	V_{meso} ($cm^3 g^{-1}$)	V_{macro} ($cm^3 g^{-1}$)	S_{ext} ($m^2 g^{-1}$)
ALi900	902	0.99	1.06	0.00	191
ACs900	758	0.80	0.07	1.20	59

* S_{BET} = BET surface area; ρ = bulk density; V_{meso} and V_{macro} = meso- and macropore volume determined by mercury porosimetry; S_{ext} = surface area due to macropores and mainly mesopores.

The mesoporous and macroporous aerogels (ALi900 and ACs900) were used to prepare supported Pt-catalysts, which after H_2-pretreatment at 300 °C showed a Pt-particle size (d_{Pt}) around 1.5 nm regardless the support used. When the catalytic activity of both catalysts pretreated at 300 °C was analyzed in the combustion of benzene (Figure 5), the only parameter affecting on their activity was the porous texture, since the surface oxygen content, pH_{PZC} (point of zero charge) and Pt-dispersion were similar. ALi900Pt was more active than ACs900Pt for the temperatures range studied, benzene being totally and selectively converted to CO_2 at only 160 °C. This larger catalytic activity of ALi900Pt is related with the role of mesoporosity and S_{ext} of the support, as previously indicated. In the mesoporous support, benzene molecules easily access to the porosity and are adsorbed throughout the large S_{ext} containing a high concentration of Pt-particles. Thus, the high benzene concentration near to reactive Pt-Particles enhanced the benzene oxidation at low temperatures.

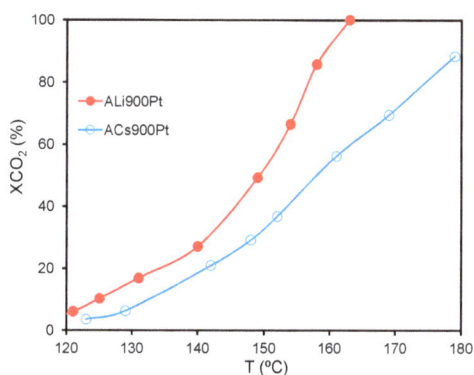

Figure 5. Influence of the macro-mesoporosity of supports on the benzene catalytic combustion ([benzene] = 1000 ppm, Q = 3.6 L h^{-1}, 0.10 $g_{catalyst}$). Figure adapted with permission from [61]. Copyright 2010, Elsevier.

The role of the support porosity was also assessed by Pd-catalysts supported on honeycomb ceramic monoliths [60], in particular three different carbon-based monoliths and their catalytic performance was studied in the total gas-phase combustion of *m*-xylene. The first monolith (HPM) was a square channel cordierite modified with α-Al$_2$O$_3$, on which a carbon layer was deposited (Figure 6c). The other two monolithic supports (WA and WB) were carbon/ceramic composites with different porosity (Figure 6b). A cross-section detailed of the channels of a monolith coated with a γ-Al$_2$O$_3$ layer is also shown in Figure 6a for comparison.

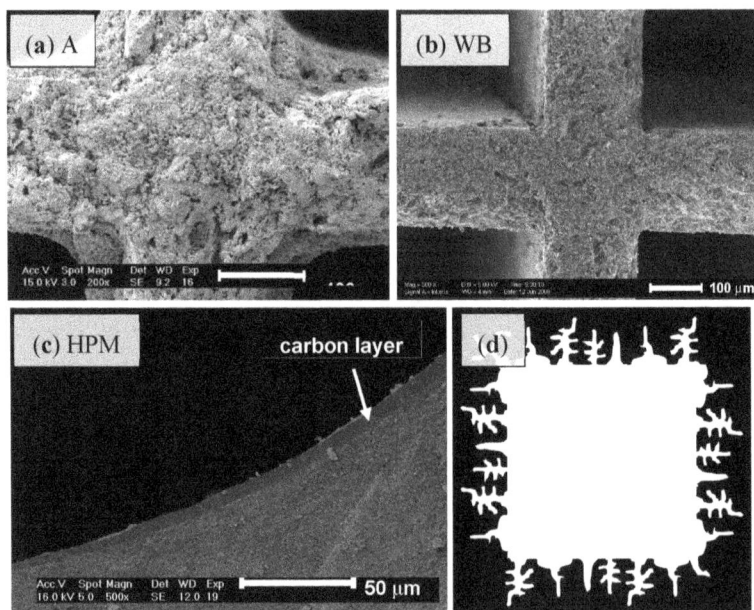

Figure 6. SEM micrographs of the monolith channels: details of (**a**) γ-Al$_2$O$_3$ layer; (**b**) carbon/ceramic composite and (**c**) carbon coating. (**d**) Model of the porosity developed for carbon-based monoliths. Figures reprinted with permission from [60,62,63]. Copyright 2008, 2009, Elsevier.

The textural properties of the carbon-based monoliths were analyzed (Table 3), different porosities being determined depending on the type of support used. In this context, the carbon layer in HPM presented only small micropores, since S_{BET} was much lower than S_{CO2}. Both WA and WB monoliths had a heterogeneous micropore size distribution (*i.e.*, S_{BET} was higher than S_{CO2}), while WA presented only macropores and WB had meso- and macropores (Figure 6d) and consequently, a larger S_{ext}.

Table 3. Surface characteristics of the monolithic supports. Data in parenthesis are given per gram of carbon.

Monolith	*S_{CO2} (m^2 g^{-1})	S_{BET} (m^2 g^{-1})	S_{ext} (m^2 g^{-1})	V_{macro} (cm^3 g^{-1})	V_{meso} (cm^3 g^{-1})
WA	329 (947)	474 (1366)	4 (12)	0.325 (0.937)	0
WB	242 (782)	460 (1489)	62 (199)	0.233 (0.754)	0.138 (0.447)
HPM	17 (269)	(2)	(<1)	(n.d.)	(n.d.)

 * By application of Dubinin-Radushkevich equation to CO$_2$ adsorption data; n.d. = not detected.

Pd-catalysts supported on the previous monoliths presented a comparable Pd loading with very similar Pd particle sizes (around 5–6 nm), although differences were found in their location. Thus, Pd was situated only in the macropores of the Pd/WA catalyst, while for Pd/WB the Pd was distributed throughout the mesoporous texture. In the case of Pd/HPM, Pd-particles were deposited on the external surface of the carbon layer, since the size of the metal particles avoided their deposition in the respective micropores.

The reaction rates of the monolithic catalysts were obtained for the gas-phase catalytic combustion of *m*-xylene (Figure 7). The total combustion of *m*-xylene was reached at 170 °C for Pd/WB, which was the most active catalyst, with a total selectivity to CO$_2$ and H$_2$O. In addition, the catalytic performance was very different, the rate of the catalysts being varied as follows: Pd/HPM < Pd/WA < Pd/WB, *i.e.*, in according to the external surface area of the corresponding support (Table 3). Therefore, the macropores and mainly mesopores, play an important role in this type of gas-phase reactions, improving the contact between Pd-particles and *m*-xylene molecules, in agreement with previous results for carbon aerogels supported catalysts.

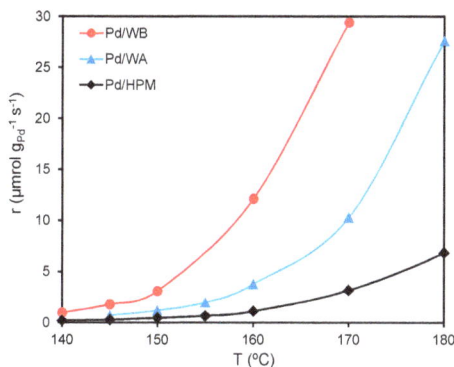

Figure 7. Catalytic performance of the monolithic catalysts normalized per gram of Pd in the *m*-xylene combustion ([*m*-xylene] = 0.10 vol.%, Q = 2000 m^3$_{gas}$ h^{-1} m^{-3}$_{monolith}$). Figure adapted with permission from [60]. Copyright 2008, Elsevier.

Since the carbon materials are highly porous solids, the preconcentration of VOCs in their surface by adsorption facilitates their combustion (as indicated), but also the synergy adsorbent/catalyst will have an important role. Therefore, the adsorptive behavior of this type of materials in the removal of VOCs will depend markedly on the operational conditions. In this context, the adsorption of different VOCs over carbon aerogel-based catalysts was studied by TG in dynamic mode (Figure 8) [59,64]. As expected, the adsorption capacity and the adsorption rate for A500-Pt decrease with higher temperatures (Figure 8a), since adsorption only takes place in progressively narrower pores and consequently, the adsorption capacity decreases. On the other hand, adsorption/desorption cycles for A500-Pt and A1000-Pt at constant temperature (Figure 8b) showed a comparable toluene adsorption rate for both aerogels regardless the catalyst and the carbonization temperature used, because the mean micropore sizes (L_0) were always around 0.57 ± 0.03 nm. However, the adsorption capacity for both samples was favored for the sample carbonized at higher temperature (A1000-Pt) due to a larger development of its microporosity, compared to A500-Pt. These results also pointed out that the toluene adsorption is reversible by thermal treatment of the corresponding support (Figure 8b).

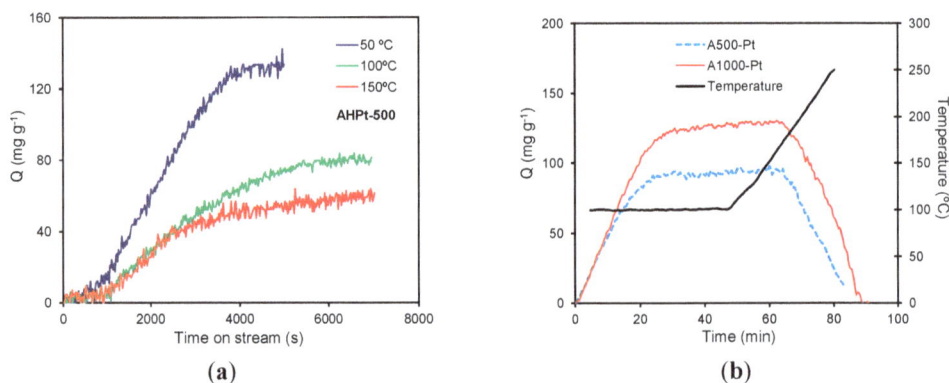

Figure 8. Dynamic toluene adsorption: (**a**) effect of the adsorption temperature and (**b**) effect of the carbonization temperature ($Q = 3.6$ L h^{-1}, 0.10 g$_{catalyst}$). Figures adapted with permission from [64]. Copyright 2011, Elsevier.

The catalytic performance of A500-Pt was assessed in the combustion of p-xylene at 170 °C and acetone at 210 °C (Figure 9a,b, respectively). The total conversion (X_T) of p-xylene was reached and maintained at 170 °C over time, however the corresponding conversion to CO_2 (X_{CO2}) showed small positive and negative oscillations around 100% (Figure 9a). For the acetone combustion (Figure 9b), the total conversion was lower than that obtained for p-xylene, *i.e.*, acetone is less reactive than p-xylene. In addition, both X_T and X_{CO2} curves showed large oscillations, in particular X_{CO2} under these experimental conditions. This oscillatory behavior was reported in the literature as a consequence of adsorbed VOCs or coke-precursor burning with Pt-catalysts supported on inorganic supports such as, zeolites or alumina [65]. However, the toluene adsorption was completely reversible for carbon aerogels with different porous textures as previously showed [59] and thereby, the VOCs interaction with the basic carbon surface should be weaker than that obtained with acidic inorganic supports. Thus, VOCs are only adsorbed and oxidized over carbon-based catalysts without any transformation and deposition of stable coke or coke precursor.

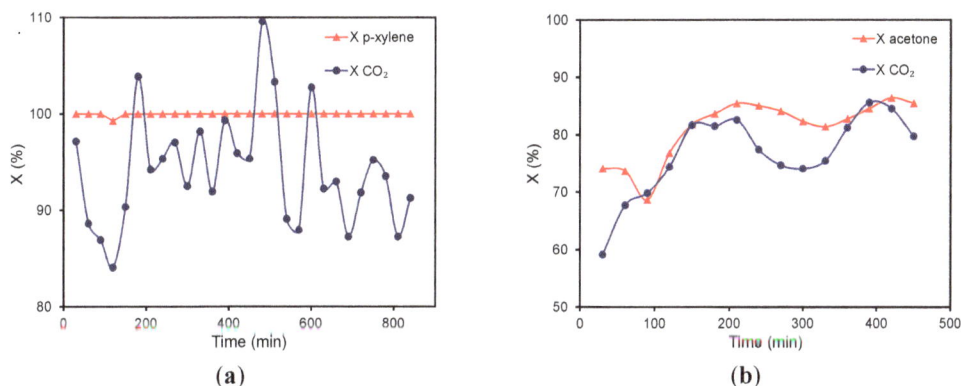

(a)   (b)

Figure 9. Catalytic stability of A500-Pt over the combustion of (**a**) p-xylene at 170 °C and (**b**) acetone at 210 °C ([VOC] = 1500 ppm, Q = 3.6 L h^{-1}, 0.10 g$_{catalyst}$). Figures adapted with permission from [64]. Copyright 2011, Elsevier.

In summary, carbon-based catalysts present a versatile behavior, since they can behave as good adsorbents at low temperatures and as excellent combustion catalysts at high reaction temperatures. The temperature range depends obviously on material properties (e.g., porosity, surface chemistry, active phase and metal dispersion) and on VOCs nature. In this context, O'Malley and Hodnett [66] defined a reactivity pattern for different Pt-inorganic supports as follows: alcohols > aromatics > ketones > carboxylic acids > alkanes, according to the weakness of C-H bonds.

The equilibrium between both adsorption and catalytic processes was studied by changing the weight of catalyst used (0.10 or 0.30 g of A1000-Pt), *i.e.*, by varying the contact time, over the m-xylene combustion at 190 °C (Figure 10). When 0.10 g of catalyst was used, the m-xylene removal reached at the beginning of the reaction was only due to adsorption, since X_{CO2} was null. After saturation (at around 2 h on stream) X_T becomes also very low, but both X_T and X_{CO2} simultaneously increased over reaction time because the m-xylene removal by catalytic combustion increased for 17 h until total conversion. The enhanced catalytic activity can be also explained by the sintering of Pt particles, as previously reported [56]. By increasing the contact time (using 0.30 g of A1000-Pt), the removal of m-xylene was also total during the first 2 h but after saturation, the X_T decay was smaller than that observed when 0.10 g of catalyst was used, because the X_{CO2} evolution was faster due to the combustion process improved by the presence of a larger amount of active sites. In this case, total removal of m-xylene to CO_2 was reached after 7 h, compared to 17 h obtained when 0.10 g of catalyst was used. Therefore, at this temperature the adsorption process occurs initially faster than the combustion reaction, although the catalytic removal of VOCs is progressively favored by the preconcentration on the catalyst surface and by the sintering of Pt-particles. The influence of this parameter will be deeply studied in the subsequent Section. Oscillations in both X_T and X_{CO2} indicate the establishment of a continuous equilibrium between different adsorption/reaction cycles, while X_{CO2} values higher than 100% demonstrated the combustion of compounds previously adsorbed.

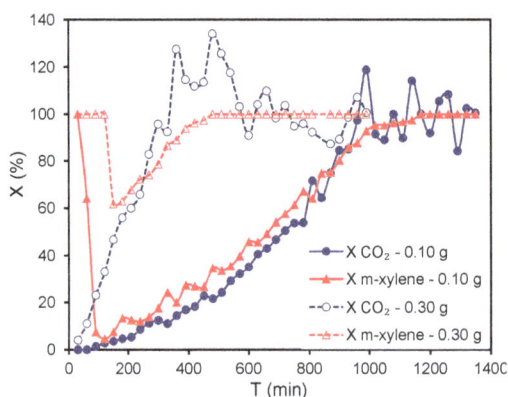

Figure 10. Influence of the contact time on the m-xylene combustion obtained for A1000-Pt at 190 °C ([m-xylene] = 1500 ppm, Q = 3.6 L h^{-1}, 0.10 or 0.30 g$_{catalyst}$). Figure adapted with permission from [64]. Copyright 2011, Elsevier.

2.3. Influence of the Metal Particle Size

The influence of this parameter on the catalytic performance in gas-phase combustion reactions has generated certain controversy. Some authors classify the combustion as a structure–sensitive reaction with the Pt particle size, having a detrimental effect on the conversion [56,67]. However, Tsou *et al.* [65] did not find any dependency on Pt-dispersion for the o-xylene combustion with Pt/HBEA zeolite. Therefore, these discrepancies suggest that the correlation between the Pt particle size and catalytic activity is influenced by the catalyst support selected.

For the sake of elucidating this issue, a Pt catalyst supported on a mesoporous carbon aerogel was pre-treated at different temperatures (400, 600 and 950 °C) in He flow (*i.e.*, APt400He, APt600He and APt950He, respectively) or at 400 °C in H_2 flow (*i.e.*, APt400H$_2$) [67]. The Pt particle size for the different catalysts was determined by H_2 chemisorption (d_{Pt}) and high resolution transmission electron microscopy ($d_{Pt-HRTEM}$), obtaining a well agreement between both techniques (Table 4). The He-pretreatments performed at 400 and 600 °C gave rise to small Pt particle sizes (between 2.3–2.8 nm). When the pre-treatment temperature increased up to 950 °C, the catalyst with the largest Pt-particle size was obtained (9.5 nm). The H_2-pretreatment at 400 °C led to a $d_{Pt-HRTEM}$ value (2.9 nm) even slightly larger than that obtained after He-pre-treatment at 600 °C (2.4 nm). Pt catalysts prepared with identical metal precursor and supported on other carbon materials [56,68] and even zeolites [69] also presented a loss of dispersion after H_2-pretreatments. This behavior is due to the formation of an intermediate hydride, [Pt(NH$_3$)$_2$H$_2$], which is unstable and mobile, leading to an agglomeration of Pt-particles. He-pretreatments give rise to a gradual decomposition of the amino complex along two intermediate reactions: (1) reaction takes place between 275 and 315 °C and (2) reaction between 315 and 370 °C [70].

$$[Pt(NH_3)_4]Cl_2 \rightarrow [Pt(NH_3)_2Cl_2] + 2NH_3 \qquad (1)$$

$$3[Pt(NH_3)_2Cl_2] \rightarrow 3Pt + N_2 + 4NH_3 + 6HCl \qquad (2)$$

Table 4. Hydrogen uptake (Q_t), dispersion (D) and Pt particle size of supported catalysts.

Catalyst	*Q_t (μmol H_2 g^{-1})	*D (%)	*d_{Pt} (nm)	$d_{Pt-HRTEM}$ (nm)
APt400He	12.0	46.8	2.3	-
APt600He	10.0	39.2	2.8	2.4
APt950He	2.9	11.3	9.5	6.6
APt400H$_2$	-	-	-	2.9

* Data obtained by H_2 chemisorption at 25 °C.

The pre-treated catalysts with different Pt particle sizes were tested in the gas-phase *o*-xylene combustion (Figure 11). The absence of CO_2 converted (X_{CO2}, Figure 11b) and the total *o*-xylene conversion ($X_{o-xylene}$, Figure 11a) at low reaction temperatures, indicated an *o*-xylene removal mainly by adsorption. The light-off curves shift to lower temperatures with increasing Pt particle size (Figure 11), APt950He and APt400He being the most and least active catalysts, respectively. Therefore, these results suggest that gas-phase *o*-xylene combustion over these Pt-catalysts is a structure-sensitive reaction.

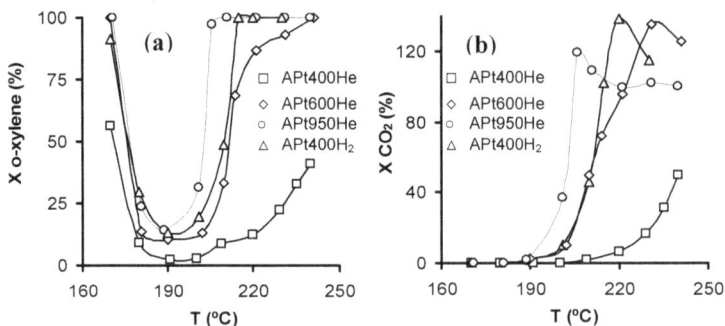

Figure 11. Light-off curves for the supported Pt catalysts: (**a**) conversion of o-xylene and (**b**) conversion to CO_2 ([o-xylene] = 0.10 vol.%, Q = 3.6 L h^{-1}, 0.10 g$_{catalyst}$). Figures reprinted with permission from [67]. Copyright 2005, Elsevier.

Furthermore, the catalyst activation during the reaction by sintering (Figure 10) was also reported by other authors [71,72]. This behavior seems to be markedly affected by the initial metal particle size of the catalyst. In this context, the CO_2 formation as a function of reaction time was monitored at 190 °C (Figure 12), in which no o-xylene combustion to CO_2 was detected (Figure 11b). As previously mentioned, Pt catalysts are activated after a certain period of time under these experimental conditions and the conversion to CO_2 increases (Figure 12). However, this behavior was not observed with APt400He, which is the catalyst with the highest Pt-dispersion.

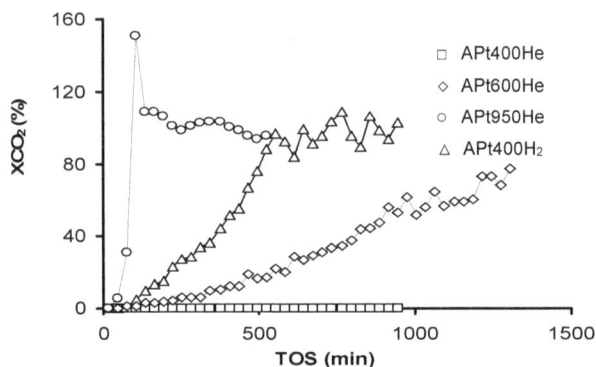

Figure 12. Conversion of o-xylene to CO_2 obtained at 190 °C in function of time on stream (TOS) ([o-xylene] = 0.10 vol.%, Q = 3.6 L h^{-1}, 0.10 g$_{catalyst}$). Figures reprinted with permission from [67]. Copyright 2005, Elsevier.

The catalyst activation during reaction time was also pointed out with two consecutive light-off curves [67]. Light-off curves in the second run were shifted

102

to a lower temperature compared to those obtained with the first run. The difference between both curves, ΔT, is a measurement of the activation degree of the catalyst after the first combustion run. These results showed that the most activated catalyst was APt950He, which was also the most active catalyst during the o-xylene combustion.

2.4. Influence of the Support Surface Chemistry

If there is a highlighted property of the carbon materials, it is their easily tuneable surface chemistry. Its influence on the catalytic performance in gas-phase VOCs combustion was pointed out by using supported Pt-catalysts on oxidized carbon aerogels. Thus, the mesoporous and macroporous carbon aerogel (ALi900 and ACs900, respectively) were oxidized with different agents: hydrogen peroxide (9.8 M H_2O_2 – AX900H samples) and ammonium peroxydisulfate $((NH_4)_2S_2O_8$–AX900S samples) [73], before being used as Pt catalyst supports.

The oxidation treatments performed on the carbon aerogels did not modify significantly their porosity, but their surface chemistry changed from basic to acid as oxygen content increased (Table 5). The porosity preservation allowed us to study accurately the influence of the surface chemistry on the Pt particle size of the corresponding catalysts, which increased for those prepared on oxidized supports (e.g., 1.6 and 2.4 nm for ALi900 and ALi900S, respectively). This loss of Pt-dispersion was more notorious when the amount of oxygen-containing surface groups increased and, in particular, for macroporous oxidized supports. In this context, the Pt particle size increased from 2.4 to 3.9 nm for ALi900SPt pre-treated at 300 and 450 °C, respectively, while a more evident increase was observed for ACs900SPt pre-treated under the same conditions, i.e., d_{Pt} increased from 3.4 to 6.0 nm, respectively. Therefore, the increase of oxygen content in the supports favors the Pt sintering, while large external surface areas (mesoporous supports) prevent in certain extent the sintering, which is more marked with high pre-treatment temperatures.

Table 5. Amounts of CO and CO_2 evolved up to 1000 °C and oxygen content (O_{TPD}) from TPD experiments, measurements of pH_{PZC} and mean Pt particle size for the derivatives Pt/C.

Support/catalyst	CO (μmol g^{-1})	CO_2 (μmol g^{-1})	O_{TPD} (%)	pH_{PZC}	*d_{Pt}(nm)
ALi900	420	61	0.9	10.22	1.6
ALi900H	1110	335	2.9	4.94	2.0
ALi900S	2846	1621	9.7	3.16	2.4 (3.9)
ACs900	534	59	1.0	10.28	1.5
ACs900H	832	364	2.5	6.89	1.8
ACs900S	2651	1522	9.1	3.85	3.4 (6.0)

* Catalysts characterized after a H_2-tretreatmen at 300 °C or 450 °C (data included in brackets).

The catalytic activity of the supported Pt catalysts was determined in the gas-phase combustion of benzene, studying the correlation with the surface chemistry once the porosity was comparable after oxidation treatments. In general, the employment of oxidized supports have two consequences: (i) a decrease of the hydrophobic character with the oxygen content of the support [74], since the oxygen-containing surface groups (called as primary centers) may interact with water molecules by hydrogen bonds; and (ii) a decrease of Pt-dispersion (as indicated). The loss of support hydrophobicity has a detrimental effect on the performance of the catalyst, because the re-adsorption of water generated during the reaction can block the active sites, namely at low reaction temperatures. Nevertheless, when the catalytic behavior for each series is analyzed, the conversion of benzene to CO_2 increased with the order: un-oxidized < H-treated < S-treated supports (Figure 13).

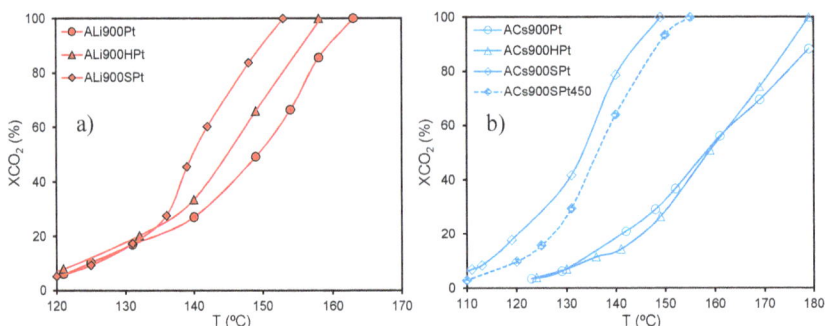

Figure 13. Light-off curves obtained in the benzene combustion for Pt catalysts supported on (**a**) mesoporous and (**b**) macroporous carbon aerogels ([benzene] = 1000 ppm, $Q = 3.6$ L h^{-1}, 0.10 g$_{catalyst}$). Figures adapted with permission from [61]. Copyright 2010, Elsevier.

As previously demonstrated, VOCs combustion with Pt/C catalysts is a sensitive–structure reaction, in which Pt-dispersion would be the main parameter controlling the catalytic activity [67]. The hydrophilic character increases therefore after oxidation, producing a weak effect on the catalytic performance of oxidation catalysts supported on carbon materials. Therefore, the possible water re-adsorption favored by the increased hydrophilicity of the oxidized support is clearly compensated by the increase of activity induced by large Pt particles formed.

2.5. Carbon versus Inorganic Supports: The Influence of the Hydrophobicity

In order to clarify the influence of hydrophobicity of supports, this study was carried out by analyzing the catalytic activity of Pd and Pt catalysts supported on both carbon and inorganic supports, looking for a deeper hydrophobicity difference taking into account that changes in the hydrophobicity of oxidized carbon aerogels did not seem to influence significantly on the catalytic activity. In this context, two types of coatings were developed on a cordierite monolithic substrate. The first one was formed by a γ-Al$_2$O$_3$ layer (A monolith) and the another contained 3.3 wt.% of carbon nanofibers (ACNF monolith) grown *in situ* on the A monolith [63].

The corresponding γ-Al$_2$O$_3$ and CNF layers were thin, homogeneous and well-adhered on the cordierite substrate (Figure 14a). The CNFs layer was developed avoiding the formation of cracks and other defects on the cordierite monolith. Figure 6a and Figure 14b show views of A and ACNF monolith channels, respectively, with different topologies being observed for each monolith. Some individual CNFs can be clearly observed through the monolith surface (Figure 14c). The different morphology obviously influenced on the textural properties of the monolithic supports. Thus, S_{BET} significantly increased with the CNF coating compared to A monolith (Table 6). The analysis of the macro- and mesoporosity indicated that both samples are mainly macroporoses, the macropore volume in ACNF being larger than in A (*i.e.*, 0.157 and 0.134 cm^3 g^{-1}, respectively) and the macropores being narrower after the CNF coating.

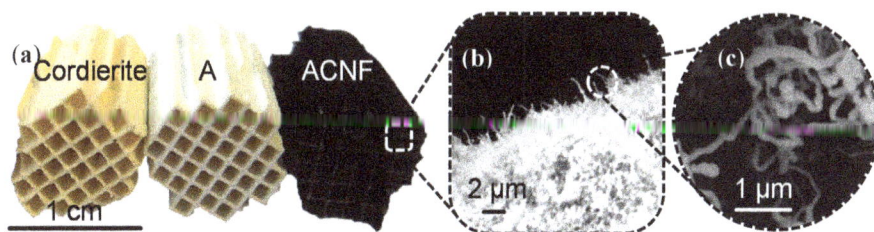

Figure 14. (**a**) Photograph of both γ-Al$_2$O$_3$ and CNF coatings on a cordierite monolith. SEM micrographs of the ACNF monolith: (**b**) cross-section of a monolith channel and (**c**) detail of CNF grown *in situ*.

Table 6. Textural characteristics of the monolithic supports.

Monolith	*S_{BET} (m^2 g^{-1})	V_{macro} (cm^3 g^{-1})	V_{meso} (cm^3 g^{-1})	*,†Enthalpy (J g^{-1})
A	164	0.134	0.027	−138.2
ACNF	371	0.157	0.028	−103.8

* Data given per gram of total coating material; † By immersion calorimetry in water.

Pt and Pd catalysts were prepared on both A and ACNF monoliths by the impregnation method and their catalytic activity was studied in the gas-phase combustion of benzene, toluene and m-xylene (Figure 15a–c, respectively). The conversion of the different VOCs tested, followed the sequence: $X_{benzene} > X_{toluene} > X_{m\text{-}xylene}$ regardless the active phase and type of support used. In addition, Pt catalysts were more active than corresponding Pd-catalysts regardless the type of aromatic compound and support employed, the most active catalysts being always those supported on CNF coated monoliths. Furthermore, catalysts supported on A monoliths seem to burn equally any BTX, while those supported on ACNF showed preference to burn benzene better than toluene or *m*-xylene (Figure 15a–c), namely the Pt/ACNF catalyst.

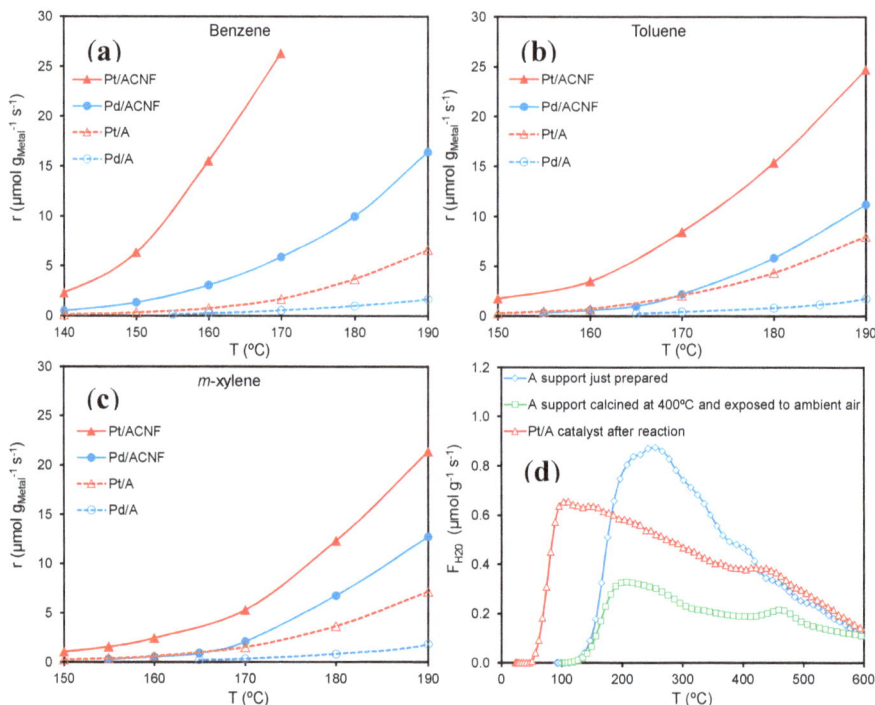

Figure 15. Catalytic activity of the catalysts (normalized per gram of Pt or Pd) as a function of the reaction temperature: (**a**) combustion of benzene, (**b**) combustion of toluene and (**c**) combustion of *m*-xylene. (**d**) Water desorption profiles obtained by TPD. Figures adapted with permission from [63]. Copyright 2006, Elsevier.

A possible explanation of these results was deduced by determining the adsorptive capacity of water for the A support and the Pt/A catalyst. The water desorption profiles obtained by TPD (Figure 15d) indicated that the A support (after calcined at 400 °C and exposed to ambient air at room temperature for a month), is able to reincorporate important amounts of water (like water itself or like hydroxyl groups by reaction with the Al atoms). Since the catalysts were pre-treated in H_2 flow at 300 °C before being used in BTX combustions, water should not appear below 300 °C during the TPD experiment. However, a large amount of water was evolved, which should have been produced during the combustion. Therefore, the water adsorption of the alumina phase could produce a decrease on the activity of the corresponding supported catalysts. These results are in agreement with the lower water immersion enthalpy (Table 6) obtained for ACNF, which corroborates its large hydrophobic character. As a summary, the support hydrophobicity allowed enhancing the activity of supported Pt or Pd catalysts in the gas-phase combustion of BTX.

3. Experimental Section

3.1. Preparation of Pt Catalysts Supported on Carbon Aerogels

Organic aerogels were synthesized by polymerization of resorcinol (R) with formaldehyde (F) in an aqueous solution (W) containing a polymerization catalyst (C) following the methodology suggested by Pekala et al. [14]. After curing and supercritical drying with CO_2, the organic RF aerogels were carbonized in N_2 flow at 500 or 900 °C yielding to the corresponding carbon aerogels, which were used as catalyst supports. Table 7 summarizes the recipe and the polymerization catalyst used: alkali carbonate (Li_2CO_3, Na_2CO_3 and Cs_2CO_3) [75] or tetraammine platinum (II) chloride ($[Pt(NH_3)_4]Cl_2$) [56]. The supports will be referred to the text by indicating the support name (i.e., A, ALi, ACs or B) and the carbonization temperature (500 or 900 °C).

Table 7. Recipes used in the preparation of organic aerogels.

Support	R/F	R/W	R/C	C
A	0.5	0.13	800	Na_2CO_3
ALi	0.5	0.07	300	Li_2CO_3
ACs	0.5	0.07	300	Cs_2CO_3
B	0.5	0.13	800	$[Pt(NH_3)_4]Cl_2$

Two carbon aerogels (ALi900 and ACs900) were oxidized with different agents: (i) concentrated hydrogen peroxide (H_2O_2, 9.8 M) and (ii) a saturated solution of ammonium peroxydisulfate ($(NH_4)_2S_2O_8$) in sulphuric acid (H_2SO_4, 1M) [73]. The oxidation treatments were carried out for 48 h at an ambient temperature. Afterwards, the materials were washed with distilled water and dried at 120 °C for 24 h. The supports oxidized with H_2O_2 and $(NH_4)_2S_2O_8$ will be labelled by adding "H" or "S" to the support name, respectively.

Supported Pt catalysts (1 wt.%) were prepared by the incipient wetness impregnation method using an aqueous solution containing $[Pt(NH_3)_4](NO_3)_2$ or $[Pt(NH_3)_4]Cl_2$ as metal precursor salts. The Pt-doped catalyst (B support) already included the active phase and thereby, no additional Pt was added. Prior to characterization or reaction, the catalysts were reduced in H_2 or He flow at different temperatures (300, 400, 600 or 950 °C) for 3 h. In determined studies (e.g., in the influence of Pt particle size), the reducing agent and temperature used will be also referred to the catalyst name.

3.2. Preparation of Pt Catalysts Supported on Carbon-Based Monoliths

For the HPM monolith, an α-Al_2O_3 layer was applied on cordierite monolithic substrates (supplied by Corning Inc.) by dip-coating method, in order to block the cordierite macroporosity, to prevent deposition of catalytic material in the substrate walls and to round the channel cross section [76]. The cylindrical (length 1.5 cm and diameter 1 cm) cordierite monoliths had square channels, a cell density of 62 cells cm^{-2} (400 cpsi) and a wall thickness of 0.18 mm. Then, a thin carbon layer was deposited by carbonization of a polyfurfuryl alcohol (PFA) coating. The total carbon content of HPM was 6.3 wt.%.

The cordierite monoliths were also coated with a γ-Al_2O_3 layer (11 wt.%) by dip-coating method to yield A monoliths. For ACNF monoliths, Ni (2 wt.%) was firstly deposited by equilibrium impregnation method of A monoliths with an aqueous solution of nickel (II) nitrate. After drying, these materials were calcined at 400 °C for 2 h and after *in situ* reduction for 1 h at 550 °C, CNF were grown on the monoliths under ethylene/N_2 atmosphere for 4 h at the same temperature. The final content of CNF in the monoliths was 3.3 wt.%.

Two composite carbon/ceramic monoliths, WA and WB, (purchased from MeadWestvaco Corporation, USA), were also used as catalyst supports. Both monoliths had a cell density of 400 cpsi, length 1.5 cm, diameter 1 cm and a total carbon content of 34.7 and 30.9 wt.%, respectively.

All monoliths were previously oxidized with an aqueous solution of H_2O_2 9.8 M for 24 h at room temperature to create anchoring sites for metal deposition. The catalysts were deposited on the monoliths by equilibrium impregnation method of the support with an aqueous solution of $[Pt(NH_3)_4](NO_3)_2$ or $[Pd(NH_3)_4](NO_3)_2$, obtaining the corresponding monolithic catalysts. The catalysts will be labelled in the text by using the support name (HPM, WA, WB, A or ACNF) and the metal deposited (Pd or Pt). Prior to characterization or reaction, the catalysts were also reduced in H_2 flow at 300 °C for 3 h.

3.3. Characterization of Supports and Catalysts

The carbonization processes were simulated by thermogravimetric (TG) analysis under N_2 flow using a Shimadzu thermobalance model TGA-50H. The morphology of the samples and adherence of the coatings deposited onto cordierite monoliths were analyzed by scanning electron microscopy (SEM) with a Leo (Carl Zeiss) Gemini-1530 microscopy

Textural characterization was carried out using mercury porosimetry (Quantachrome Autoscan-60) and physical adsorption of N_2 at -196 °C and CO_2

at 0 °C (Quantachrome Autosorb-1). From mercury porosimetry data, V_{meso} and V_{macro} were calculated, which correspond to the volume of pores with diameter between 3.7–50 nm and with diameter larger than 50 nm, respectively. The external surface area (S_{ext}) is due to macropores and mainly mesopores, previously estimated from mercury porosimetry. The BET and Dubinin-Radushkevich equations were applied to the corresponding adsorption isotherms, S_{BET} and S_{CO2} being determined, respectively [77].

The immersion enthalpies into water of the previously outgassed samples (110 °C for 12 h) were measured at 30 °C with an isothermal calorimeter of Tian-Calvet type (Setaram C-80). Corrections corresponding to the bulb breaking energy and to the liquid vaporization energy were also taken into account [78].

The surface chemistry of the supports was characterized by temperature programmed desorption (TPD) and measurements of pH_{PZC} (point of zero charge). TPD experiments were carried out by heating the samples to 1000 °C in He flow at a heating rate of 5 °C min^{-1}. The amount of evolved gases was recorded as a function of temperature using a quadrupole mass spectrometer (Balzers, model Thermocube), as described elsewhere [61]. The oxygen content (O_{TPD}) was calculated from the amounts of CO and CO_2 evolved during the TPD experiments.

Pt dispersion (D) and average Pt particle size (d_{Pt}) were obtained by H_2 chemisorption at 25 °C and high resolution transmission microscopy ($d_{Pt, HRTEM}$). Pt dispersion was obtained from the amount of H_2 chemisorbed assuming a stoichiometric ratio of H_2:Pt = 1:2 (dissociative chemisorption) and the average particle sizes were calculated as $d_{Pt} = 1.08/D$ (nm). HRTEM images were acquired with a Phillips CM-20 electron microscope.

3.4. Gas-Phase Catalytic Combustion of BTX

The catalytic combustion of BTX and acetone was performed in a glass, U-shaped, reactor operated in continuous mode at atmospheric pressure. In a typical run, certain amount of catalyst was placed in the reactor and a feed stream containing He/O_2 mixture (80/20%) was saturated with the hydrocarbon by bubbling through a saturator filled with VOCs cooled at a determined temperature. The total flow of the feed stream was controlled by a flow controller and the reaction was studied at temperatures ranging from 110 to 190 °C.

Analysis of the reaction products was carried out by using a Perkin Elmer (mod. 8500) gas chromatograph (GC) with a thermal conductivity detector and Paraplot Q capillary column. The only primary products found were CO_2 and H_2O, which were also confirmed by mass spectrometry. In all cases, conversion was calculated by VOCs consumption and formation to CO_2.

4. Conclusions

This review has established that carbon supports have shown high efficiency in the development of combustion catalysts. Different types of carbon materials, including carbon aerogels, carbon nanofibers and carbon coatings or structured carbon monoliths with different porosity and surface chemistry were used to develop specific combustion catalysts, when used as support of noble metals (Pt and Pd). Different series of catalysts were prepared, characterized and tested in the combustion of aromatics BTX and acetone. In all cases, total combustion is reached below 190 °C and selectively to CO_2, in spite that the VOCs conversion depends on both VOCs nature and concentration.

The gas-phase catalytic combustion of VOCs is clearly a structure–sensitive reaction, being the active phase dispersion the most influencing parameter. Therefore, the temperature and atmosphere of catalyst pretreatment should be carefully fitted, since its activity increases as the Pt particle size increases, at least up to 10 nm (minimum dispersion studied in the present work). The catalyst activation occurs along the combustion reaction and due to the increasing sintering. However, very large particles should lead to a decrease of activity.

Support mesoporosity favors markedly the active phase dispersion and the concentration of adsorbed molecules. The hydrophobicity changes induced by oxidation treatments of carbon materials have a weak effect on the catalyst performance. Oxidized supports seem also to increase the activity of the supported catalyst, but it is mainly due to the loss of metal dispersion induced by the release of less stable oxygen-containing surface groups acting as Pt anchoring sites. Nevertheless, the influence of the support hydrophobicity on the catalyst performance was clearly pointed out by comparing carbon and inorganic supports, since inorganic supports (e.g., alumina) presented a lower activity due to the high water adsorption capacity, leading to the deactivation of active sites regarding to those supported on carbon materials.

Author Contributions: All the authors of this manuscript participate in the development of the experimental works, including materials and equipment management, analysis and redaction of results for publication.

Conflicts of Interest: The authors declare no conflict of interest.

References

1. Berenjian, A.; Chan, N.; Malmiri, H.J. Volatile Organic Compounds removal methods: A review. *Am. J. Biochem. Biotechnol.* **2012**, *8*, 220–229.
2. Huang, H.; Xu, Y.; Feng, Q.; Leung, D.Y.C. Low temperature catalytic oxidation of volatile organic compounds: A review. *Catal. Sci. Technol.* **2015**.

3. Spivey, J.J. Complete catalytic oxidation of volatile organics. *Ind. Eng. Chem. Res.* **1987**, *26*, 2165–2180.

4. García, T.; Solsona, B.; Taylor, S.H. Naphthalene total oxidation over metal oxide catalysts. *Appl. Catal. B* **2006**, *66*, 92–99.

5. Kim, S.C.; Park, C.Y. The complete oxidation of a volatile organic compound (toluene) over supported metal oxide catalysts. *Res. Chem. Intermediates* **2002**, *28*, 441–449.

6. Kulkarni, D.; Wachs, I.E. Isopropanol oxidation by pure metal oxide catalysts: Number of active surface sites and turnover frequencies. *Appl. Catal. A* **2002**, *237*, 121–137.

7. Li, W.B.; Chu, W.B.; Zhuang, M.; Hua, J. Catalytic oxidation of toluene on Mn-containing mixed oxides prepared in reverse microemulsions. *Catal. Today* **2004**, *93–95*, 205–209.

8. Barbero, B.P.; Costa-Almeida, L.; Sanz, O.; Morales, M.R.; Cadus, L.E.; Montes, M. Washcoating of metallic monoliths with a MnCu catalyst for catalytic combustion of volatile organic compounds. *Chem. Eng. J.* **2008**, *139*, 430–435.

9. Hu, C.; Zhu, Q.; Jiang, Z. Nanosized $CuO-Zr_xCe_{1-x}O_y$ aerogel catalysts prepared by ethanol supercritical drying for catalytic deep oxidation of benzene. *Powder Technol.* **2009**, *194*, 109–114.

10. Kim, S.C.; Shim, W.G. Catalytic combustion of VOCs over a series of manganese oxide catalysts. *Appl. Catal. B* **2010**, *98*, 180–185.

11. Santos, V.P.; Pereira, M.F.R.; Órfão, J.J.M.; Figueiredo, J.L. The role of lattice oxygen on the activity of manganese oxides towards the oxidation of volatile organic compounds. *Appl. Catal. B* **2010**, *99*, 353–363.

12. Yang, P.; Yang, S.; Shi, Z.; Meng, Z.; Zhou, R. Deep oxidation of chlorinated VOCs over CeO_2-based transition metal mixed oxide catalysts. *Appl. Catal. B* **2015**, *162*, 227–235.

13. Sinha, A.K.; Suzuki, K. Three-dimensional mesoporous chromium oxide: A highly efficient material for the elimination of volatile organic compounds. *Angew. Chemie Int. Ed.* **2005**, *44*, 271–273.

14. Jones, J.; Ross, J.R.H. The development of supported vanadia catalysts for the combined catalytic removal of the oxides of nitrogen and of chlorinated hydrocarbons from flue gases. *Catal. Today* **1997**, *35*, 97–105.

15. Liotta, L.F. Catalytic oxidation of volatile organic compounds on supported noble metals. *Appl. Catal. B* **2010**, *100*, 403–412.

16. Li, W.B.; Wang, J.X.; Gong, H. Catalytic combustion of VOCs on non-noble metal catalysts. *Catal. Today* **2009**, *148*, 81–87.

17. Papaefthimiou, P.; Ioannides, T.; Verykios, X.E. Performance of doped Pt/TiO_2 (W^{6+}) catalysts for combustion of volatile organic compounds (VOCS). *Appl. Catal. B* **1998**, *15*, 75–92.

18. Gélin, P.; Primet, M. Complete oxidation of methane at low temperature over noble metal based catalysts: A review. *Appl. Catal. B* **2002**, *39*, 1–37.

19. Patterson, M.J.; Angove, D.E.; Cant, N.W. The effect of carbon monoxide on the oxidation of four C6 to C8 hydrocarbons over platinum, palladium and rhodium. *Appl. Catal. B* **2000**, *26*, 47–57.

20. Li, N.; Gaillard, F.; Boréave, A. Electrochemical promotion of Ag catalyst for the low temperature combustion of toluene. *Catal. Commun.* **2008**, *9*, 1439–1442.

21. Scirè, S.; Liotta, L.F. Supported gold catalysts for the total oxidation of volatile organic compounds. *Appl. Catal. B* **2012**, *125*, 222–246.

22. Carabineiro, S.A.C.; Chen, X.; Martynyuk, O.; Bogdanchikova, N.; Avalos-Borja, M.; Pestryakov, A.; Tavares, P.B.; Órfão, J.J.M.; Pereira, M.F.R.; Figueiredo, J.L. Gold supported on metal oxides for volatile organic compounds total oxidation. *Catal. Today* **2015**, *244*, 103–114.

23. Kim, K.J.; Boo, S.I.; Ahn, H.G. Preparation and characterization of the bimetallic Pt-Au/ZnO/Al$_2$O$_3$ catalysts: Influence of Pt-Au molar ratio on the catalytic activity for toluene oxidation. *J. Ind. Eng. Chem.* **2009**, *15*, 92–97.

24. Dos Santos, A.A.; Lima, K.M.N.; Figueiredo, R.T.; Egues, S.M.D.S.; Ramos, A.L.D. Toluene deep oxidation over noble metals, Copper and Vanadium Oxides. *Catal. Lett.* **2007**, *114*, 59–63.

25. Santos, V.P.; Carabineiro, S.A.C.; Tavares, P.B.; Pereira, M.F.R.; Órfão, J.J.M.; Figueiredo, J.L. Oxidation of CO, ethanol and toluene over TiO$_2$ supported noble metal catalysts. *Appl. Catal. B* **2010**, *99*, 198–205.

26. Escandón, L.S.; Ordóñez, S.; Vega, A.; Díez, F.V. Sulphur poisoning of palladium catalysts used for methane combustion: Effect of the support. *J. Hazard. Mater.* **2008**, *153*, 742–750.

27. Debecker, D.P.; Farin, B.; Gaigneaux, E.M.; Sanchez, C.; Sassoye, C. Total oxidation of propane with a nano-RuO$_2$/TiO$_2$ catalyst. *Appl. Catal. A* **2014**, *481*, 11–18.

28. Yazawa, Y.; Takagi, N.; Yoshida, H.; Komai, S.-i.; Satsuma, A.; Tanaka, T.; Yoshida, S.; Hattori, T. The support effect on propane combustion over platinum catalyst: Control of the oxidation-resistance of platinum by the acid strength of support materials. *Appl. Catal. A* **2002**, *233*, 103–112.

29. Yazawa, Y.; Yoshida, H.; Takagi, N.; Komai, S.-i.; Satsuma, A.; Hattori, T. Acid Strength of Support Materials as a Factor Controlling Oxidation State of Palladium Catalyst for Propane Combustion. *J. Catal.* **1999**, *187*, 15–23.

30. Zhang, C.; He, H. A comparative study of TiO$_2$ supported noble metal catalysts for the oxidation of formaldehyde at room temperature. *Catal. Today* **2007**, *126*, 345–350.

31. An, N.; Wu, P.; Li, S.; Jia, M.; Zhang, W. Catalytic oxidation of formaldehyde over Pt/Fe$_2$O$_3$ catalysts prepared by different method. *Appl. Surface Sci.* **2013**, *285*(Part B), 805–809.

32. Yu, X.; He, J.; Wang, D.; Hu, Y.; Tian, H.; He, Z. Facile Controlled Synthesis of Pt/MnO$_2$ Nanostructured Catalysts and Their Catalytic Performance for Oxidative Decomposition of Formaldehyde. *J. Phys. Chem. C* **2012**, *116*, 851–860.

33. Minicò, S.; Scirè, S.; Crisafulli, C.; Maggiore, R.; Galvagno, S. Catalytic combustion of volatile organic compounds on gold/iron oxide catalysts. *Appl. Catal. B* **2000**, *28*, 245–251.

34. Haruta, M. Size- and support-dependency in the catalysis of gold. *Catal. Today* **1997**, *36*, 153–166.

35. Centeno, M.A.; Paulis, M.; Montes, M.; Odriozola, J.A. Catalytic combustion of volatile organic compounds on $Au/CeO_2/Al_2O_3$ and Au/Al_2O_3 catalysts. *Appl. Catal. A* **2002**, *234*, 65–78.

36. Ousmane, M.; Liotta, L.F.; Carlo, G.D.; Pantaleo, G.; Venezia, A.M.; Deganello, G.; Retailleau, L.; Boreave, A.; Giroir-Fendler, A. Supported Au catalysts for low-temperature abatement of propene and toluene, as model VOCs: Support effect. *Appl. Catal. B* **2011**, *101*, 629–637.

37. Wu, H.; Wang, L.; Zhang, J.; Shen, Z.; Zhao, J. Catalytic oxidation of benzene, toluene and p-xylene over colloidal gold supported on zinc oxide catalyst. *Catal. Commun.* **2011**, *12*, 859–865.

38. Scirè, S.; Minicò, S.; Crisafulli, C. Pt catalysts supported on H-type zeolites for the catalytic combustion of chlorobenzene. *Appl. Catal. B* **2003**, *45*, 117–125.

39. Xia, Q.H.; Hidajat, K.; Kawi, S. Adsorption and catalytic combustion of aromatics on platinum-supported MCM-41 materials. *Catal. Today* **2001**, *68*, 255–262.

40. Aznárez, A.; Delaigle, R.; Eloy, P.; Gaigneaux, E.M.; Korili, S.A.; Gil, A. Catalysts based on pillared clays for the oxidation of chlorobenzene. *Catal. Today* **2015**, *246*, 15–27.

41. Gil, A.; Vicente, M.A.; Korili, S.A. Effect of the nature and structure of pillared clays in the catalytic behaviour of supported manganese oxide. *Catal. Today* **2006**, *112*, 117–120.

42. Giraudon, J.M.; Elhachimi, A.; Leclercq, G. Catalytic oxidation of chlorobenzene over Pd/perovskites. *Appl. Catal. B* **2008**, *84*, 251–261.

43. Arzamendi, G.; de la Peña O'Shea, V.A.; Álvarez-Galván, M.C.; Fierro, J.L.G.; Arias, P.L.; Gandía, L.M. Kinetics and selectivity of methyl-ethyl-ketone combustion in air over alumina-supported PdO_x-MnO_x catalysts. *J. Catal.* **2009**, *261*, 50–59.

44. Takeguchi, T.; Aoyama, S.; Ueda, J.; Kikuchi, R.; Eguchi, K. Catalytic combustion of volatile organic compounds on supported precious metal catalysts. *Top. Catal.* **2003**, *23*, 159–162.

45. Marécot, P.; Fakche, A.; Kellali, B.; Mabilon, G.; Prigent, P.; Barbier, J. Propane and propene oxidation over platinum and palladium on alumina: Effects of chloride and water. *Appl. Catal. B* **1994**, *3*, 283–294.

46. Papaefthimiou, P.; Ioannides, T.; Verykios, X.E. Catalytic incineration of volatile organic compounds Present in industrial waste streams. *Appl. Therm. Eng.* **1998**, *18*, 1005–1012.

47. Wu, J.C.S.; Chang, T.Y. VOC deep oxidation over Pt catalysts using hydrophobic supports. *Catal. Today* **1998**, *44*, 111–118.

48. Wu, J.C.S.; Lin, Z.A.; Tsai, F.M.; Pan, J.W. Low-temperature complete oxidation of BTX on Pt/activated carbon catalysts. *Catal. Today* **2000**, *63*, 419–426.

49. Shim, W.G.; Kim, S.C. Heterogeneous adsorption and catalytic oxidation of benzene, toluene and xylene over spent and chemically regenerated platinum catalyst supported on activated carbon. *Appl. Surf. Sci.* **2010**, *256*, 5566–5571.

50. Bedia, J.; Rosas, J.M.; Rodríguez-Mirasol, J.; Cordero, T. Pd supported on mesoporous activated carbons with high oxidation resistance as catalysts for toluene oxidation. *Appl. Catal. B* **2010**, *94*, 8–18.

51. Huang, S.; Zhang, C.; He, H. *In situ* adsorption-catalysis system for the removal of o-xylene over an activated carbon supported Pd catalyst. *J. Environ. Sci.* **2009**, *21*, 985–990.

52. Jamalzadeh, Z.; Haghighi, M.; Asgari, N. Synthesis and physicochemical characterizations of nanostructured Pd/carbon-clinoptilolite-CeO$_2$ catalyst for abatement of xylene from waste gas streams at low temperature. *J. Ind. Eng. Chem.* **2014**, *20*, 2735–2744.

53. Liu, Z.-S.; Chen, J.-Y.; Peng, Y.-H. Activated carbon fibers impregnated with Pd and Pt catalysts for toluene removal. *J. Hazard. Mater.* **2013**, *256–257*, 49–55.

54. Joung, H.-J.; Kim, J.-H.; Oh, J.-S.; You, D.-W.; Park, H.-O.; Jung, K.-W. Catalytic oxidation of VOCs over CNT-supported platinum nanoparticles. *Appl. Surf. Sci.* **2014**, *290*, 267–273.

55. Lu, C.-Y.; Wei, M.-C.; Chang, S.-H.; Wey, M.-Y. Study of the activity and backscattered electron image of Pt/CNTs prepared by the polyol process for flue gas purification. *Appl. Catal. A* **2009**, *354*, 57–62.

56. Maldonado-Hodar, F.J.; Moreno-Castilla, C.; Perez-Cadenas, A.F. Catalytic combustion of toluene on platinum-containing monolithic carbon aerogels. *Appl. Catal. B* **2004**, *54*, 217–224.

57. Maldonado-Hódar, F.J.; Ferro-García, M.A.; Rivera-Utrilla, J.; Moreno-Castilla, C. Synthesis and textural characteristics of organic aerogels, transition-metal-containing organic aerogels and their carbonized derivatives. *Carbon* **1999**, *37*, 1199–1205.

58. Veser, G.; Ziauddin, M.; Schmidt, L.D. Ignition in alkane oxidation on noble-metal catalysts. *Catal. Today* **1999**, *47*, 219–228.

59. Maldonado-Hodar, F.J.; Moreno-Castilla, C.; Carrasco-Marin, F.; Perez-Cadenas, A.F. Reversible toluene adsorption on monolithic carbon aerogels. *J. Hazard. Mater.* **2007**, *148*, 548–552.

60. Pérez-Cadenas, A.F.; Morales-Torres, S.; Kapteijn, F.; Maldonado-Hódar, F.J.; Carrasco-Marín, F.; Moreno-Castilla, C.; Moulijn, J.A. Carbon-based monolithic supports for palladium catalysts: The role of the porosity in the gas-phase total combustion of m-xylene. *Appl. Catal. B* **2008**, *77*, 272–277.

61. Morales-Torres, S.; Maldonado-Hódar, F.J.; Pérez-Cadenas, A.F.; Carrasco-Marín, F. Design of low-temperature Pt-carbon combustion catalysts for VOC's treatments. *J. Hazard. Mater.* **2010**, *183*, 814–822.

62. Pérez-Cadenas, A.F.; Morales-Torres, S.; Maldonado-Hódar, F.J.; Carrasco-Marín, F. Carbon-based monoliths for the catalytic elimination of benzene, toluene and m-xylene. *Appl. Catal. A* **2009**, *366*, 282–287.

63. Morales-Torres, S.; Pérez-Cadenas, A.F.; Kapteijn, F.; Carrasco-Marín, F.; Maldonado-Hódar, F.J.; Moulijn, J.A. Palladium and platinum catalysts supported on carbon nanofiber coated monoliths for low-temperature combustion of BTX. *Appl. Catal. B* **2009**, *89*, 411–419.

114

64. Maldonado-Hódar, F.J. Removing aromatic and oxygenated VOCs from polluted air stream using Pt–carbon aerogels: Assessment of their performance as adsorbents and combustion catalysts. *J. Hazard. Mater.* **2011**, *194*, 216–222.

65. Tsou, J.; Magnoux, P.; Guisnet, M.; Órfão, J.J.M.; Figueiredo, J.L. Oscillations in the catalytic oxidation of volatile organic compounds. *J. Catal.* **2004**, *225*, 147–154.

66. O'Malley, A.; Hodnett, B.K. The influence of volatile organic compound structure on conditions required for total oxidation. *Catal. Today* **1999**, *54*, 31–38.

67. Padilla-Serrano, M.N.; Maldonado-Hodar, F.J.; Moreno-Castilla, C. Influence of Pt particle size on catalytic combustion of xylenes on carbon aerogel-supported Pt catalysts. *Appl. Catal. B* **2005**, *61*, 253–258.

68. Ubago-Pérez, R.; Carrasco-Marín, F.; Moreno-Castilla, C. Carbon-supported Pt as catalysts for low-temperature methanol decomposition to carbon monoxide and hydrogen. *Appl. Catal. A* **2004**, *275*, 119–126.

69. Dalla Betta, R.A.; Boudart, M. Well-Dispersed platinum on Y Zeolite: Preparation and Catalytic Activity. In Proceedings of 5th International Congress on Catalysis, Palm Beach, FL, USA, 21–25 August 1972; pp. 1329–1341.

70. Richard, M.A.; Pancirov, R.J. Thermal decomposition of tetraamineplatinum(II) chloride by simultaneous TG/DTG/DTA/MS and direct insertion probe mass spectrometry. *J. Therm. Anal.* **1987**, *32*, 825–834.

71. Garetto, T.F.; Apesteguía, C.R. Oxidative catalytic removal of hydrocarbons over Pt/Al2O3 catalysts. *Catal. Today* **2000**, *62*, 189–199.

72. Garetto, T.F.; Apesteguía, C.R. Structure sensitivity and *in situ* activation of benzene combustion on Pt/Al2O3 catalysts. *Appl. Catal. B* **2001**, *32*, 83–94.

73. Moreno-Castilla, C.; Ferro-Garcia, M.A.; Joly, J.P.; Bautista-Toledo, I.; Carrasco-Marin, F.; Rivera-Utrilla, J. Activated Carbon Surface Modifications by Nitric Acid, Hydrogen Peroxide, and Ammonium Peroxydisulfate Treatments. *Langmuir* **1995**, *11*, 4386–4392.

74. Slasli, A.M.; Jorge, M.; Stoeckli, F.; Seaton, N.A. Modelling of water adsorption by activated carbons: Effects of microporous structure and oxygen content. *Carbon* **2004**, *42*, 1947–1952.

75. Morales-Torres, S.; Maldonado-Hódar, F.J.; Pérez-Cadenas, A.F.; Carrasco-Marín, F. Textural and mechanical characteristics of carbon aerogels synthesized by polymerization of resorcinol and formaldehyde using alkali carbonates as basification agents. *Phys. Chem. Chem. Phys.* **2010**, *12*, 10365–10372.

76. Perez-Cadenas, A.F.; Kapteijn, F.; Moulijn, J.A. Tuning the morphology of monolith coatings. *Appl. Catal. A* **2007**, *319*, 267–271.

77. Bansal, R.C.; Donnet, J.B.; Stoeckli, F. *Active Carbon*; Marcel Dekker: New York, NY, USA, 1988.

78. Denoyel, R.; Fernandez-Colinas, J.; Grillet, Y.; Rouquerol, J. Assessment of the surface area and microporosity of activated charcoals from immersion calorimetry and nitrogen adsorption data. *Langmuir* **1993**, *9*, 515–518.

Non-Thermal Plasma Combined with Cordierite-Supported Mn and Fe Based Catalysts for the Decomposition of Diethylether

Quang Hung Trinh and Young Sun Mok

Abstract: The removal of dilute diethylether (DEE, concentration: 150 ppm) from an air stream (flow rate: 1.0 L min^{-1}) using non-thermal plasma combined with different cordierite-supported catalysts, including Mn, Fe, and mixed Mn-Fe oxides, was investigated. The experimental results showed that the decomposition of DEE occurred in a one-stage reactor without the positive synergy of plasma and supported catalysts, by which *ca.* 96% of DEE was removed at a specific input energy (SIE) of *ca.* 600 J L^{-1}, except when the mixed Mn-Fe/cordierite was used. Among the catalysts that were examined, Mn-Fe/cordierite, the catalyst that was the most efficient at decomposing ozone was found to negatively affect the decomposition of DEE in the one-stage reactor. However, when it was utilized as a catalyst in the post-plasma stage of a two-part hybrid reactor, in which Mn/cordierite was directly exposed to the plasma, the reactor performance in terms of DEE decomposition efficiency was improved by more than 10% at low values of SIE compared to the efficiency that was achieved without Mn-Fe/cordierite. The ozone that was formed during the plasma stage and its subsequent catalytic dissociation during the post-plasma stage to produce atomic oxygen therefore played important roles in the removal of DEE.

Reprinted from *Catalysts*. Cite as: Trinh, Q.H.; Mok, Y.S. Non-Thermal Plasma Combined with Cordierite-Supported Mn and Fe Based Catalysts for the Decomposition of Diethylether. *Catalysts* **2015**, *5*, 800–814.

1. Introduction

The use of non-thermal plasma (NTP) for the abatement of volatile organic compounds (VOC) has become increasingly attractive to many researchers in the field of air puification [1–6]. NTP offers a number of advantages over conventional methods for air pollution control, namely usability at ambient temperature and pressure, system compactness, and flexible power adjustment [7,8]. However, the use of NTP alone results in a poor selectivity towards the target compounds and low energy efficiency, which may present barriers to practical applications.

The combination of non-thermal plasma with catalysis is a feasible way to overcome the abovementioned issues as they complement each other. NTP is nonselective in chemical reactions, but it can be ignited under normal conditions.

On the other hand, catalysis is featured as a selective process, but most catalysts are only activated at high temperatures. The most important characteristic of this complementary combination is that plasma enables catalysts to function at low temperatures, offering advantages for the treatment of VOCs in terms of energy efficiency, product selectivity and carbon balance [9,10]. In a plasma-catalysis system, a catalyst can either be included in the same reactor in a one-stage arrangement or it can be located separately in a post-plasma stage. In a one-stage arrangement, the catalyst is in direct contact with the electrical discharge plasma, which contains a variety of reactive species, which can be both short-lived and long-lived, such as excited-state atoms and molecules, radicals, photons, and energetic electrons. In comparison, in a two-stage arrangement, the catalyst is only exposed to long-lived reactive species, namely ozone and, possibly, vibrationally excited species [11].

Among the catalysts that are widely used, iron oxide (Fe_2O_3) has been reported to be superior to the other transition metal oxides for the plasma-driven catalysis of the oxidation of diesel particulate matter [12,13]. Besides, manganese oxides (e.g., MnO_2) were also widely investigated in combination with NTP in both one- and two-stage arrangements for VOC abatement [14–16]. Manganese-based catalysts were proven capable of effectively decomposing ozone to form atomic oxygen, which plays a key role in the oxidation of VOCs [17–19]. Recently, the use of bimetallic Mn-Fe catalysts has gained attraction due to their excellent activity for ozone decomposition [20,21]. Hence, the one-stage combination of NTP with a catalyst that either possesses good oxidative activity or superior ozone destruction capabilities is expected to promote VOC abatement. In this regard, the synergistic effects of plasma and supported catalysts, such as Pt, TiO_2, MnO_x, and CoO_x, have been observed in previous studies [22–25].

In this work, diethylether (DEE), an odorous and highly volatile organic compound, was chosen as a model VOC. The study aimed to investigate the decomposition of DEE using NTP combined with different cordierite honeycomb-supported catalysts, including Fe, Mn, and mixed Mn-Fe oxides. The use of cordierite honeycomb as a catalyst support does not cause a significant pressure drop across the reactor, which is meaningful from a practical viewpoint. A one-stage arrangement was used to examine the performance of the catalysts in terms of the decomposition of DEE in relation to their catalytic activities toward ozone decomposition. Based on the results that were obtained, a hybrid reactor combining one- and two-stage arrangements was proposed with the aim of enhancing catalyst performance with respect to DEE decomposition and optimizing energy efficiency.

2. Results and Discussion

2.1. X-ray Diffraction (XRD) Characterization of Prepared Catalysts

The powder XRD patterns for both the bare cordierite and cordierite-supported catalysts are shown in Figure 1. As can be seen, the separate introduction of the manganese and iron oxides into cordierite led to the appearance of new diffraction peaks. For Fe/cordierite, several peaks were observed at 2θ, namely at $24.0°$, $33.0°$, and $35.6°$, and were assigned to the formation of α-Fe_2O_3 [26]. A new peak that appeared at $32.8°$ for Mn/cordierite was attributed to the presence of Mn_2O_3. Other peaks of this phase (e.g., at $38.3°$ and $55.2°$) overlapped with those of the support [27]. Interestingly, the XRD pattern of cordierite did not exhibit any changes after co-impregnation of the two metal oxides. This observation suggested a strong interaction between the Mn and Fe species, which likely existed as an amorphous mixed oxide. A similar result was reported by Lian *et al.* for a bimetallic Mn-Fe catalyst prepared by a hydrothermal method [20].

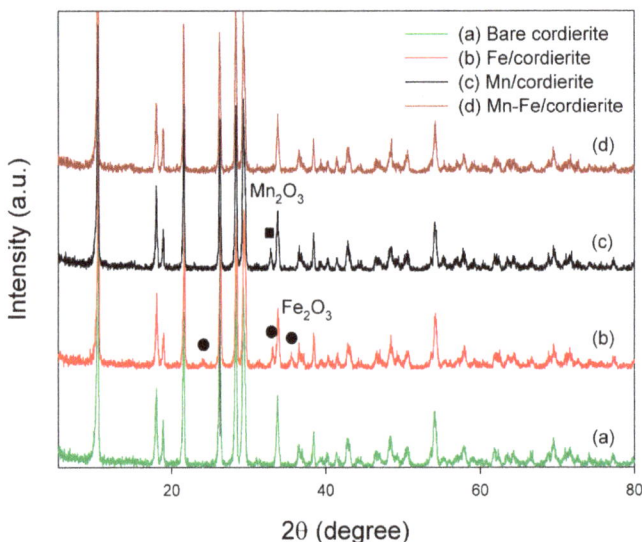

Figure 1. Powder XRD patterns of bare and coated cordierites.

2.2. Catalytic Activities of Prepared Catalysts for Ozone Decomposition

The catalytic activities of the prepared catalysts toward the decomposition of ozone generated in the air plasma were investigated and shown in Figure 2a. The preliminary experiment showed that the bare cordierite support did not show any activity for ozone decomposition. Meanwhile, all the prepared catalysts were observed to achieve a complete destruction of ozone early in the test. However, the performance of the catalysts subsequently decreased to different extents as a function of time-on-stream. Both the Mn and Fe-coated cordierites exhibited a drastic attenuation in their respective catalytic activities. Indeed, the ozone decomposition efficiency of Fe/cordierite reached zero after 6 h, whereas that of Mn/cordierite was reduced to only *ca.* 20% during the same period of time. As expected, the behavior exhibited by Mn-Fe/cordierite differed from those of the other catalysts. The ozone decomposition efficiency for the mixed oxide catalyst gently decreased to *ca.* 89% after 3 h, which exceeded the simple sum of those of the monometallic catalysts. This result clearly showed the synergistic effect of Mn and Fe on the decomposition of ozone. The low crystallinity of the Mn-Fe mixed oxide, as observed by XRD, which indicated more defects and possibly an enhanced surface area, is believed to favor the decomposition of ozone [20].

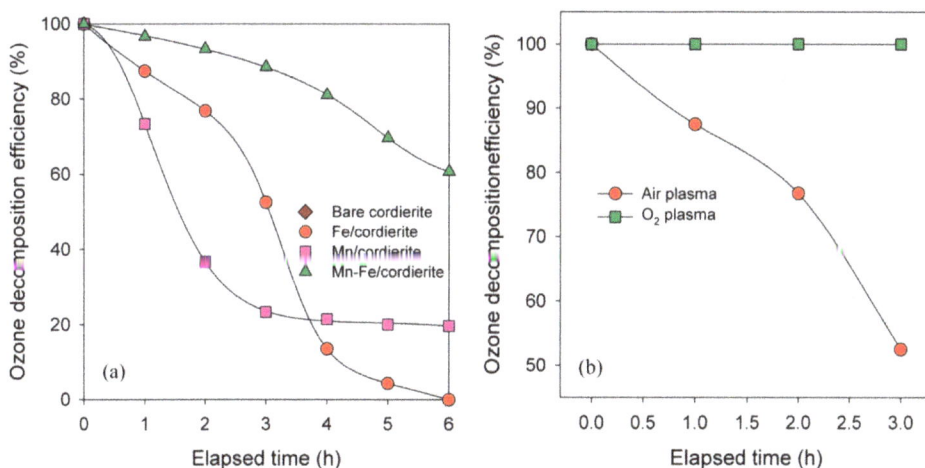

Figure 2. (a) Ozone decomposition efficiency of prepared catalysts with air plasma (initial ozone concentration: 300 ppm); (b) Ozone decomposition efficiency of Fe/cordierite with oxygen plasma (initial ozone concentration: 300 ppm).

The deactivated Fe/cordierite was then regenerated at 250 °C for 1.0 h in ambient air to continue investigating the catalyst performance toward ozone decomposition in the presence of oxygen plasma. As shown in Figure 2b, Fe/cordierite recovered its catalytic activity after the thermal treatment. In contrast to its behavior in the presence of air-generated ozone, catalyst deactivation was not observed during 3.0 h with oxygen plasma. The nitrogen oxides (NO_x) that formed in the air plasma were thus considered to be responsible for the degradation of the catalysts. The strong adsorption of NO_x resulted in a blockage of the active sites that were used for ozone decomposition. In our preliminary experiment, the Fe/cordierite could not be regenerated below a temperature of 150 °C. Hence, the selection of an

119

ozone decomposition catalyst with a long-term stability at low temperature should be considered, especially for the two-stage arrangement. The use of oxygen plasma is ineffective unless a cyclic treatment (*i.e.*, VOC adsorption followed by oxygen plasma oxidation) is applied [28,29].

Samples of Fe/cordierite were collected before and after the ozone decomposition test (with air plasma), after which they were powdered, thoroughly mixed with KBr, and then pelletized for Fourier transform infrared (FTIR) characterization. As shown in Figure 3, the FTIR spectrum of the supported catalyst did not display any characteristic changes after the ozone decomposition reaction, except for a peak that appeared at a wavenumber of 1385 cm^{-1}, which confirmed the presence of nitrate species (NO_3^-) that formed according to the following reactions [30]:

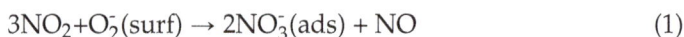

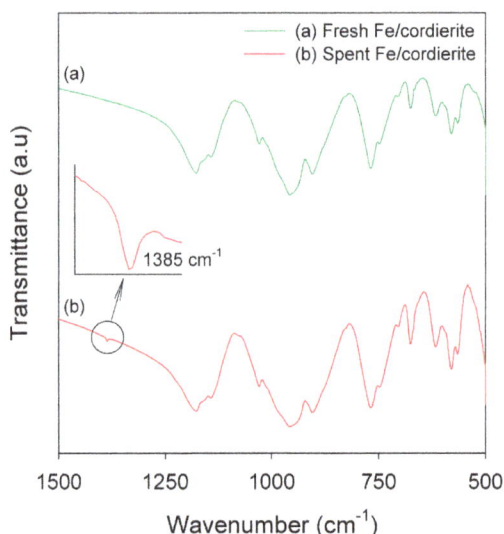

$$3NO_2 + O_2^-(surf) \rightarrow 2NO_3^-(ads) + NO \tag{1}$$

$$2NO_2 + O_2^-(surf) \rightarrow NO_3^-(ads) + NO_2^-(ads) \tag{2}$$

Figure 3. FTIR spectra of Fe/cordierite before and after ozone decomposition with air plasma.

2.3. DEE Decomposition in a One-Stage Reactor

The decomposition efficiency of DEE in the one-stage reactor containing different catalysts is shown in Figure 4. For comparison purposes, the experiment was also performed with the blank dielectric barrier discharge (DBD) reactor. Generally, in all of the cases, the DEE removal rate increased as the specific input energy (SIE) was increased. Here, SIE was defined by the discharge power determined from the Lissajous figure divided by the flow rate. For the blank reactor, *ca.* 71% of DEE conversion was obtained at an SIE of *ca.* 618 J L^{-1}, a result which was in good agreement with a previous report, which described the energy efficiency for DEE decomposition using a double dielectric barrier discharge (DDBD) reactor [8]. The introduction of cordierite honeycomb into the plasma appeared to improve the destruction of DEE, especially for low amounts of SIE. The improvement was probably due to the modification of plasma discharge and the extension of the DEE residence time. Unexpectedly, the separate incorporation of Fe and Mn into the cordierite did not lead to any further improvement in the DEE decomposition, indicating that, in this case, most of the DEE was destroyed by the plasma. Therefore, in the three particular cases, *ca.* 96% of DEE was removed from the gas stream at an SIE of *ca.* 600 J L^{-1}. Jo *et al.* also reported a similar result for the plasma-catalytic decomposition of ethylene using bare and MnO$_2$-coated alumina ceramic membranes [31]. Even worse, the bimetallic Mn-Fe/cordierite catalyst lowered the DEE decomposition efficiency by more than 15% across the whole range of SIEs that were used. Raising the SIE beyond this range to 900 J L^{-1} resulted in only a *ca.* 90% of DEE conversion.

Figure 5 shows the outlet ozone concentration during the decomposition of DEE in the one-stage reactor as a function of the SIE for the different catalysts. For comparison, the ozone concentration of the blank reactor is also included. At low SIE, the presence of the bare and Fe coated cordierites resulted in the formation of more ozone probably by enhancing the local electric field near the catalyst surfaces. As opposed to its reduced ability to affect DEE decomposition, the catalytic activity of the Mn-Fe/cordierite was observed to exceed those obtained using the other methods for ozone decomposition under plasma activation. The ozone concentration of Mn-Fe/cordierite was saturated at only 220 ppm, while the saturation values obtained otherwise ranged to above 600 ppm. This behavior was similar to that observed in the ozone decomposition test (shown in section 2.2). Thus, the one-stage combination of plasma with a well-performing ozone decomposition catalyst (*i.e.*, Mn-Fe/cordierite) did not prove advantageous for the removal of DEE. It seemed that under plasma activation, Mn-Fe/cordierite accelerated the decomposition of O$_3$ to O$_2$, in that the adsorbed O· radicals that formed were not mainly consumed during the oxidation of DEE. As proposed by Li and Oyama [32], the decomposition of ozone is initiated by the dissociative adsorption of ozone to form an oxygen molecule and an atomic oxygen species, which in turn reacts with ozone to finally form molecular oxygen. The mechanism is summarized as follows:

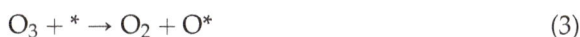

$$O_3 + {}^* \rightarrow O_2 + O^* \tag{3}$$

$$O_3 + O^* \rightarrow O_2 + O_2{}^* \tag{4}$$

$$O_2{}^* \rightarrow O_2 + {}^* \tag{5}$$

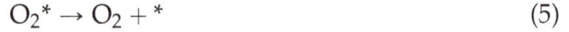

where the * symbol represents an active site. Through reaction (4), the ozone that effectively formed near the catalyst surface might compete with the DEE for atomic oxygen. The low DEE decomposition efficiency with Mn-Fe/cordierite suggested that in plasma, besides active short-lived species, ozone itself played a significant role in the reaction of DEE.

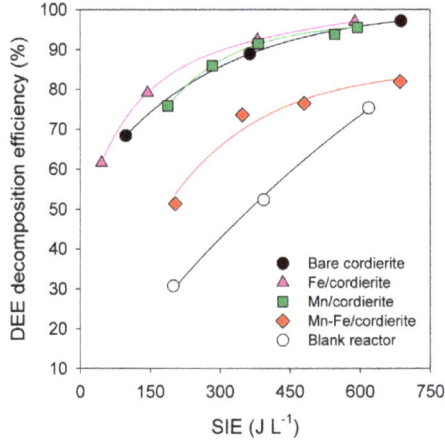

Figure 4. DEE decomposition efficiency in a one-stage reactor with different catalysts.

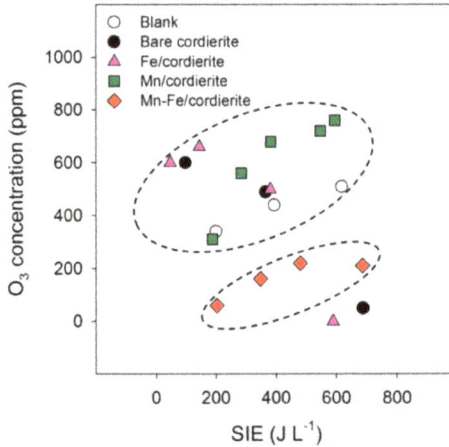

Figure 5. Outlet ozone concentration during decomposition of DEE in one-stage reactor.

In an earlier study, Ogata *et al.* [33] also reported that the use of a Cu-Cr catalyst to dissociate ozone into active atomic oxygen species was not effective for the reaction under plasma discharge, for which the extent of the decomposition of toluene and dichloromethane in a one-stage reactor was almost the same or even worse (for an increasing amount of catalyst) than in a conventional plasma reactor. These authors concluded that, for the single-stage combination (*i.e.*, the one-stage arrangement), the ozone decomposition property of the catalysts was not important. Similarly, Van Durme *et al.* [34] found that when MnO_2-CuO/TiO_2 was subjected to a plasma, it resulted in a low toluene removal efficiency compared to TiO_2, which does not have the ability to decompose ozone. However, a positive effect was observed for the in-plasma supported MnO_x catalyst when dealing with acetone, a persistent compound [16,30], in which case the formation of ozone-induced atomic oxygen therefore becomes important. It is noted that, unlike acetone, the above VOCs (e.g., ethylene, diethyl ether, and toluene) are more reactive with ozone. The gas-phase reaction of ozone with toluene, for example, is slow; however, in the presence of plasma discharge, ozone could be excited, whereby its activity increased [35,36]. The destruction of vibrationally excited ozone, generated by a three-body reaction, was reported to be about 1600 times faster than ground-state ozone due to collisions with oxygen atoms [37,38].

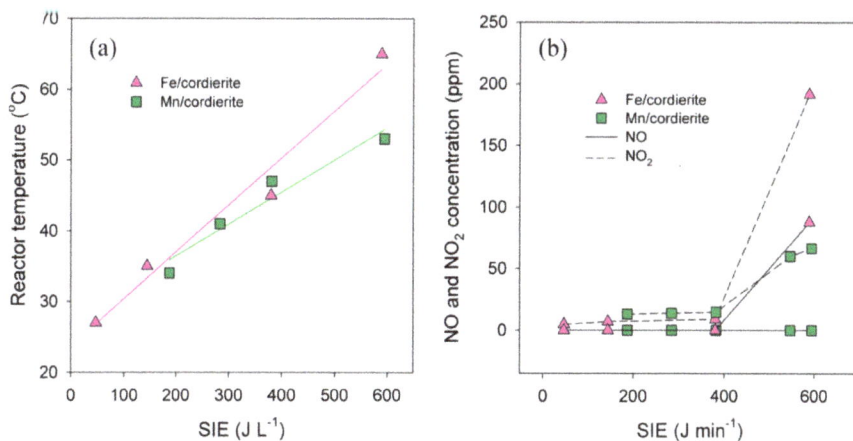

Figure 6. (a) Temperature and (b) NO and NO_2 concentrations of one-stage reactor with Fe and Mn coated cordierites.

Interestingly, the ozone concentrations of the bare and Fe-coated cordierites reached almost zero at a high SIE of *ca.* 600 J L^{-1}. In our preliminary experiment, ozone was only significantly decomposed in the gas phase above 150 °C; however, as seen in Figure 6a, the reactor temperature did not exceed 70 °C, indicating the negligible thermal decomposition of ozone. The occurrence of a zero ozone concentration at high input energies was also reported elsewhere [33,39]. It was proposed that O_3 is consumed by NO, which is favorably formed at high SIE values to produce NO_2, which subsequently reacts with atomic oxygen to regenerate NO [40,41]. The catalytic cycle for O_3 destruction by NO_x is described as follows:

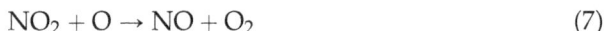

$$NO + O_3 \rightarrow NO_2 + O_2 \tag{6}$$

$$NO_2 + O \rightarrow NO + O_2 \tag{7}$$

The net reaction can be written as

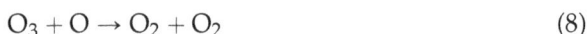

$$O_3 + O \rightarrow O_2 + O_2 \tag{8}$$

The concentrations of NO_x at the outlet of the one-stage reactor for the Mn and Fe coated cordierites were measured during plasma-catalytic oxidation of DEE. As can be seen in Figure 6b, NO was not detected within the investigated range of SIE for Mn/cordierite but appeared at above 400 J L^{-1} for the Fe/cordierite. Meanwhile, the formation of NO_2 slowly took place at SIE below 400 J L^{-1} for both catalysts and then steeply increased as further raising SIE, especially for the Fe/cordierite. At the SIE of *ca.* 590 J L^{-1}, the NO_2 concentration obtained with the Fe/cordierite reached *ca.* 190 ppm, while that of the Mn/cordierite was only *ca.* 60 ppm. The concentration of NO was much less than that of NO_2 because NO rapidly reacted with ozone (reaction (6)) and atomic oxygen (reaction (9)) to form NO_2.

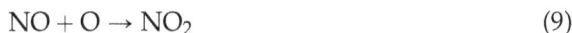

$$NO + O \rightarrow NO_2 \tag{9}$$

The regeneration of NO from NO_2 through reaction (7) was relatively slow. The behaviors of the two catalysts for the formation of NO_x were in contrast to their effects on the O_3 concentration at high SIE shown in Figure 5, indicating that NO_x played significant roles in the suppression of O_3 formation.

Our results indicated that it is possible to improve the performance of the reactor by using a hybrid reactor containing Mn/cordierite in the plasma and Mn-Fe/cordierite in the post-plasma stage. This reactor, henceforth denoted as the Mn+(Mn-Fe) reactor, was proposed for further experiments, in which the post-plasma catalyst bed was retained at room temperature.

2.4. DEE Decomposition in the Mn+(Mn-Fe) Reactor

As seen in Figure 7, the efficiency of the DEE decomposition increased by more than 10% compared to that without Mn-Fe/cordierite at SIE values below 300 J L^{-1}, after which saturation occurred at around 99%. Thus, an appropriate combination of plasma with the two Mn-based catalysts resulted in an obvious improvement of the reactor performance in terms of both DEE conversion and energy consumption.

The ozone that formed during plasma discharge and its catalytic dissociation to produce atomic oxygen in the post-plasma was therefore effective for achieving the oxidation of DEE. The hybrid combination also substantially diminished the emission of ozone, a toxic gas, to levels below 200 ppm (Figure 8). Additionally, it is shown in Figure 7 that the hybrid reactor was also more efficient than the two-stage reactor with Mn-Fe/cordierite as the post-plasma catalyst, especially at low values of SIE. The result came from the advantage of the in-plasma catalysts, which probably altered the plasma discharge mode and enhanced the residence time of gaseous species.

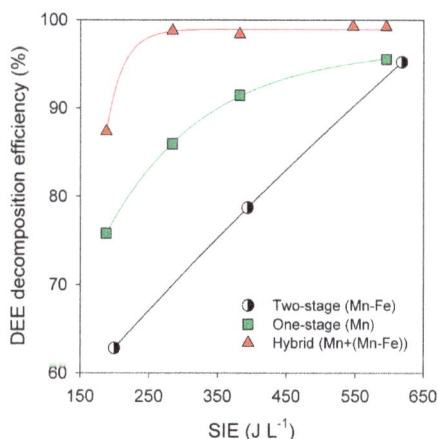

Figure 7. DEE decomposition efficiency of hybrid Mn+(Mn-Fe) reactor, one-stage reactor with Mn/cordierite, and two-stage reactor with Mn-Fe/cordierite.

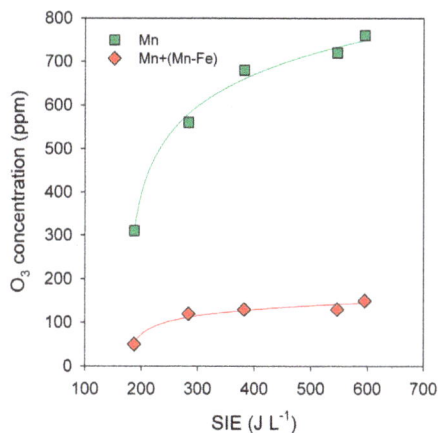

Figure 8. Outlet ozone concentration of hybrid Mn+(Mn-Fe) reactor and one-stage reactor with Mn/cordierite catalyst.

The concentrations of carbon oxides (CO and CO_2) that were detected during DEE decomposition in the Mn+(Mn-Fe) reactor are shown in Figure 9. The amount of CO_2 increased almost linearly as the SIE was increased, whereas that of CO tended to rise exponentially to a maximum within the SIE range. These trends were also observed in previous studies, thereby implying a high degree of oxidation of VOCs to CO_2, which is favored at high SIEs [16,31]; that is, *ca.* 89% of DEE was oxidatively transformed to carbon oxides at around 600 J L^{-1}.

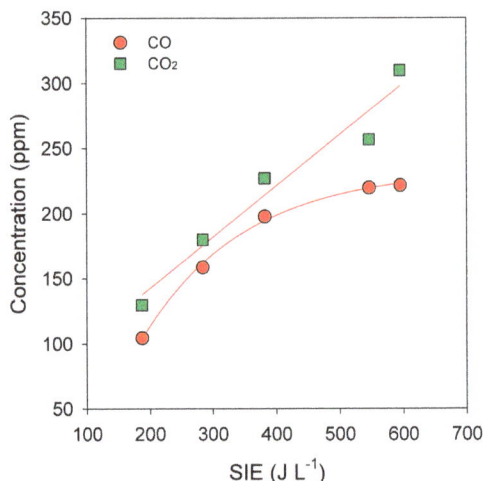

Figure 9. Concentrations of carbon oxides during DEE decomposition in hybrid Mn+(Mn-Fe) reactor.

The stabilities of catalysts used in the Mn+(Mn-Fe) reactor were judged in terms of DEE decomposition efficiency and outlet ozone concentration at SIEs of *ca.* 382 J L^{-1}. In Figure 10, it can be seen that there was no decline in the DEE decomposition efficiency during the course of the experiment. The outlet ozone concentration, however, increased with time-on-stream and reached a steady-state at 180 ppm after 5 h. This indicated that the unreacted DEE and oxidation products, such as CO_x, H_2O, and the intermediates that had reversibly adsorbed on Mn-Fe/cordierite, could partially inhibit NO_x adsorption, thereby slowing down the catalyst deactivation.

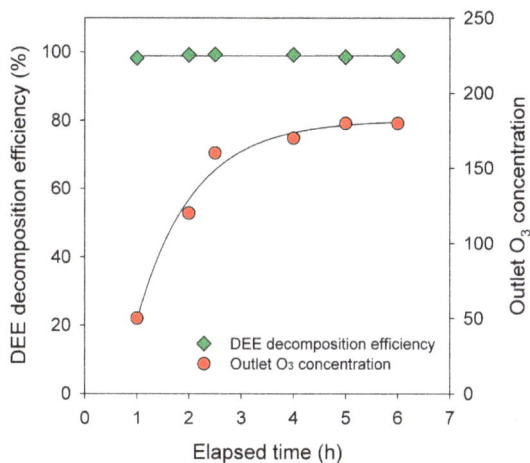

Figure 10. DEE decomposition efficiency and outlet ozone concentration of hybrid Mn+(Mn-Fe) reactor at a SIE of 382 J L^{-1}.

3. Experimental Section

3.1. Apparatus and Materials

The experimental reactor system is schematically shown in Figure 11. The DBD reactor consisted of a dielectric quartz tube (inner diameter: 26 mm, thickness: 4 mm) wrapped in aluminum foil (length: 70 mm) to act as a ground electrode. A copper wire (diameter: 1 mm) coaxially inserted into the quartz tube served as a high voltage (HV) electrode, resulting in a 12.5 mm discharge gap and a 70 mm discharge length. In preparation for the DEE removal test, a one-stage reactor was constructed by placing the cordierite-supported catalysts directly inside the discharge region of the DBD reactor (Figure 11a). The plasma was ignited by using an alternating current (AC) power source at a fixed frequency of 400 Hz. In preparation for the examination of the catalytic activity toward ozone decomposition, the catalysts under investigation were placed downstream from the blank DBD reactor (*i.e.*, similar to a two-stage arrangement, Figure 11b). This arrangement allowed for the initial generation of ozone in the plasma reactor, after which it would be transferred with either an air or an oxygen stream to be subsequently decomposed over the catalyst bed at room temperature. The hybrid Mn+(Mn-Fe) reactor containing Mn/cordierite in plasma region and Mn-Fe/cordierite in post-plasma region for decomposition of DEE is schematically shown in Figure 11c.

The catalyst support had a cylindrical shape (diameter: 26 mm, length: 70 mm, weight: 15.5 g) and was cut from a commercial cordierite monolith (diameter: 144 mm, high: 152 mm, 400 cells per square inch). Different catalysts including Mn, Fe, and mixed Mn-Fe oxides were prepared by a dip-coating method using $Mn(NO_3)_2.6H_2O$

(Junsei Chemical, Tokyo, Japan) and $Fe(NO_3)_3.9H_2O$ (Daejung Chemicals & Metals, Shiheung, Korea) as metal precursors. The Mn-Fe catalyst was prepared by using a mixture of the two nitrates. The cordierite support was dipped into the respective aqueous nitrate solution at room temperature and then dried overnight at 110 °C, before it was calcinated at 500 °C for 3 h. The dipping step was repeated several times before calcination. Finally, cordierite-supported Mn (1.7 wt.%), Fe (1.6 wt.%), and Mn-Fe (1.5–1.5 wt.%) catalysts, denoted as Mn/cordierite, Fe/cordierite, and Mn-Fe/cordierite, respectively, were obtained. The weight percentages of the active components were estimated from the amount of solution that was adsorbed on the cordierite supports.

The concentration of the DEE in the gas (synthetic dry air: 21% O_2 and 79% N_2, by volume) and the flow rate were controlled at 150 ppm and 1.0 L min^{-1}, respectively, by using mass flow controllers (MFC). For the ozone decomposition test, either synthetic dry air or pure oxygen (without DEE) was fed into the reactor at a flow rate of 1.0 L min^{-1} and the ozone concentration was maintained at 300 ppm in both of these cases. All catalysts were thermally treated at 450 °C for 1 h in ambient air before each experiment, unless otherwise noted.

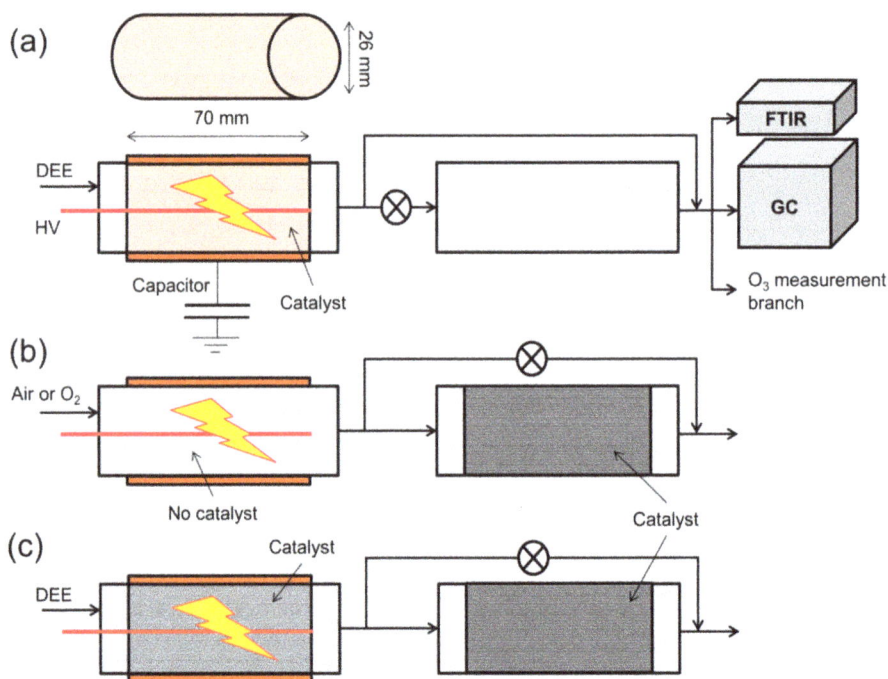

Figure 11. Experimental reactor system. (**a**) One-stage reactor for DEE decomposition, (**b**) Reactor for investigating ozone decomposition on different catalysts and (**c**) Hybrid Mn+(Mn-Fe) reactor for DEE decomposition.

3.2. Measurement Methods

The concentration of DEE was monitored using a gas chromatograph (Bruker 450-GC) equipped with a flame ionization detector (FID) and a 60 m × 0.32 mm BR-624ms capillary column. The oven temperature was ramped from 50 to 60 °C at a rate of 1.0 °C min^{-1}. FTIR spectroscopy (IRPrestige-21, Shimadzu) with a 76 mm-long gas cell was used to measure concentrations of carbon oxides and nitrogen oxides formed during the oxidation of DEE. All samples were collected at a resolution of 2.0 cm^{-1} with 20 scans. The ozone concentration was measured by using a portable gas analyzer (PortaSens II, Analytical Technology). Reactor temperature measurements were performed by using an ethanol thermometer positioned 5 mm downstream from the discharge region. The experimental data were recorded 20 min after plasma ignition, unless otherwise noted.

Charge dissipation in the DBD reactor was measured by measuring the voltage drop on a capacitor (capacitance: 1.0μF) that was connected in series to the reactor. Details of the voltage measurement are described elsewhere [29]. The discharge power was calculated from the so-called voltage-charge Lissajous figures.

The prepared catalysts were characterized by FTIR spectroscopy and an XRD system (D/MAX2200H, Bede 200, Rigaku instruments) equipped with Cu Kα radiation (40 kV, 40 mA).

4. Conclusions

The decomposition of DEE using NTP combined with cordierite-supported Mn, Fe, and mixed Mn-Fe oxides was investigated in this work. It was found that, for the one-stage experimental arrangement, most of the DEE was destroyed by the plasma, because the incorporation of either Mn oxide or Fe oxide in the cordierite did not lead to a significant improvement in the removal of the DEE. Of all the catalysts that were tested, Mn-Fe/cordierite, the catalyst that was most efficient at decomposing ozone, was found to lower the DEE decomposition efficiency. Thus, the catalytic activity for ozone decomposition under plasma discharge conditions was such that DEE removal could not be achieved. However, Mn-Fe/cordierite was able to catalyze the ozonation of unreacted DEE when it was placed downstream from the Mn/cordierite reactor in which the plasma discharge occurred, enhancing the reactor performance in terms of DEE decomposition and energy efficiencies as well as reducing ozone emission.

Acknowledgments: This work was supported by the R&D Program of "Plasma Advanced Technology for Agriculture and Food (Plasma Farming)" through the National Fusion Research Institute of Korea (NFRI), and by the Cooperative Research Program for Agriculture, Science & Technology Development (Project No. PJ008508032012), Rural Development Administration, Korea.

Author Contributions: Quang Hung Trinh carried out the experimental work and analyzed the data; Young Sun Mok supervised all of the study.

Conflicts of Interest: The authors declare no conflict of interest.

References

1. Yamamoto, T. VOC decomposition by nonthermal plasma processing-a new approach. *J. Electrostat.* **1997**, *42*, 227–238.

2. Roland, U.; Holzer, F.; Kopinke, F.-D. Improved oxidation of air pollutants in a non-thermal plasma. *Catal. Today* **2002**, *73*, 315–323.

3. Mista, W.; Kacprzyk, R. Decomposition of toluene using non-thermal plasma reactor at room temperature. *Catal. Today* **2008**, *137*, 345–349.

4. Schmid, S.; Jecklin, M.C.; Zenobi, R. Degradation of volatile organic compounds in a non-thermal plasma air purifier. *Chemosphere* **2010**, *79*, 124–130.

5. Abd Allah, Z.; Whitehead, J.C.; Martin, P. Remediation of dichloromethane (CH_2C_{l2}) using non-thermal, atmospheric pressure plasma generated in a packed-bed reactor. *Environ. Sci. Technol.* **2013**, *48*, 558–565.

6. Demidiouk, V.; Chae, J.O. Decomposition of volatile organic compounds in plasma-catalytic system. *IEEE Trans. Plasma Sci.* **2005**, *33*, 157–161.

7. Zhu, T.; Li, J.; Jin, Y.; Liang, Y.; Ma, G. Decomposition of benzene by non-thermal plasma processing: Photocatalyst and ozone effect. *Int. J. Environ. Sci. Tech.* **2008**, *5*, 375–384.

8. Chae, J.; Moon, S.; Sun, H.; Kim, K.; Vassiliev, V.A.; Mikholap, E.M. A study of volatile organic compounds decomposition with the use of non-thermal plasma. *KSME Int. J.* **1999**, *13*, 647–655.

9. Kim, H.H.; Ogata, A. Nonthermal plasma activates catalyst: from current understanding and future prospects. *Eur. Phys. J. Appl. Phys.* **2011**, *55*, 13806.

10. Kim, H.H.; Ogata, A. Interaction of nonthermal plasma with catalyst for the air pollution control. *Int. J. Plasma Environ. Sci. Technol.* **2012**, *6*, 43–48.

11. Whitehead, J.C. Plasma catalysis: A solution for environmental problems. *Pure Appl. Chem.* **2010**, *82*, 1329–1336.

12. Yamamoto, S.; Yao, S.; Kodama, S.; Mine, C.; Fujioka, Y. Investigation of transition metal oxide catalysts for diesel PM removal under plasma discharge conditions. *Open Catal. J.* **2008**, *1*, 11–16.

13. Yao, S.; Yamamoto, S.; Kodama, S.; Mine, C.; Fujioka, Y. Characterization of catalyst-supported dielectric barrier discharge reactor. *Open Catal. J.* **2009**, *2*, 79–85.

14. Zhu, T.; Wan, Y.D.; Li, J.; He, X.W.; Xu, D.Y.; Shu, X.Q.; Liang, W.J.; Jin, Y.Q. Volatile organic compounds decomposition using nonthermal plasma coupled with a combination of catalysts. *Int. J. Environ. Sci. Technol.* **2011**, *8*, 621–630.

15. Chen, H.L.; Lee, H.M.; Chen, S.H.; Chang, M.B.; Yu, S.J.; Li, S.N. Removal of volatile organic compounds by single-stage and two-stage plasma catalysis systems: A review of the performance enhancement mechanisms, current status, and suitable applications. *Environ. Sci. Technol.* **2009**, *43*, 2216–2227.

16. Trinh, H.Q.; Mok, Y.S. Plasma-catalytic oxidation of acetone in annular porous monolithic ceramic-supported catalysts. *Chem. Eng. J.* **2014**, *251*, 199–206.

17. Einaga, H.; Ogata, A. Benzene oxidation with ozone over supported manganese oxide catalysts: effect of catalyst support and reaction conditions. *J. Hazard. Mater.* **2009**, *164*, 1236–1241.

18. Ye, L.; Feng, F.; Liu, J.; Tang, X.; Zhang, X.; Huang, Y.; Liu, Z.; Yan, K. Toluene decomposition by a two-stage hybrid plasma catalyst system in dry air. *IEEE Trans. Plasma Sci.* **2014**, *42*, 3529–3538.

19. Jin, M.; Kim, J.H.; Kim, J.M.; Jeon, J.-K.; Jurng, J.; Bae, G.-N.; Park, Y.-K. Benzene oxidation with ozone over MnO_x/SBA-15 catalysts. *Catal. Today* **2013**, *204*, 108–113.

20. Lian, Z.; Ma, J.; He, H. Decomposition of high-level ozone under high humidity over Mn–Fe catalyst: The influence of iron precursors. *Catal. Commun.* **2015**, *59*, 156–160.

21. Zaloznaya, L.A.; Tkachenko, S.N.; Egorova, G.V.; Tkachenko, I.S.; Sobolev, A.V.; Golosman, E.Z.; Troshina, V.A.; Lunin, V.V. Ozone decomposition and benzene oxidation catalysts based on iron and manganese oxides as industrial wastes from water decontamination by ozone treatment. *Catal. Ind.* **2009**, *1*, 224–228.

22. Demidiouk, V.; Moon, S.; Chae, J.; Lee, D. Application of a plasma-catalytic system for decomposition of volatile organic compounds. *J. Korean Phys. Soc.* **2003**, *42*, 966–970.

23. Futamura, S.; Einaga, H.; Kabashima, H.; Hwan, L.Y. Synergistic effect of silent discharge plasma and catalysts on benzene decomposition. *Catal. Today* **2004**, *89*, 89–95.

24. Ayrault, C.; Barrault, J.; Blin-Simiand, N.; Jorand, F.; Pasquiers, S.; Rousseau, A.; Tatibouët, J.M. Oxidation of 2-heptanone in air by a DBD-type plasma generated within a honeycomb monolith supported Pt-based catalyst. *Catal. Today* **2004**, *89*, 75–81.

25. Subrahmanyam, C.; Renken, A.; Kiwi-Minsker, L. Novel catalytic dielectric barrier discharge reactor for gas-phase abatement of isopropanol. *Plasma Chem. Plasma Process.* **2006**, *27*, 13–22.

26. Jaafar, N.F.; Abdul Jalil, A.; Triwahyono, S.; Muhd Muhid, M.N.; Sapawe, N.; Satar, M.A.H.; Asaari, H. Photodecolorization of methyl orange over α-Fe_2O_3-supported HY catalysts: The effects of catalyst preparation and dealumination. *Chem. Eng. J.* **2012**, *191*, 112–122.

27. Boxiong, S.; Yan, Y.A. O.; Hongqing, M.A.; Ting, L.I.U. Ceria modified MnO_x/TiO_2-pillared clays catalysts for selective catalytic reduction of NO with NH_3 at low temperature. *Chin. J. Catal.* **2011**, *32*, 1803–1811.

28. Kim, H.H.; Kim, J.H.; Ogata, A. Adsorption and oxygen plasma-driven catalysis for total oxidation of VOCs. *Int. J. Plasma Environ. Sci. Technol.* **2008**, *2*, 106–112.

29. Trinh, Q.H.; Gandhi, M.S.; Mok, Y.S. Adsorption and plasma-catalytic oxidation of acetone over zeolite-supported silver catalyst. *Jpn. J. Appl. Phys.* **2015**, *54*, 01AG04.

30. Zhu, X.; Gao, X.; Zheng, C.; Wang, Z.; Ni, M.; Tu, X. Plasma-catalytic removal of a low concentration of acetone in humid conditions. *RSC Adv.* **2014**, *4*, 37796.

31. Jo, J.-O.; Lee, S.B.; Jang, D.L.; Mok, Y.S. Plasma–catalytic ceramic membrane reactor for volatile organic compound control. *IEEE Trans. Plasma Sci.* **2013**, *41*, 3021–3029.

32. Li, W.; Oyama, S.T. Mechanism of ozone decomposition on a manganese oxide catalyst. 2. Steady-state and transient kinetic studies. *J. Am. Chem. Soc.* **1998**, *120*, 9047–9052.

33. Ogata, A.; Saito, K.; Kim, H.-H.; Sugasawa, M.; Aritani, H.; Einaga, H. Performance of an ozone decomposition catalyst in hybrid plasma reactors for volatile organic compound removal. *Plasma Chem. Plasma Process.* **2009**, *30*, 33–42.

34. Van Durme, J.; Dewulf, J.; Sysmans, W.; Leys, C.; van Langenhove, H. Efficient toluene abatement in indoor air by a plasma catalytic hybrid system. *Appl. Catal. B* **2007**, *74*, 161–169.

35. Vandenbroucke, A.M.; Morent, R.; De Geyter, N.; Leys, C. Non-thermal plasmas for non-catalytic and catalytic VOC abatement. *J. Hazard. Mater.* **2011**, *195*, 30–54.

36. Lopaev, D.V.; Malykhin, E.M.; Namiot, V.A. UV absorption of vibrationally excited ozone. *J. Phys. B* **2008**, *41*, 085104.

37. Marinov, D.; Guerra, V.; Guaitella, O.; Booth, J.-P.; Rousseau, A. Ozone kinetics in low-pressure discharges: vibrationally excited ozone and molecule formation on surfaces. *Plasma Sources Sci. Technol.* **2013**, *22*, 055018.

38. Eliasson, B.; Hirth, M.; Kogelschatz, U. Ozone synthesis from oxygen in dielectric barrier discharges. *J. Phys. D* **1987**, *20*, 1421.

39. Yamamoto, T.; Ramanathan, K.; Lawless, P.A.; Ensor, D.S.; Newsome, J.R.; Plaks, N.; Ramsey, G.H. Control of volatile organic compounds by an AC energized ferroelectric pellet reactor and a pulsed corona reactor. *IEEE Trans. Ind. Appl.* **1992**, *28*, 528–534.

40. Yagi, S.; Tanaka, M. Mechanism of ozone generation in air-fed ozonisers. *J. Phys. D* **1979**, *12*, 1509–1520.

41. Mok, Y.S.; Nam, I.-S. Role of organic chemical additives in pulsed corona discharge process for conversion of NO. *J. Chem. Eng. Japan* **1998**, *31*, 391–397.

The Role of Ozone in the Reaction Mechanism of a Bare Zeolite-Plasma Hybrid System

Yoshiyuki Teramoto, Hyun-Ha Kim, Nobuaki Negishi and Atsushi Ogata

Abstract: We investigated the reaction mechanism of a metal-unloaded zeolite-plasma hybrid system for decomposition of toluene at room temperature. Short-lived radicals and fast electrons did not contribute substantially to the reaction mechanism of toluene decomposition in the zeolite-plasma hybrid system. The main factor enhancing the reaction mechanism was gas-phase O_3 directly reacting with toluene adsorbed onto the zeolite (the Eley-Rideal mechanism). CO_2 selectivity was not improved by using H-Y zeolite due to its low ability to retain the active oxygen species formed by O_3. The gas-phase O_3 reacted with adsorbed toluene to form a ring cleavage intermediate that was slowly converted into formic acid. The decomposition rate of formic acid was much lower than that of toluene on the H-Y surface.

Reprinted from *Catalysts*. Cite as: Teramoto, Y.; Kim, H.-H.; Negishi, N.; Ogata, A. The Role of Ozone in the Reaction Mechanism of a Bare Zeolite-Plasma Hybrid System. *Catalysts* **2015**, *5*, 838–850.

1. Introduction

Non-thermal plasma processes have attracted significant interest as promising technologies for decomposing volatile organic compounds (VOCs) [1]. Over the last decade, research has focused on improving the process efficiency of decomposition of VOCs using various plasma reactors such as pulsed corona [2–4], ferroelectric pellet packed-bed discharge [5], silent (dielectric barrier) discharge [6], and surface discharge [7,8] reactors. The feasibility of plasma processes is limited by energy efficiency and formation of toxic by-products. To improve energy efficiency and CO_2 selectivity, catalyzed hybrid plasma reactors have been designed [9–11]. For example, it has been reported that the combination of dielectric barrier discharge and a metal loaded catalyst enhances trichloroethylene decomposition and CO_2 selectivity [12,13]. Loading of active metals not only enhances CO_2 selectivity for various VOCs [9,14,15], but also improves plasma coverage on the surface [16]. We have also proposed a non-thermal plasma reactor combined with a catalyst [8,17] and cyclic plasma operation [18]. Plasma induced surface oxygen species on metal nanoparticles play a significant role in improvement of the process [19,20]. In particular, O_3 enhances formation of active oxygen species on the catalyst surface. Manganese oxides (MnO_x)

are well known as promising catalysts for decomposition of VOCs by O_3 at low temperatures through the Langmuir-Hinshelwood (L-H) mechanism, as follows [21]:

$$O_3 + * \rightarrow O_2 + *O_{ads} \qquad (1)$$

$$*O_{ads} + VOC_{ads} \rightarrow CO_x + H_2O + * \qquad (2)$$

where * is an active sites on the catalyst and $*O_{ads}$ is the active oxygen species on the catalyst. Previous studies have reported that metal nanoparticles on zeolite strongly retain the active oxygen species formed by plasma compared with bare zeolite. We previously investigated active oxygen species formed by plasma on a silver nanoparticle-loaded zeolite [22]. The amount of active oxygen fixed onto the catalyst surface by O_2 plasma was approximately proportional to the square of the amount of supported silver. Thus, most of the active oxygen was present on the silver surface. Interestingly, the ozone demand factor ($DF_{O3} = \Delta[O_3]/\Delta[toluene]$), *i.e.*, the amount of ozone required for oxidation of toluene of bare zeolite (ZSM-5), was half that of silver-loaded zeolite [23]. Smaller DF_{O3} values correlate with better performance during ozone-assisted catalysis of VOCs. These findings suggest that the supported metal does not necessarily improve the efficiency of the plasma hybrid system and that performance improvements in that system cannot be explained solely by surface reactions of active oxygen species formed by O_3 (the L-H mechanism).

In this study, we focused on reaction mechanisms of a metal-unloaded zeolite in a plasma hybrid system for toluene decomposition. The role of metal nanoparticles has been studied by several research groups and has been found to be important in VOC removal, affecting decomposition efficiency, the carbon balance, and CO_2 selectivity. However, the role of supports (e.g., without active metals) has not been well studied in terms of interaction with plasma. The role of zeolite in the toluene decomposition process was investigated by changing the position of the zeolite in the plasma reactor. The reaction mechanisms of toluene decomposition in the gas phase and on the zeolite surface were studied using gaseous and surface FT-IR analyses.

2. Results and Discussion

2.1. Role of Zeolites in the Plasma Hybrid System

Energy efficiency in the decomposition of VOCs was enhanced by using a zeolite-hybrid plasma reactor [18]. Improvements in efficiency differed widely depending on the type of zeolite and its position in the reactor. Table 1 summarizes the performance of the zeolites in toluene oxidation, and their basic properties. Decomposition efficiency in the downstream position was 3.7 times higher than that in a conventional reactor, while that in the upstream position was 1.2–1.5 times higher. Additionally, decomposition efficiency in the downstream position was significantly affected by the type of zeolite, compared with upstream. The dependence of the decomposition efficiency on the position of the zeolite suggests that improvements in performance cannot be explained solely by activation of the zeolite surface by plasma.

Table 1. Decomposition efficiency in hybrid plasma reactors and the properties of zeolites.

Zeolite	C₇H₈ decomposition (%) Upstream:Downstream		CO		CO₂		HCOOH		Carbon balance (%)		Surface area (m²/g)	SiO₂/Al₂O₃	Channel (Å)
None	15		35		48		17		97		-	-	-
Na-Y	21	52	36	33	56	63	8	4	95	92	750	3.1	7.4
H-Y(1)	18	55	41	45	52	47	7	8	97	99	650	1.9	7.4
H-Y(2)	20	50	47	42	46	51	7	7	98	96	520	5.6	7.4
MOR	23	35	46	42	45	52	9	6	97	96	460	15.2	6.7 × 7.0
FER	19	20	38	33	49	65	13	2	96	99	270	17.9	4.3 × 5.5

Figure 1 shows the relationship between the amount of toluene adsorbed and decomposition efficiency (normalized to that of a reactor without zeolites). In general, the surface area greatly affects the adsorption capacity and decomposition of VOCs. In a previously reported ozone-assisted catalyst system, the rate of benzene oxidation linearly increased with the specific surface area of the MnO_x-loading catalyst, regardless of the type of catalyst support [24]. In the present study, the amount of toluene (~6 Å) adsorbed by ferrierite (FER) was small and limited to the surface due to the micropore size (4.3×5.5 Å^2) of FER.

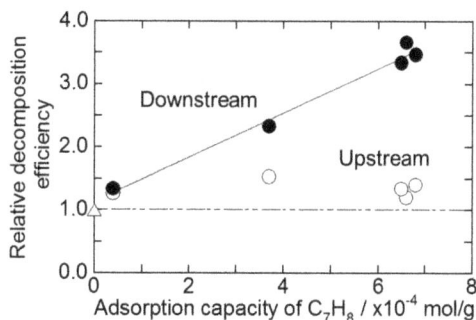

Figure 1. Relationship between the decomposition efficiency and adsorption capacity for toluene.

The decomposition efficiency of the downstream configuration was linearly proportional to the adsorption capacity (Figure 1). On the other hand, that of the upstream configuration was near constant. If the toluene adsorbed in the internal areas of the zeolite was decomposed by short-lived radicals and fast electrons, the decomposition efficiency would not depend on the zeolite position. We conclude that reactions between toluene adsorbed onto the zeolite and short-lived radicals or fast electrons did not contribute to improvements in decomposition efficiency. One possible explanation is the limited area of plasma for bare zeolite compared to metal-loaded zeolites; plasma mainly formed at the points of contact with the zeolite pellets [16]. This limited plasma area reduced effective interactions at the surface of the zeolite. These results confirm the important contribution of metal nanoparticles to both plasma properties and catalytic performance. In addition, the diffusion lengths of short-lived radicals (O and OH) and fast electrons are <100 µm at atmospheric pressure. Thus, short-lived radicals and fast electrons had limited effect in this system.

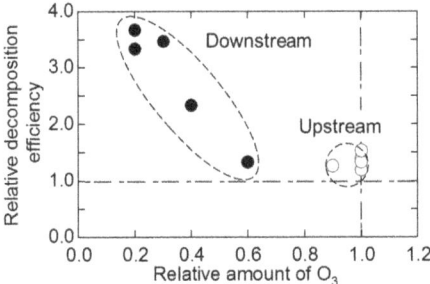

Figure 2. Relationship between the relative decomposition efficiency and relative amount of ozone.

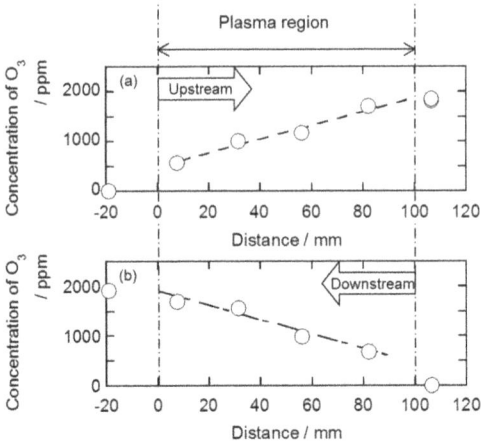

Figure 3. Ozone distribution in a conventional plasma reactor: (**a**) upstream configuration and (**b**) downstream configuration.

Ozone has a long diffusion length (>10 mm) due to its long lifetime, and can reach the inner area of micropores. Figure 2 shows the decomposition efficiency as a function of the outlet O_3 concentration, normalized to that of a reactor without zeolites. In the downstream position, the O_3 concentration was affected by the type of zeolite and decreased with increasing decomposition efficiency. Therefore, O_3 was consumed approximately linearly with toluene oxidation. In the upstream position, there were no significant differences in O_3 concentration or decomposition efficiency among the zeolite types. This can be explained by the O_3 concentration at the entrance of the plasma reactor. The distribution of O_3 concentrations in the reactor tube were measured using a dedicated reactor (shown in section 3.2.). Figure 3a,b shows the distribution of O_3 in an unfilled reactor under dry air flow (without toluene or H_2O) in the upstream and downstream configurations, respectively. The O_3 concentration increased linearly with distance from the entrance of the reactor in both directions. When 1.0 g of H-Y was placed in the upstream or downstream position, the O_3 concentration remained constant in the zeolite-filled region (Figure 4a,b)). These results were also observed for other zeolite types. Thus, formation of O_3 was strongly suppressed in the zeolites-filled region. Based on these results, the low decomposition efficiency of the upstream configuration was likely due to a low O_3 concentration, while the high efficiency of the downstream configuration was due to a high O_3 concentration. Therefore, adsorbed toluene in the internal area of the zeolite was manly decomposed by O_3. In addition to reactions (1) and (2), the following reaction represents this mechanism:

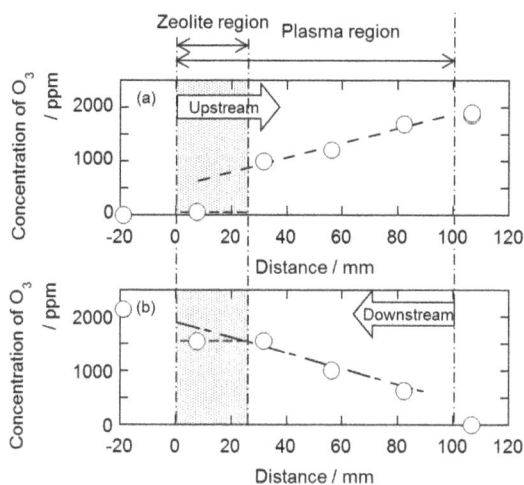

$$*VOC_{ads} + O_3 \rightarrow CO_x + H_2O + * \qquad (3)$$

Figure 4. Relationship between the decomposition efficiency and the adsorption of ozone: (**a**) upstream configuration and (**b**) downstream configuration.

In reactions (1) and (2), O_3 initially decomposes at the zeolite surface. Figure 5 shows the outlet concentration of O_3 in the downstream reactor filled with 1.0–3.0 g of H-Y. No decomposition of O_3 was observed for any case. Thus, H-Y has little ability to decompose O_3 and form active oxygen species on its surface. Therefore, the main process involved in toluene decomposition is likely direct reaction of gas phase O_3 with toluene adsorbed onto H-Y, *i.e.*, the Eley-Rideal (E-R) mechanism. This process is different from that for other types of zeolites and metal-loaded zeolites, *i.e.*, the L-H mechanism. For example, active oxygen species formed by O_3 on the MCM-41 and MS-13X surfaces significantly contributed to VOC decomposition [25,26]. These active oxygen species on the catalyst surface are consumed in reactions with VOCs, O_3, O, *etc.* [27–29]. In other words, some of the active oxygen species (=O_3) form unwanted byproducts in the L-H process, unlike in E-R. Therefore, in the E-R process, the H-Y zeolite has the potential for achieving low DF_{O3} in VOC decomposition.

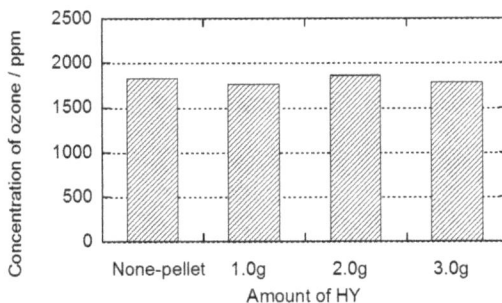

Figure 5. Concentration of ozone in the downstream reactor with amount of H-Y(1).

CO_2 selectivity was similar between the H-Y filled and unfilled reactors (Table 1). On the other hand, production of formic acid was lower for the H-Y filled reactors. Considering the role of O_3 in this system (reaction (3)), it may be difficult for gaseous O_3 to oxidize the adsorbed CO to CO_2, unlike for toluene and formic acid. Generally, in a plasma-catalyst hybrid system, bare zeolite contributes little to oxidization of CO. Metal-loaded zeolites significantly improves CO_2 selectivity due to surface reaction with the active oxygen species formed by O_3, most of which are present on the silver surface of the catalyst [22]. Thus, active oxygen species on the metal surface contribute greatly to improvements in CO_2 selectivity. The H-Y zeolite cannot easily retain active oxygen species on its surface through reactions with O_3, leading to low CO_2 selectivity.

Higher concentrations of VOCs result in higher energy efficiencies for removal [30]. Kuroki *et al.* [31] reported that adsorbed toluene desorbed easily by DC pulsed plasma discharge, resulting in an increase in the concentration of VOCs. Although this experimental result using standard adsorbents cannot be directly compared with zeolites, if such an effect occurred in our system, the efficiency in the upstream reactor should be nearly the same as that in the downstream one. However, efficiency in the downstream configuration was much higher than that in the upstream reactor. This suggested that enhancement due to desorption of toluene from the zeolite was negligible.

The hydrophilicity of zeolite depends on the SiO_2/Al_2O_3 ratio and affects the performance of thermal catalytic processes. Generally, hydrophilicity decreases with increasing SiO_2/Al_2O_3 ratio. However, the difference in hydrophilicity between H-Y(1) and H-Y(2) did not affect decomposition efficiency in our system (Table 1), unlike for thermal catalytic processes.

2.2. Decomposition of Toluene Adsorbed onto the Zeolites by O_3

Figure 6a,b shows spectra for toluene adsorbed onto H-Y and FER, respectively, measured 10 min after introducing 200 ppm of toluene in N_2. Figure 6c also shows the spectrum for toluene in the gas phase as a reference. IR peaks assigned to the aromatic ring were observed between 2000 and 1750 cm^{-1} in both Figure 6a,b. These peak positions were shifted compared with those in the gas phase. Table 2 summarizes the IR peaks for toluene in both the gas and liquid phases. The adsorption peaks near 3000 cm^{-1} were assigned to the CH_3 group of toluene and those at 2000–1700 cm^{-1} were assigned to the C–H bonds in the aromatic ring. The shift in peaks for toluene adsorbed onto the zeolite is due to the perturbation of the π-electric system by adsorption onto the H-Y zeolite. These peaks were assigned to toluene physically adsorbed onto the solid surface, because they completely disappeared upon heating the sample to 200 °C. The peak signal for toluene on FER was much smaller than that on H-Y, because the adsorption capacity of FER for toluene is limited due to its small micropore size.

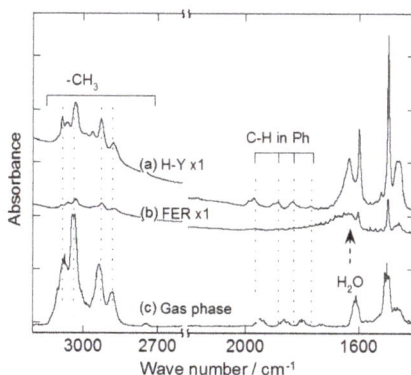

Figure 6. Infrared spectra for toluene (**a**) on H-Y(1) zeolite, (**b**) on FER, and (**c**) in the gas phase.

Table 2. Wave number of IR peaks for toluene in gas phase, in the liquid film, and on H-Y.

Bond	Gas phase	Liquid film	on H-Y
C–H in Ph	3101	3087	3083
	3076	3062	3065
	3037	3026	3031
C–H in CH$_3$	-	2948	2963
	2938	2920	2928
	2881	2873	2879
C–H in Ph	1942	1942	1969
	1856	1858	1884
	1801	1803	1832
	1732	-	1772
C–C in Ph	1611	1605	1599
	-	1524	-
	1500	1496	1495
	1463	1451	1457
	Infra-red DB	SDBS	this work

The reactivity of toluene adsorbed onto H-Y was then investigated. Initially, toluene was adsorbed onto the H-Y sample for 10 min (Figure 7a). N$_2$ gas was then supplied to remove the gas phase toluene from the reactor. A mixture of 400 ppm O$_3$ in O$_2$ was then introduced and the FT-IR spectrum was recorded (Figure 7b–f). The IR peaks for toluene disappeared immediately after O$_3$ introduction, in less than 1 min. In contrast, IR peaks at 1390, 1630 and 1725 cm^{-1} appeared and their intensities increased for 40 min after O$_3$ introduction. These peaks were assigned to formic acid (HCOOH). These results suggest that gas-phase O$_3$ reacted with adsorbed toluene to form ring cleavage intermediate that was slowly converted to formic acid. We anticipate that the performance of the zeolite-plasma hybrid system could be improved by increasing the operating temperature, because the decomposition rate of formic acid is significantly affected by temperature [18].

A distinctive peak for the active oxygen species formed by O$_3$ on MCM-41 and MS-13X was observed at around 1380 cm^{-1} [25,26]. In the present experiment, this peak was not observed. This result supports the observation that the H-Y zeolite does not retain active oxygen species on its surface through reaction with O$_3$.

Figure 8 shows IR spectra for formic acid adsorbed to H-Y for 10 min under HCOOH(200 ppm)/N$_2$ flow. The observed peaks were assigned to the carboxylic acid group (-COOH). The peak positions in the spectra and the profiles for formic acid adsorbed onto H-Y (Figure 8b) were in good agreement with those for toluene adsorbed onto H-Y treated with O$_3$ (Figure 7c–f). Therefore, the peaks in Figure 7c–f can be assigned to formic acid. When a mixture of 400 ppm O$_3$ in O$_2$ was introduced for 10 min, the peak at 1725 cm^{-1} decreased by approximately half of its initial intensity, and the peak at 1630 cm^{-1} increased significantly in intensity. The decomposition rate of formic acid was much lower than that of toluene under the same conditions. Therefore, the benzene ring was relatively easily decomposed on the H-Y surface compared to carboxylic acid. With reference to the IR spectra of H_2O (Figure 8d), the peak at 1630 cm^{-1} with increased intensity was assigned to H_2O the final product of formic acid oxidation.

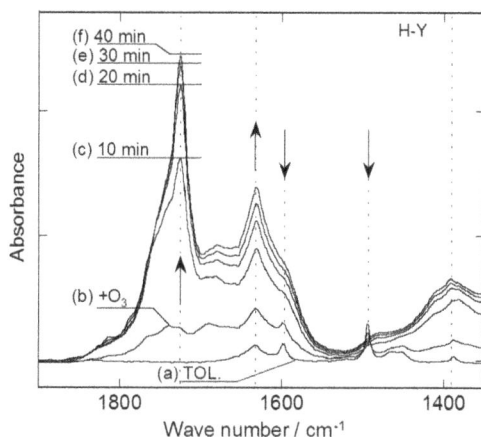

Figure 7. Infrared spectra for toluene decomposition on H-Y(1): (**a**) toluene adsorbed onto H-Y; after addition of O_3 for (**b**) 0, (**c**) 10, (**d**) 20, (**e**) 30, and (**f**) 40 min.

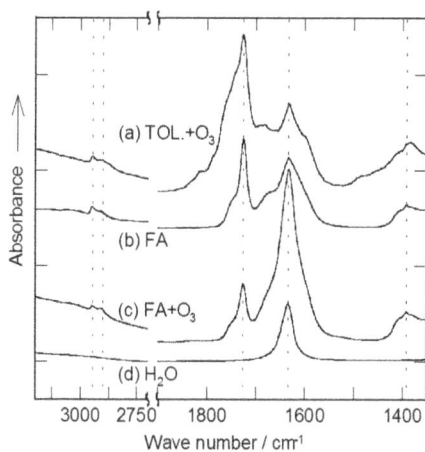

Figure 8. Infra-red spectra of (**a**) toluene adsorbed onto H-Y(1) after introducing O_3, (**b**) formic acid adsorbed onto H-Y, (**c**) after adducting O_3, and (**d**) H_2O adsorbed onto H-Y.

3. Experimental Section

3.1. Decomposition of Toluene (C_7H_8)

The configuration of the surface discharge reactor is shown in Figure 9a. A quartz tube (i.d.: 10 mm, thickness: 1.5 mm) was used as the dielectric barrier. A spiral platinum wire coil (70 turns/100 mm) with a 0.3 mm diameter was placed in contact with the inner wall of the barrier tube. A piece of copper foil was wrapped around the outside of the barrier tube as the ground electrode; the outer electrode length was 100 mm.

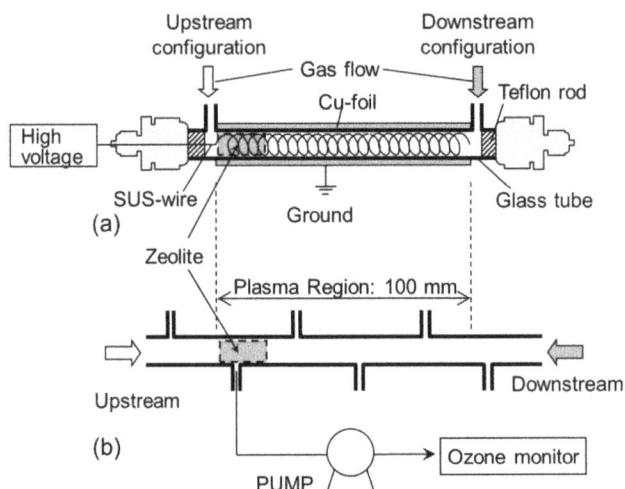

Figure 9. Configurations of (**a**) the zeolite hybrid reactor and (**b**) the reactor for monitoring the ozone distribution in the plasma region.

Zeolites (1 g) were packed into one end of the reactor. The reaction conditions were adjusted based on the flow direction and the position of the zeolites. The position of the zeolites was near the gas inlet in the upstream configuration and near the gas outlet in the downstream configuration.

The reactant gas consisted of 200 ppm toluene (C_7H_8), 0.5% H_2O, and 20.0% O_2, with N_2 making up the remaining fraction. It was fed at a flow rate of 500 cm^3/min into the reactor, where a high AC voltage was applied between the two electrodes. The plasma energy was supplied using an AC high-voltage neon transformer (24 kHz, NEON M-5; Lecip Co., Gifu, Japan) and the plug-in power was measured using a digital power meter (WT110; Yokogawa Electric Co., Tokyo, Japan). The input power was 5 W.

The plasma was cycled four times in 160 min (off for 30 min → on for 10 min → off for 30 min → etc.) (Figure 10), because cyclic operation improves energy efficiency by inducing periods of adsorption followed by plasma decomposition [8,13]. The concentrations of products such as COx were consistent during each plasma period. The experiment was carried out at room temperature. Under plasma operation, the reactor temperature was increased to approximately 60 °C.

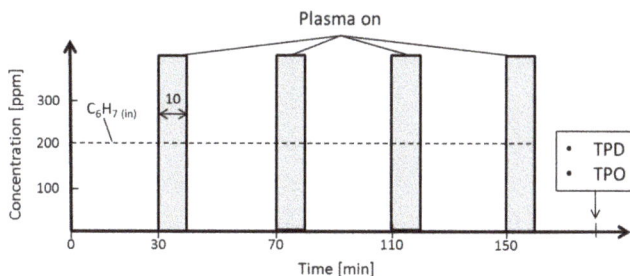

Figure 10. Plasma cyclic operation.

The effluent gas was monitored continuously using an FT-IR spectrometer (FTS135; Bio-Rad Laboratories, Hercules, CA, USA) equipped with a gas cell (2.4 m path, 125 cm^3 volume; Infrared Analysis Inc., Anaheim, CA, USA) and the total amounts of the products were calculated from their integrated spectra. Under plasma operation, production of N_2O and HNO_3 was confirmed. The concentration of N_2O was approximately 60 ppm. An accurate concentration for HNO_3 could not be determined due to overlap of the FT-IR spectrum with that for H_2O. The amounts of adsorbed components on the zeolite were measured using temperature-programmed desorption and oxidation (TPD and TPO, respectively) after cyclical operation. For the TPD process, the zeolite was heated to 500 °C at a rate of 10 °C/min and held at that temperature under N_2 atmosphere. After the TPD process, the residual solid carbon was oxidized at 600 °C under synthetic air. Decomposition efficiency was calculated as follows:

$$C_7H_8 \text{ decomposition } (\%) = (C_7H_{8in} - C_7H_{8out} - C_7H_{8TPD})/(C_7H_{8in}) \times 100 \quad (4)$$

where C_7H_{8in} and C_7H_{8out} are the total input and output of toluene, respectively. $(C_7H_8)_{TPD}$ is the amount of toluene desorbed from the TPD reactor, where the adsorbed toluene could be easily desorbed from the zeolite surface without noticeable decomposition. The carbon balance was estimated as follows:

$$\text{Carbon balance } (\%) = (CO_2 + CO + HCOOH)/7C_7H_{8decomposition} \times 100 \quad (5)$$

Based on the relatively high carbon balance (Table 1), the influence of products other than those above, such as carbon deposition on the zeolite surface, was negligible.

143

Thus, the product selectivity, where $X = CO$, CO_2, or HCOOH, was calculated as follows:

$$\text{Selectivity for } X\,(\%) \;=\; (X)/(CO_2 + CO + HCOOH) \times 100 \qquad (6)$$

Four types of zeolites were used in this study: A Na-forms of faujasite (Na-Y), two H-forms of faujasite (H-Y), an H-form of mordenite (MOR), and an H-forms of ferrierite (FER). These samples molded into pellets were purchased from Tosoh Co. (Tokyo, Japan) and Nikki-Universal Co. Ltd. (Tokyo, Japan). Quite similar V-I patterns were observed for all zeolites.

3.2. Measurement of the Ozone Concentration in the Reactor

A reactor with six branch pipes (Figure 9b) was used to measure the O_3 distribution in the plasma region at room temperature. The gas was sampled from each branch pipe using a diaphragm pump and O_3 was measured using an ozone-monitor (PG-620H; Ebara Jitsugyo Co., Ltd., Tokyo, Japan). The zeolite pellets (1.0–3.0 g) were filled in the area indicated by the dotted line. Dry air (20% O_2 and 80% N_2) was introduced into the reactor at a flow rate of 500 cm^3/min, and an input power of 2.0 W was applied.

3.3. Infrared Measurement of the Chemical Species Adsorbed onto the Zeolites

The IR spectra for chemical species adsorbed onto H-Y and FER were recorded using an FT-IR spectrometer (Spectrum 100; Perkin-Elmer Co. Ltd., Waltham, MA, USA) equipped with an IR cell for solid samples (Figure 11). The pellets were ground to a fine powder with an agate mortar and pestle, and the sample (0.043–0.044 g) was pressed into a thin disc 20 mm in diameter. The disc was inserted in the sample holder (Figure 11a) and the spectrum was measured using the transmission method. The sample could be moved to a heated position under desired gas flow conditions and to the measuring position at room temperature. The spectrum was measured at a resolution of 4.0 cm^{-1} and scan numbers of 200 using mercury cadmium telluride detector.

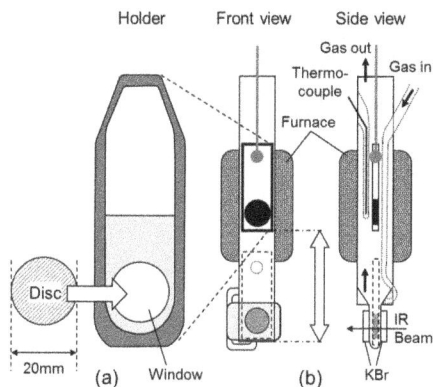

Figure 11. Configurations of (**a**) the sample holder and (**b**) the IR cell equipped with a furnace.

4. Conclusions

The reaction mechanism of the zeolite-plasma hybrid system for toluene decomposition was studied using gas and surface FT-IR analyses. The contributions of short-lived radicals and fast electrons to decomposition of adsorbed toluene were low. On the other hand, O_3 directly decomposed toluene adsorbed in the inner areas of the zeolite micropore because it has a long life time. O_3 provided the main mechanism of toluene decomposition in the zeolite-plasma hybrid system, unlike in metal-loaded zeolite-plasma systems. CO_2 selectivity was not improved by use of the bare zeolite, because H-Y zeolite does not retain active oxygen species on its surface through surface reactions of O_3. Gas-phase O_3 reacted with adsorbed toluene to form a ring cleavage intermediate that was slowly converted to formic acid. The decomposition rate of formic acid was much slower than that of toluene on the H-Y surface.

Author Contributions: Y.T., H.H.K., N.N. and A.O. designed the research; A.O. performed research; Y.T., H.H.K. and A.O. analysed data; Y.T. wrote, and all authors commented on the manuscript.

Conflicts of Interest: The authors declare no conflict of interest.

References

1.   Oda, T. Non-thermal plasma processing for environmental protection: Decomposition of dilute VOCs in air. *J. Electrostat.* **2003**, *57*, 293–311.
2.   Nunez, C.M.; Ramsey, G.H.; Ponder, W.H.; Abbott, J.H.; Hamel, L.E.; Kariher, P.H. Corona destruction—An innovative control technology for VOCs and air toxics. *J. Air Waste Manag.* **1993**, *43*, 242–247.

3. Yamamoto, T.; Futamura, S. Nonthermal plasma processing for controlling volatile organic compounds. *Combust. Sci. Technol.* **1998**, *133*, 117–133.

4. Sobacchi, M.G.; Saveliev, A.V.; Fridman, A.A.; Gutsol, A.F.; Kennedy, L.A. Experimental assessment of pulsed corona discharge for treatment of VOC emissions. *Plasma Chem. Plasma Process.* **2003**, *23*, 347–370.

5. Ogata, A.; Ito, D.; Mizuno, K.; Kushiyama, S.; Gal, A.; Yamamoto, T. Effect of coexisting components on aromatic decomposition in a packed-bed plasma reactor. *Appl. Catal. A* **2002**, *236*, 9–15.

6. Park, D.W.; Yoon, S.H.; Kim, G.J.; Sekiguchi, H. The effect of catalyst on the decomposition of dilute benzene using dielectric barrier discharge. *J. Ind. Eng. Chem.* **2002**, *8*, 393–398.

7. Oda, T.; Yamashita, R.; Haga, I.; Takahashi, T.; Masuda, S. Decomposition of gaseous organic contaminants by surface discharge induced plasma chemical processing spcp. *IEEE Trans. Ind. Appl.* **1996**, *32*, 118–124.

8. Oh, S.M.; Kim, H.H.; Ogata, A.; Einaga, H.; Futamura, S.; Park, D.W. Effect of zeolite in surface discharge plasma on the decomposition of toluene. *Catal. Lett.* **2005**, *99*, 101–104.

9. Kim, H.H.; Oh, S.M.; Ogata, A.; Futamura, S. Decomposition of benzene using Ag/TiO$_2$ packed plasma-driven catalyst reactor: Influence of electrode configuration and Ag-loading amount. *Catal. Lett.* **2004**, *96*, 189–194.

10. Futamura, S.; Einaga, H.; Kabashima, H.; Lee, Y.H. Synergistic effect of silent discharge plasma and catalysts on benzene decomposition. *Catal. Today* **2004**, *89*, 89–95.

11. Ogata, A.; Einaga, H.; Kabashima, H.; Futamura, S.; Kushiyama, S.; Kim, H.H. Effective combination of nonthermal plasma and catalysts for decomposition of benzene in air. *Appl. Catal. B* **2003**, *46*, 87–95.

12. Magureanu, M.; Mandache, N.B.; Parvulescu, V.I.; Subrahmanyam, C.; Renken, A.; Kiwi-Minsker, L. Improved performance of non-thermal plasma reactor during decomposition of trichloroethylene: Optimization of the reactor geometry and introduction of catalytic electrode. *Appl. Catal. B* **2007**, *74*, 270–277.

13. Han, S.B.; Oda, T.; Ono, R. Improvement of the energy efficiency in the decomposition of dilute trichloroethylene by the barrier discharge plasma process. *IEEE Trans. Ind. Appl.* **2005**, *41*, 1343–1349.

14. Hakoda, T.; Matsumoto, K.; Mizuno, A.; Kojima, T.; Hirota, K. Catalytic oxidation of xylene in air using TiO$_2$ under electron beam irradiation. *Plasma Chem. Plasma Process.* **2008**, *28*, 25–37.

15. Rousseau, A.; Meshchanov, A.V.; Roepcke, J. Evidence of plasma-catalyst synergy in a low-pressure discharge. *Appl. Phys. Lett.* **2006**, *88*.

16. Kim, H.H.; Kim, J.H.; Ogata, A. Microscopic observation of discharge plasma on the surface of zeolites supported metal nanoparticles. *J. Phys. D* **2009**, *42*, 135210.

17. Ogata, A.; Ito, D.; Mizuno, K.; Kushiyama, S.; Yamamoto, T. Removal of dilute benzene using a zeolite-hybrid plasma reactor. *IEEE Trans. Ind. Appl.* **2001**, *37*, 959–964.

18. Oh, S.M.; Kim, H.H.; Einaga, H.; Ogata, A.; Futamura, S.; Park, D.W. Zeolite-combined plasma reactor for decomposition of toluene. *Thin Solid Films* **2006**, *506*, 418–422.

19. Sano, T.; Negishi, N.; Sakai, E.; Matsuzawa, S. Contributions of photocatalytic/catalytic activities of TiO_2 and g-Al_2O_3 in nonthermal plasma on oxidation of acetaldehyde and co. *J. Mol. Catal. A Chem.* **2006**, *245*, 235–241.

20. Bulanin, K.M.; Lavalley, J.C.; Lamotte, J.; Mariey, L.; Tsyganenko, N.M.; Tsyganenko, A.A. Infrared study of ozone adsorption on CeO_2. *J. Phys. Chem. B* **1998**, *102*, 6809–6816.

21. Einaga, H.; Futamura, S. Catalytic oxidation of benzene with ozone over mn ion-exchanged zeolites. *Catal. Commun.* **2007**, *8*, 557–560.

22. Teramoto, Y.; Kim, H.H.; Ogata, A.; Negishi, N. Study of plasma-induced surface active oxygen on zeolite-supported silver nanoparticles. *Catal. Lett.* **2013**, *143*, 1374–1378.

23. Kim, H.H.; Sugasawa, M.; Hirata, H.; Teramoto, Y.; Kosuge, K.; Negishi, N.; Ogata, A. Ozone-assisted catalysis of toluene with layered ZSM-5 and Ag/ZSM-5 zeolites. *Plasma Chem. Plasma Process.* **2013**, *33*, 1083–1098.

24. Einaga, H.; Ogata, A. Benzene oxidation with ozone over supported manganese oxide catalysts: Effect of catalyst support and reaction conditions. *J. Hazard Mater.* **2009**, *164*, 1236–1241.

25. Kwong, C.W.; Chao, C.Y.H.; Hui, K.S.; Wan, M.P. Catalytic ozonation of toluene using zeolite and MCM-41 materials. *Environ. Sci. Technol.* **2008**, *42*, 8504–8509.

26. Chao, C.Y.H.; Kwong, C.W.; Hui, K.S. Potential use of a combined ozone and zeolite system for gaseous toluene elimination. *J. Hazard Mater.* **2007**, *143*, 118–127.

27. Sullivan, R.C.; Thornberry, T.; Abbatt, J.P.D. Ozone decomposition kinetics on alumina: Effects of ozone partial pressure, relative humidity and repeated oxidation cycles. *Atmos. Chem. Phys.* **2004**, *4*, 1301–1310.

28. Li, W.; Gibbs, G.V.; Oyama, S.T. Mechanism of ozone decomposition on a manganese oxide catalyst. I. In situ raman spectroscopy and ab initio molecular orbital calculations. *J. Am. Chem. Soc.* **1998**, *120*, 9041–9046.

29. Li, W.; Oyama, S.T. Mechanism of ozone decomposition on a manganese oxide catalyst. 2. Steady-state and transient kinetic studies. *J. Am. Chem. Soc.* **1998**, *120*, 9047–9052.

30. Ogata, A.; Shintani, N.; Mizuno, K.; Kushiyama, S.; Yamamoto, T. Decomposition of benzene using a nonthermal plasma reactor packed with ferroelectric pellets. *IEEE Trans. Ind. Appl.* **1999**, *35*, 753–759.

31. Kuroki, T.; Fujioka, T.; Okubo, M.; Yamamoto, T. Toluene concentration using honeycomb nonthermal plasma desorption. *Thin Solid Films* **2007**, *515*, 4272–4277.

Co-Al Mixed Oxides Prepared via LDH Route Using Microwaves or Ultrasound: Application for Catalytic Toluene Total Oxidation

Eric Genty, Julien Brunet, Christophe Poupin, Sandra Casale, Sylvie Capelle, Pascale Massiani, Stéphane Siffert and Renaud Cousin

Abstract: Co_6Al_2HT hydrotalcite-like compounds were synthesized by three different methods: co-precipitation, microwaves-assisted and ultrasound-assisted methods. The mixed oxides obtained after calcination were studied by several techniques: XRD, TEM, H_2-TPR and XPS. They were also tested as catalysts in the reaction of total oxidation of toluene. The physico-chemical studies revealed a modification of the structural characteristics (surface area, morphology) as well as of the reducibility of the formed mixed oxides. The solid prepared by microwaves-assisted synthesis was the most active. Furthermore, a relationship between the ratio of Co^{2+} on the surface, the reducibility of the Co-Al mixed oxide and the T_{50} in toluene oxidation was demonstrated. This suggests a Mars Van Krevelen mechanism for toluene total oxidation on these catalysts.

Reprinted from *Catalysts*. Cite as: Genty, E.; Brunet, J.; Poupin, C.; Casale, S.; Capelle, S.; Massiani, P.; Siffert, S.; Cousin, R. Co-Al Mixed Oxides Prepared via LDH Route Using Microwaves or Ultrasound: Application for Catalytic Toluene Total Oxidation. *Catalysts* **2015**, *5*, 851–867.

1. Introduction

The emissions of Volatile Organic Compounds (VOCs) in industrial activities are now strictly regulated due to their harmfulness towards public health and the atmospheric environment. Among the many technologies available for VOCs control, catalytic oxidation of these pollutants to carbon dioxide and water has been recognized as one of the most promising ones. Supported noble metals are generally considered as efficient catalysts for VOCs oxidation [1–5] and their efficiency at low temperature can sometimes be considered as a key advantage favoring the use of such catalysts compared with mixed oxide catalysts. However, due to the high cost of noble metals, many researchers have recently devoted strong efforts to the development of suitable catalysts containing only transition metal oxides [6–10]. In this domain, special attention has been focused on cobalt materials in the spinel form [10], on oxide catalysts [11,12] and on mixed oxides issued from hydrotalcite precursors [13,14]. For the generation of well-dispersed, active and very stable mixed oxides, the Layered Double Hydroxides (LDH) route shows much potential. Indeed, LDH, also referred to as hydrotalcite-like compounds, constitute a large group of natural and synthetic minerals whose physico-chemical properties have strong analogies with clay minerals and more particularly with cationic clay minerals. This class of layered double hydroxides consists of positively charged metal hydroxide layers separated from each other by anions and water molecules. The wide range of possible cations and anions that can be incorporated in the Hydrotalcite (HT) structure gives rise to a large number of materials [7,13,14]. In addition, a high surface area, a good thermal stability, a good mixed oxides homogeneity, basic properties and a high metal dispersion are obtained after thermal treatment [10,15].

The interest in applying calcined LDH as catalysts in the environmental field has already been demonstrated. For example, mixed oxides have been shown to be very active and selective in the decomposition of nitrogen oxides [16] or in the total oxidation of Volatile Organic Compounds [13,14]. Also, mixed oxides obtained after calcination were used equally as catalysts [13,17] or as catalytic support [18–20] in redox reactions such as the selective oxidation of hydrocarbons or the total oxidation of VOCs [21,22].

The conventional and most widely used method to prepare LDH is co-precipitation [7,8,13,14]. However, alternative methods providing well-crystallized materials have also been studied [23,24]. In particular, the use of microwaves during the synthesis of catalytic materials presented several advantages over conventional methods. Thus, Ran et $al.$ [24] have synthesized ternary oxides of $La_{0.7}Sr_{0.3}MnO_{3+\lambda}$ using microwaves and they have highlighted the benefits of this synthesis method, compared to conventional sol-gel synthesis, in terms of rapidity (3–5 min compared to several hours), energy efficiency, uniformity of heating (no hot spot in the solution), homogeneity of the obtained solid (larger specific surface area and small particle sizes). Kaddouri et $al.$ [25] showed that a $La_{1-x}B_xMnO_{3.15}$ perovskite type solid synthesized under microwaves exhibit better catalytic properties than the same solid synthesized conventionally. Similarly, the use of ultrasounds in the synthesis of material was reported to be beneficial due to the phenomenon of acoustic cavitation that corresponds to the formation, growth and collapse of bubbles generated in the solution. Zhou et $al.$ [26] performed the impregnation of zirconium and cobalt supported on SiO_2 ($Co/Zr/SiO_2$) with sonication and they noted that ultrasounds increase the specific surface area and lead to a better dispersion of the cobalt particles on the substrate surface due to such cavitation effects. Similar cavitation phenomena have allowed obtaining a better catalytic reactivity of solids in the Fischer-Tropsch reaction. Recently, microwaves radiation heating [27–29] and ultrasound [27,30,31] have been applied to the synthesis of hydrotalcites with different chemical compositions as an alternative to the conventional hydrothermal treatments. Several of the advantages listed above have been observed, especially the shorter treatment times needed to achieve enhanced crystallinity degrees, the higher specific surface area and the improved metallic dispersions [28,32,33].

A previous study in our laboratory has shown that Co-Al mixed oxide catalysts present interesting catalytic activity for total oxidation of toluene [13]. In order to improve this type of catalytic materials, the aim of the present work was to investigate the effect of treatment by microwaves or by ultrasound during the preparation of the LDH samples. Thus, mixed oxides were prepared by calcination of hydrotalcite like compounds precursors (Co_6Al_2HT and Mg_6Al_2HT) using three different routes (microwaves (MW), ultrasound (US) and conventional treatment (CT)) and the catalytic properties of the samples were tested in the total oxidation of toluene. The mixed oxide properties were also characterized by several techniques: powder X-ray Diffraction (XRD), Transmission Electron Microscopy (TEM) Hydrogen Temperature Programmed Reduction (H_2-TPR) and X-ray photoelectron spectrometry (XPS).

2. Results and Discussion

Table 1 indicates the code names and the specific surface areas of the mixed oxides obtained after calcination of the hydrotalcite precursors. With the aim to study the effect of the preparation method, the cobalt-based materials are compared in this table to reference Mg-Al-O materials having more conventional hydrotalcite compositions but synthesized following the same protocols as those used for the preparation of the Co-Al mixed oxides. Compared to the samples prepared by coprecipitation, a significant increase of the specific surface area is observed when the mixed oxides were prepared using either the US or the MW method. This result was already reported in the literature [28,30,31]. Indeed, Neto *et al.* [31] have shown on LDH synthesis by sonication that ultrasounds not only accelerate crystal formation but also better disperse small particles, thus reducing their aggregation during nucleation and crystal growth and giving rise to increased specific surface area. Moreover, concerning the effect of microwaves during the LDH preparation, it was observed that MW radiation for short time leads to faster nucleation, which produces a larger number of small crystals and therefore an increase of the surface area [28]. Thus, the treatment by microwaves or by ultrasounds during synthesis improves the specific area of the obtained materials.

Table 1 also reports the chemical compositions of the solids. It can be clearly seen that the ratio $M^{2+}/Al^{3+} = 3$ is well respected for each solid. The T_{50} values reported as well in Table 1 (temperature at which 50% of the toluene is converted)

150

show that the catalytic performances of the samples in toluene total oxidation follow the order:

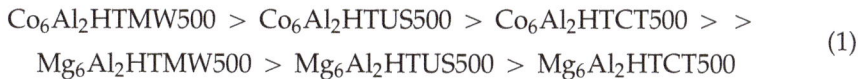

$$Co_6Al_2HTMW500 > Co_6Al_2HTUS500 > Co_6Al_2HTCT500 > >$$
$$Mg_6Al_2HTMW500 > Mg_6Al_2HTUS500 > Mg_6Al_2HTCT500 \tag{1}$$

This order is also illustrated in Figure 1 that shows the evolution of the conversion levels as a function of the temperature. From these data, it can be concluded that the replacement of the bivalent cation Mg^{2+} by Co^{2+} in the chemical composition of sample has a significant effect. This observation is in line with a previous work [13] that indicated the importance of the nature of the bivalent cation in the sample. Moreover, an increase of the catalytic activity is clearly observed for the solids prepared by microwaves or ultrasounds methods.

Table 1. Code name, BET specific area, chemical composition, crystallite size and temperature of 50% toluene conversion (T_{50}) of catalysts.

Solids	Specific surface area $(m^2 \cdot g^{-1})$	Chemical composition M^{2+}/Al^{3+}		T_{50} (°C)	Crystallite size of oxide * (nm)
		Theoretical	Experimental		
$Mg_6Al_2HTCT500$	216	3	2.79	477	11.6
$Mg_6Al_2HTUS500$	272	3	2.84	421	4.5
$Mg_6Al_2HTMW500$	342	3	2.95	295	3.7
$Co_6Al_2HTCT500$	123	3	2.85	287	9.4
$Co_6Al_2HTUS500$	157	3	3.15	278	7.0
$Co_6Al_2HTMW500$	167	3	3.00	271	7.6

* Estimated by the XRD measurement.

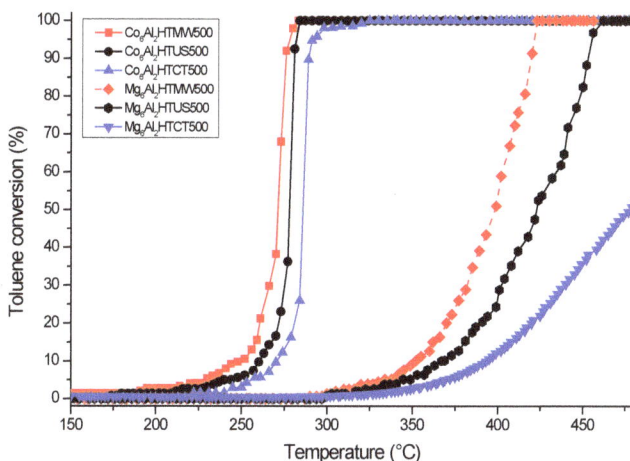

Figure 1. Conversion of toluene (%) on mixed oxide *vs.* reaction temperature (°C).

When toluene conversion was complete, H_2O and CO_2 were the only products observed. However, during toluene conversion at lower temperatures on fall samples, few ppm of benzene and CO were detected. Different reaction paths have been already recognized [34]. The most important reaction route is the one initiated by attack on the methyl group with subsequent oxidation steps; however, direct dealkylation of toluene into benzene is possible.

For a better understanding of the effect of these preparation methods on the catalysts, physico-chemicals characterizations of the Co-Al mixed oxides were performed. Firstly, XRD was done to investigate the structure of the samples. The X-ray diffraction patterns (Figure 2) reveal several crystalline spinel phases. These spinel phases are divided into two groups: the spinel phase of cobalt (II) and cobalt (III) (Co_3O_4 (JCPDS-ICDD 42-1467)) and cobalt aluminum spinel $CoAl_2O_4$ phases (JCPDS-ICDD 44-0160) and Co_2AlO_4 (JCPDS-ICDD 38-0814). The presence of the first phase (Co_3O_4) is due to the easy oxidation of Co^{2+} to Co^{3+} in contact with the air and the greater thermodynamic stability of this phase compared to CoO [35]. However, the three diffraction patterns of these three spinel phases are very similar in position and in intensity [36–38]. For this reason, it is not possible to differentiate the phases by XRD. The Co-Al mixed oxide leads to the normal spinel, *i.e.*, $CoAl_2O_4$ corresponding to the Co^{2+} in the tetrahedral positions and the Al^{3+} in the octahedral position. However, a slight decrease in the crystallite size of the oxide is observed for solids prepared using the microwaves or ultrasound methods (Table 1).

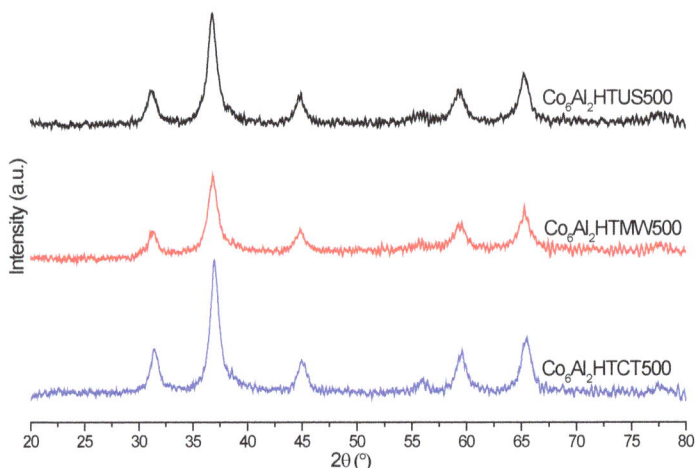

Figure 2. X-ray diffraction patterns of the mixed oxides.

In order to investigate the morphology of the samples prepared via the three different routes, TEM analysis was performed. From the representative micrographs shown in Figure 3, some morphological differences can be observed depending on the cobalt based mixed oxide. Firstly, the SEM images show that the three samples are formed of agglomerates (with sizes in the mm range), themselves composed of an aggregation of very small particles. As seen by both SEM (Figure 3a–c) and TEM (Figure 3a'–c') images, the size of these individual particles is bigger and less regular in the solid resulting from the conventional synthesis. By contrast, the other two samples have a sharper size distribution of grains with more uniform particle sizes, less than 10 nm. Moreover, the grains appear much thinner, especially in the case of $Co_6Al_2HTMW500$ (see insets in SEM images). These observations are consistent with the assessment of the crystallite size made during the XRD analysis (Table 1). Moreover, the HR-TEM images confirm that all three samples are composed of grains with crystalline structure. Finally, the various TEM images suggest a difference in porosity among the three samples, being more important on the solid resulting from both the Ultrasonic and microwave assisted syntheses compared to the solid issued from the conventional route.

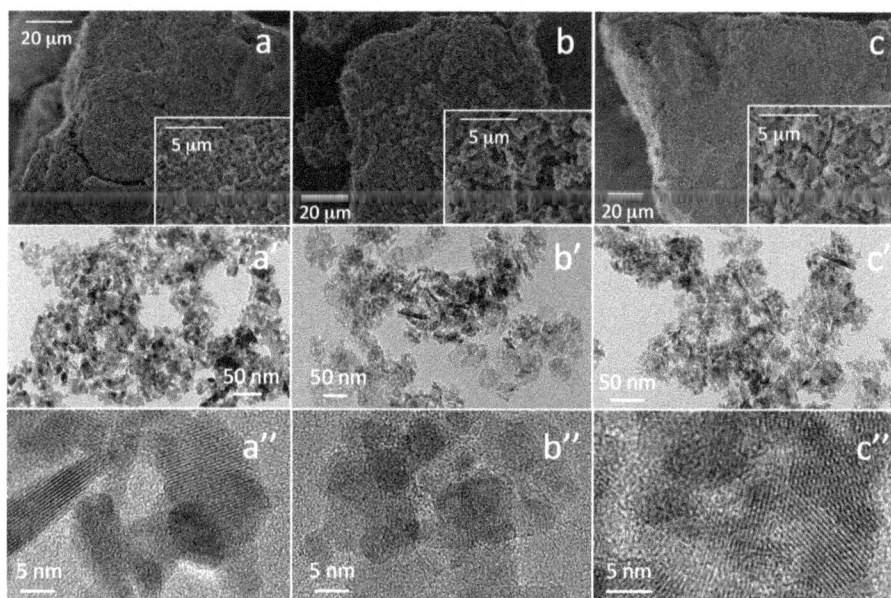

Figure 3. (a–c) SEM; (a'–c') TEM and (a''–c'') HR-TEM images of samples of (a,a',a'') $Co_6Al_2HTCT500$, (b,b',b'') $Co_6Al_2HTUS500$ and (c,c',c'') $Co_6Al_2HTMW500$.

Regarding the grains morphology, each grain has solid elliptical shapes of larger or smaller size. However, in the case of solids issued from unconventional syntheses, the presence of elongated grains forms ("rods") is observed. This form of grain may be due to two factors: the orientation of the grain in the analysis is perpendicular to the plane or the grain has "rod" type morphology. Such rod shape morphology has been already observed in the literature for solid cobalt oxide (Co_3O_4) [39,40] and it has been interestingly related to samples having good performances in the oxidation of carbon monoxide at low temperatures.

Analysis of the samples by Temperature-Programmed Reduction (H_2-TPR) was next performed (Figure 4). The reduction profile has a similar shape for the three samples but the reduction temperature of samples slightly differs depending on the preparation method of the solid. Indeed, the solid resulting from the synthesis with microwaves is reduced at lower temperatures. In order to identify more specifically the reduction peaks, deconvolution of these peaks using Gaussian-type components was done. This peak deconvolution (Figure 4), with a correlation factor greater than 0.999, reveal that the reduction of Co_6Al_2HT500 consists of two steps at low temperatures ($T < 400\,°C$) for the Co_3O_4 species and three steps at high temperature ($T > 400\,°C$) for cobalt/aluminum spinel species [39,41–43].

At $T < 400\,°C$

$$Co_3O_4 + H_2 \rightarrow 3CoO + H_2O \tag{2}$$

$$3CoO + 3H_2 \rightarrow 3Co + 3H_2O \tag{3}$$

At $T > 400\,°C$

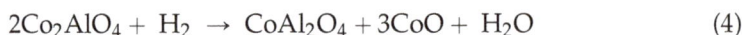

$$2Co_2AlO_4 + H_2 \rightarrow CoAl_2O_4 + 3CoO + H_2O \tag{4}$$

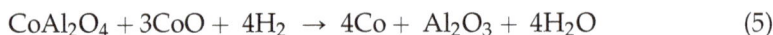

$$CoAl_2O_4 + 3CoO + 4H_2 \rightarrow 4Co + Al_2O_3 + 4H_2O \tag{5}$$

$$CoAl_2O_4 + H_2 \rightarrow Co + Al_2O_3 + H_2O \tag{6}$$

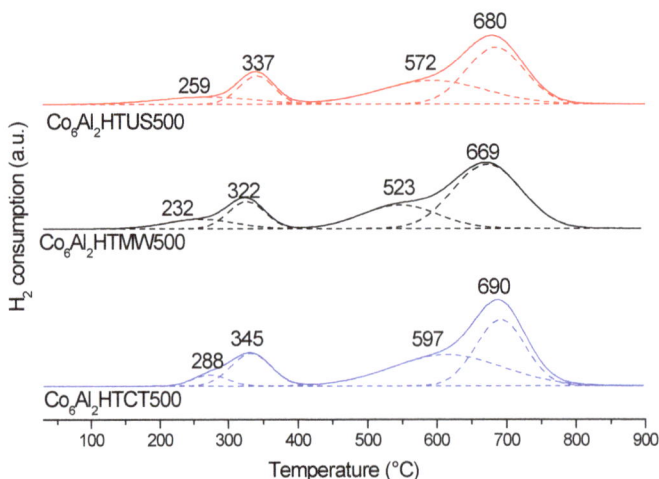

Figure 4. H_2-TPR profiles of the mixed oxides.

For the three solids, a distinction between the two reduction stages of Co_3O_4 is possible. Similar discrimination was done earlier by Moulijn and Arnoldy [41] and explained by the reduction of Co_3O_4 into CoO phase at around 240 °C. Then, the second peak (at 330 °C) corresponds to the reduction of either the CoO phase into Co^0 or of Co_3O_4 into Co^0 [39,41–43]. Regarding the reduction of species at higher temperatures ($T > 400$ °C), only a limited change in the shape of the profiles is observed. The observation of three stages of reduction is more difficult because they are in the same temperature range. Indeed, a smaller reduction temperature is observed in the case of the solids synthesized using microwaves or ultrasound. Cook et $al.$ [44] showed the importance of the crystallite size for the reduction step. Thus, when the crystallite size decreases, the reducibility takes place at lower temperature. The consumption for all peaks is reported into Table 2. The stoichiometry of the first steps ($T < 400$ °C) predicts a ratio equal to 3 for the consumption ratio peak 2/1. This ratio is respected for all samples, which confirms the hypothesis for the chemical equation. Concerning the second step, it is difficult to distinguish the three steps due to the position in the same temperature range. So, for the quantitative study, the first step ($Co_2AlO_4 \rightarrow CoAl_2O_4$ and CoO) corresponds to Peak 3 and the last two correspond to Peak 4. The ratio between Peaks 4 and 3 is equal to 5 which is in accordance with the literature [39,41–44].

Table 2. Temperatures for H_2 consumptions by TPR experiments Co_6Al_2HT500 issued from different preparation methods.

Catalyst	Temperature (°C) H_2 consumption ($\mu mol\cdot g^{-1}$)				Consumption ratio peak 2/1	Consumption ratio peak 4/3	H_2 consumption ($\mu mol\cdot g^{-1}$)
	Peak 1	Peak 2	Peak 3	Peak 4			
$Co_6Al_2HTCT500$	288 °C 644	345 °C 2012	597 °C 1458	690 °C 7410	3.12	5.08	11715
$Co_6Al_2HTUS500$	259 °C 651	337 °C 2192	572 °C 1557	680 °C 7444	3.36	4.78	11845
$Co_6Al_2HTMW500$	232 °C 777	322 °C 2242	523 °C 1229	669 °C 7711	2.88	4.69	11961

The cobalt based catalysts were also characterized by XPS in order to examine the effect of the preparation method on the nature and the oxidation degrees of the surface species. The XPS spectra of O 1s and Co 2p are shown in Figure 5, and the XPS data are summarized in Table 3.

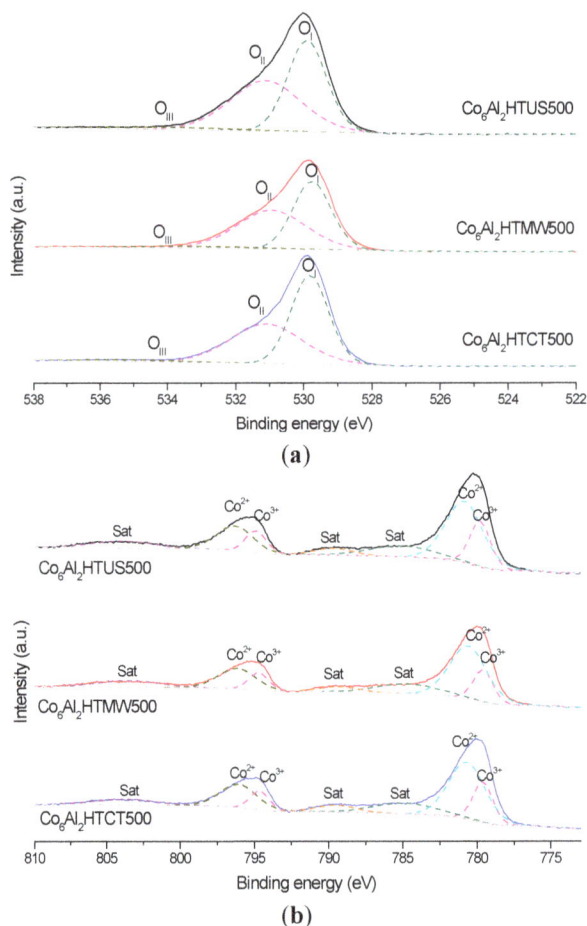

Figure 5. XPS spectra of (**a**) O 1s and (**b**) Co 2p from the mixed oxides.

Table 3. Binding energy (E_b) for O_{1s} and Co_{2p} levels for the catalysts.

Catalyst	E_b (O_{1s}) (eV)			E_b (Co_{2p}) (eV)			
				Co $2p_{1/2}$		Co $2p_{3/2}$	
	O_I	O_{II}	O_{III}	Co^{2+}	Co^{3+}	Co^{2+}	Co^{3+}
$Co_6Al_2HTCT500$	529.8	531.1	533.9	796.1	794.7	780.7	779.6
$Co_6Al_2HTUS500$	529.9	531.1	533.8	796.1	794.7	780.8	779.7
$Co_6Al_2HTMW500$	529.8	531.0	533.5	796.2	794.8	780.7	779.6

Regarding the photopeak of oxygen 1s (O 1s), three components are observed. The first component (O_{III}) at 533–534 eV corresponds to the oxygen present in the form of carbonates or of water molecules [45]. The second component (O_{II}) at 531 eV corresponds to adsorbed surface oxygen (O_2^- or O^-) or hydroxyl groups

(HO$^-$) [46,47]. The last component (O$_I$) which is located at a lower binding energy is consistent with lattice oxygen O^{2-}. The oxygen composition of the surface of the solid plays an important role in the catalytic activity in the oxidation reactions. Indeed, O$_{II}$ species exhibit greater mobility than the lattice oxygen. In addition, several authors [46,47] have shown that the catalytic activity of the solid could be related to the presence of a relationship between oxygen O$_{II}$ and the largest O$_I$ oxygen. From the data, a ratio between the O$_I$ and O$_{II}$ species was calculated for all samples (Table 4). This revealed that for solids synthetized under ultrasound or microwaves, the ratio O$_{II}$/O$_I$ increased.

Table 4. XPS data of Co$_6$Al$_2$HT500 issued from different preparation methods and catalytic intrinsic activity calculated at 10% toluene conversion.

Catalyst	Surface Content (%) by XPS			T_{50} (°C)	Catalytic intrinsic activity (mol·m^{-2}·h^{-1})
	O$_{II}$/O$_I$	Co^{2+}/Co^{3+}	Co/Al		
Co$_6$Al$_2$HTCT500	0.86	2.38	2.9	287	2.59 × 10^{-10}
Co$_6$Al$_2$HTUS500	1.01	2.69	2.9	278	2.80 × 10^{-10}
Co$_6$Al$_2$HTMW500	1.05	2.98	3.3	271	3.05 × 10^{-10}

Concerning the cobalt species (Table 3), an overlay of the photopeaks attributed to Co 2p is represented in Figure 5b. The observed profiles are characteristic of a mixture of Co^{2+} and Co^{3+} belonging to CoAl$_2$O$_4$ or Co$_3$O$_4$ species [48,49]. A relationship between the Co^{2+} and Co^{3+} species was calculated as the ratio of the area of the Co^{2+} signal component Co 2p$_{3/2}$ and the Co^{3+} signal component Co 2p$_{3/2}$ [6] (Table 4). This ratio suggests that the surface amount of Co^{2+} is higher in the Co$_6$Al$_2$HTMW500 solid. It can be observed as well that the presence of oxygen defects is at a higher quantity in the case of the Co$_6$Al$_2$HTMW500 mixed oxide, which could explain the best catalytic reactivity of the solid [6,48].

For the VOC oxidation on metal oxide based catalysts, several catalyst parameters, such as the surface area, the reducibility of the active species, the oxidation degree of the metal species or the oxygen species present on the surface, can affect the catalytic performances [6,50]. Thus, the better activity of the mixed oxides synthesized using microwaves can be related to a larger specific surface area. Indeed, this increased surface area can provide greater availability of the active phase for the reaction.

Moreover, a relationship between the H$_2$ consumption of the first temperature peak (corresponding to the reduction of the phase Co$_3$O$_4$ into Co^0) between 0 and 400 °C and the catalytic intrinsic activity is obtained (Figure 6). This relationship between the low-temperature reducibility and the toluene conversion suggests that Co$_3$O$_4$ is the active phase. Then, a redox mechanism for this reaction can be proposed, corresponding to a Mars Van Krevelen mechanism [6,51]. This mechanism involves the participation of the lattice oxygen by a redox cycle. For the combustion of alkane with cobalt oxide, Solsona *et al.* [51] reported that O^{2-} anions participate in the total oxidation according to the redox mechanism. Moreover, Bahlawaue [52] showed that the reactivity depends on the fast migration of oxygen ions through the lattice cobalt oxide.

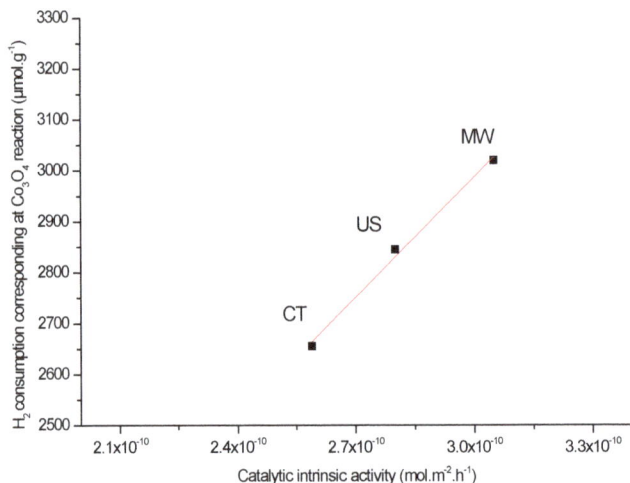

Figure 6. Relationship between catalytic intrinsic activity and H_2 consumption during the H_2-TPR corresponding at Co_3O_4 species.

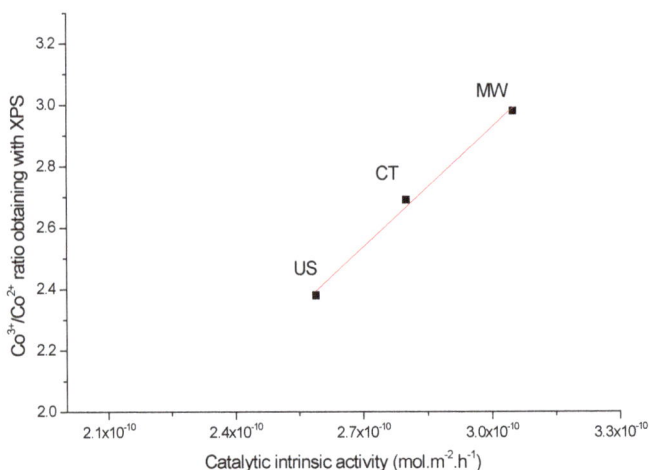

Figure 7. Relationship between catalytic intrinsic activity and the ratio Co^{2+} and Co^{3+} species calculated with XPS analysis.

Moreover, another relationship could be identified as well, involving XPS results. Indeed, Figure 7 shows that the Co^{2+}/Co^{3+} ratio evaluated by XPS is also linearly related to the catalytic intrinsic activity. The relationship shows that the proportion of solid in the Co^{2+} is an important factor for achieving a good activity of the solid. Indeed, Figure 7 shows that the most active solid in the oxidation of toluene has the highest Co^{2+} proportion on the surface of the solid. This can also be linked to the increased presence of O_{II} species (adsorbed oxygen O_2^- or O^- type and oxygen vacancies in the surface of the solid). The greater presence of such a type of oxygen in the case of solids prepared by microwaves and ultrasound is an important feature that provides a higher availability of oxygen to effect the reaction (and the reduction of the solid) at low temperatures. Indeed, this oxygen is more mobile and therefore more available for reaction than in the case of the solid prepared conventionally.

3. Experimental Section

3.1. Preparation of Catalysts

Conventional Hydrotalcite (HT) with Co:Al molar ratio 6:2 was prepared by the coprecipitation method. An aqueous solution containing appropriate amounts of $Co(NO_3)_2$, $6H_2O$ and $Al(NO_3)_3$, $9H_2O$ ions (solution A) was added drop-wise under stirring into an aqueous solution containing Na_2CO_3 and $NaOH$ (solution B). During the synthesis, the temperature and the pH were maintained at 60 °C and 10.5, respectively. The solution was aged at 60 °C during 18 h. Then, the precipitate was filtered, washed several times with hot deionized water (50 °C) and dried at 60 °C for 64 h. The obtained sample was called Co_6Al_2HTCT (where HTCT corresponds to the HT synthetized using the conventional method).

Two other samples were prepared using non-conventional methods involving either microwaves (MW) or ultrasounds (US). The first one was prepared by adding solution A (containing Co^{2+} and Al^{3+} ions, see above) into solution B (containing Na_2CO_3 and $NaOH$). The obtained gel was aged at 80 °C for 1 h, in a monomode reactor Synthewave Prolabo (Fontenay sous bois, France) 402 (300 W) equipped with an infrared pyrometer control and mechanical stirring. After the precipitate was filtered and washed until pH 7, the solid was dried at 60 °C for 12 h. The sample was labeled Co_6Al_2HTMW (where HTMW corresponds to the hydrotalcite preparation route with the microwaves procedure). The preparation of the last sample was carried out as described above, but adding the aqueous solution (A) containing the metal cations (Co^{2+} and Al^{3+}) onto solution (B) containing the base while simultaneously submitting the mixture to ultrasounds irradiation (Branson sonifier (Danbury, CT, USA) 450 (20 kHz)). This procedure was carried out at atmospheric pressure and at ambient temperature. The precipitate was filtered, washed to eliminate the alkali metals and the nitrate ions until the pH 7, and then dried at 60 °C for 12 h. The sample was labelled Co_6Al_2HTUS (where HTUS corresponds to the HT synthetized using the ultrasound method).

159

Thermal treatment of all three samples was performed under flow of air ($4\,\mathrm{L\cdot h^{-1}}$, $1\,\mathrm{^\circ C\cdot min^{-1}}$, 4 h at 500 °C). The calcined samples were named $Co_6Al_2HTCT500$, $Co_6Al_2HTMW500$ and $Co_6Al_2HTUS500$.

Reference Mg-Al-O materials were also prepared to allow comparisons. These reference solids, derived from classic hydrotalcite compositions, were synthesized and calcined following the same protocols as the ones presented above for the preparation of Co-Al mixed oxides. These samples were respectively denominated $Mg_6Al_2HTCT500$, $Mg_6Al_2HTMW500$ and $Mg_6Al_2HTUS500$.

3.2. Characterization Techniques

Specific surface areas were evaluated by the BET method using a QSurf M1 apparatus (Thermo Electron, Waltham, MA, USA). The gas adsorbed at $-196\,\mathrm{^\circ C}$ was pure nitrogen.

Crystalline structures were determined at room temperature from X-ray Diffractograms (XRD) recorded on a D8 Advance from Bruker AXS (Champs-Sur–Marne, France) diffractometer equipped with a copper anode ($\lambda = 1.5406$ Å) and a LynxEye Detector. The scattering intensities were measured over an angular range of $10° \leqslant 2\theta \leqslant 80°$ for all samples with a step-size of $\Delta(2\theta) = 0.02°$ and a count time of 4 s per step. The diffraction patterns were indexed by comparison with the "Joint Committee on Powder Diffraction Standards" (JCPDS) files. Crystallite sizes were determined (Scherrer equation) using a graphic based profile analysis program (TOPAS from Bruker AXS).

Elemental compositions of samples were analyzed using Inductively Coupled Plasma-Optical Emission Spectrometer (ICP OES, Thermo Electron (Waltham, MA, USA) ICAP 6300 DUO).

Samples morphologies were investigated by Transmission Electron Microscopy observations carried out in scanning (SEM) and transmission (TEM) modes. The SEM images were recorded on a Hitachi (Tokyo, Japan) SU-70 SEM-FEG microscope whereas both TEM and HR-TEM (high resolution) images were obtained on a JEOL (Akishima, Japan) JEM-2010 operating at 200 kV).

X-ray photoelectron spectroscopy (XPS) analyses were conducted on a Kratos (Manchester, UK) Axis Ultra DLD spectrometer with a monochromatic Al $K\alpha$ ($h\nu = 1486.6$ eV) radiation source operated at 15 kV and 15 mA. The binding energy (BE) was calibrated based on the line position of C 1s (285 eV). The CasaXPS processing software (Casa Software Ltd. Teignmouth, UK) was used to estimate the relative abundance of the different species. The surface atomic ratios were calculated by correcting the intensity with theoretical sensibility factors based on the Scofield cross section and using a nonlinear Shirley background.

The atomic composition, N_A/N_B, between A and B was derived from the equation:

$$\frac{N_A}{N_B} = \frac{I_A \, \sigma_A}{I_B \, \sigma_B} \left(\frac{E_{CA}}{E_{CB}}\right)^{-0.23} \qquad (7)$$

where σ is the cross section, E_C kinetic energy and I is the intensity of the peak.

Temperature-Programmed Reduction (H_2-TPR) experiments were carried out in an Altamira AMI-200 apparatus (Labor und Analysen Technik GmbH, Garbsen, Germany). Prior to the experiment, the sample (30 mg) was activated under argon at 150 °C for 1 h. The sample was then heated from ambient temperature to 850 °C under H_2 flow (5 vol% in argon 30 mL·min^{-1}) with a heating rate of 5 °C·min^{-1}.

3.3. Catalytic Tests

The activity for toluene total oxidation of the catalysts (100 mg) was measured in a continuous flow system in a fixed bed reactor at atmospheric pressure. Before each test, the catalyst was reactivated *in situ* in flowing air (2 L·h^{-1}) at 500 °C for 4 h. The flow of reactant gases (100 mL·min^{-1} with 1000 ppm of C_7H_8 and 20% O_2 in He) was adjusted using a instrument apparatus constituted of a saturator and mass flow controllers. After reaching a stable flow, the reactants were passed through the catalyst bed and the temperature was increased from room temperature to 500 °C (1 °C·min^{-1}). The feed and the reactor outflow gases were analyzed on line by a 490 Micro gas chromatography from Agilent Technologies France (les Ulis, France). Model 4400IR infrared analyzers from ADEV srl (Cesano Maderno, Italy) were also used to perform the analysis of both CO and CO_2. The catalysts performance was assessed in terms of T_{50} temperature, defined as the temperature at which 50% conversion was obtained.

Catalytic intrinsic activities were evaluated at 10% of toluene conversion and considering a plug-flow reactor:

$$A = \frac{Q \times 273.15 \times [C_7H_8]_0 \times X}{V_M \times T_{10} \times 10^6 \times m \times S_{BET}} \qquad (8)$$

where:

A is the catalytic intrinsic activity (mol·m^{-2}·h^{-1});
Q is the volume flow (L·h^{-1});
V_M is the molar volume (L·mol^{-1});
T_{10} is the catalyst temperature for 10% toluene conversion (K);
$[C_7H_8]_0$ is the toluene initial concentration (ppm);
X is the toluene conversion (%);
m is the catalyst mass (g);
S_{BET} is the specific surface area of the catalyst (m^2·g^{-1}).

4. Conclusions

Co$_6$Al$_2$HT hydrotalcite-like compounds were synthesized by three different methods: conventional co-precipitation, methods assisted by microwaves or ultrasounds. The mixed oxides obtained after calcination were studied. The specific surface areas of solids prepared under sonication method or using microwaves radiations are much higher than that obtained by coprecipitation. The H$_2$-TPR analysis shows a reduction at lower temperatures of the Co$_3$O$_4$ species. The XPS study shows that the solid resulting from the synthesis by microwaves leads to a higher content in Co^{2+} species on the sample. These Co^{2+} species induce a higher presence of oxygen vacancies that could explain the higher activity of the catalyst. Moreover, correlations between the T50 of the toluene oxidation and the Co^{2+}/Co^{3+} ratio as well as the H$_2$ consumption of the solids are evidenced, suggesting a Mars Van-Krevelen mechanism. Thus, the preparation of cobalt-based mixed oxide via a Layer Double Hydroxide synthesis route using treatment by microwaves seems to be efficient to obtain a performing catalyst for VOCs oxidation.

Acknowledgments: The authors thank the Industrial Environmental Research Institute (IRENI) and the European Community (Interreg IV France-Wallonie-Flandres project, REDUGAZ) for financial support. The authors would like to thank Francine Cazier, François Delattre and Pierre-Edouard Danjou for their precious advices about the microwaves and ultrasound synthesis.

Author Contributions: Eric Genty and Julien Brunet performed the synthesis of catalysts, carried out the TPR measurements, contributed to the XPS characterization and catalytic tests. Christophe Poupin achieved the XRD experiments. Sandra Casale and Pascale Massiani realized the TEM analysis. All the authors contributed equally to the data interpretation and discussion. Renaud Cousin and Stéphane Siffert coordinate the manuscript. Eric Genty and Renaud Cousin write the manuscript. Sylvie Capelle and Pascale Massiani improved the manuscript.

Conflicts of Interest: The authors declare no conflict of interest.

References

1. Santos, V.P.; Carabineiro, S.A.C.; Tavares, P.B.; Pereira, M.F.R.; Órfão, J.J.M.; Figueiredo, J.L. Applied Catalysis B: Environmental Oxidation of CO, ethanol and toluene over TiO$_2$ supported noble metal catalysts. *Appl. Catal. B* **2010**, *99*, 198–205.
2. Liotta, L.F. Catalytic oxidation of volatile organic compounds on supported noble metals. *Appl. Catal. B* **2010**, *100*, 403–412.
3. Barakat, T.; Rooke, J.C.; Genty, E.; Cousin, R.; Siffert, S.; Su, B.-L. Gold catalysts in environmental remediation and water-gas shift technologies. *Energy Environ. Sci.* **2013**, *6*, 371–391.
4. Ordóñez, S.; Bello, L.; Sastre, H.; Rosal, R.; Fernando, V.D. Kinetics of the deep oxidation of benzene , toluene, *n*-hexane and their binary mixtures over a platinum on γ-alumina catalyst. *Appl. Catal. B* **2002**, *38*, 139–149.

5. Scirè, S.; Liotta, L.F. Supported gold catalysts for the total oxidation of volatile organic compounds. *Appl. Catal. B* **2012**, *125*, 222–246.

6. Gao, W.; Zhao, Y.; Liu, J.; Huang, Q.; He, S.; Li, C.; Zhao, J.; Wei, M. Catalytic conversion of syngas to mixed alcohols over CuFe-based catalysts derived from layered double hydroxides. *Catal. Sci. Technol.* **2013**, *3*, 1324–1332.

7. Jiratova, K.; Cuba, P.; Kovanda, F.; Hilaire, L.; Pitchon, V. Preparation and characterisation of activated Ni(Mn)/Mg/Al hydrotalcites for combustion catalysis. *Catal. Today* **2002**, *76*, 43–53.

8. Kovanda, F.; Jiratova, K.; Rymes, J.; Kolousek, D. Characterization of activated CuMgAl hydrotalcites and their catalytic activity in toluene combustion. *Appl. Clay Sci.* **2001**, *18*, 71–80.

9. Zhu, Z.; Lu, G.; Zhang, Z.; Guo, Y.; Guo, Y.; Wang, Y. Highly Active and Stable Co_3O_4/ZSM-5 Catalyst for Propane Oxidation: Effect of the preparation Method. *ACS Catal.* **2013**, *3*, 1154–1164.

10. Tang, C.-W.; Kuo, M.-C.; Lin, C.-J.; Wang, C.-B.; Chien, S.-H. Evaluation of carbon monoxide oxidation over CeO_2/Co_3O_4 catalysts: Effect of ceria loading. *Catal. Today* **2008**, *131*, 520–525.

11. Liotta, L.F.; Ousmane, M.; di Carlo, G.; Pantaleo, G.; Deganello, G.; Boreave, A.; Giroir-Fendler, A. Catalytic Removal of Toluene over Co_3O_4–CeO_2 Mixed Oxide Catalysts: Comparison with Pt/Al_2O_3. *Catal. Lett.* **2008**, *127*, 270–276.

12. Liotta, L.F.; Wu, H.; Pantaleo, G.; Venezia, A.M. Co_3O_4 nanocrystals and Co_3O_4–MO_x binary oxides for CO, CH_4 and VOC oxidation at low temperatures: A review. *Catal. Sci. Technol.* **2013**, *3*, 3085–3102.

13. Genty, E.; Cousin, R.; Capelle, S.; Gennequin, C.; Siffert, S. Catalytic Oxidation of Toluene and CO over Nanocatalysts Derived from Hydrotalcite-Like Compounds ($X_6^{2+}Al_2^{3+}$): Effect of the Bivalent Cation. *Eur. J. Inorg. Chem.* **2012**, *2012*, 2802–2811.

14. Gennequin, C.; Siffert, S.; Cousin, R.; Aboukaïs, A. Co–Mg–Al Hydrotalcite Precursors for Catalytic Total Oxidation of Volatile Organic Compounds. *Top. Catal.* **2009**, *52*, 482–491.

15. Debecker, D.P.; Gaigneaux, E.M.; Busca, G. Exploring, tuning, and exploiting the basicity of hydrotalcites for applications in heterogeneous catalysis. *Chemistry* **2009**, *15*, 3920–3935.

16. Chang, K.S.; Song, H.; Park, Y.-S.; Woo, J.-W. Analysis of N_2O decomposition over fixed bed mixed metal oxide catalysts made from hydrotalcite-type precursors. *Appl. Catal. A* **2004**, *273*, 223–231.

17. Palacio, L.A.; Velásquez, J.; Echavarría, A.; Faro, A.; Ribeiro, F.R.; Ribeiro, M.F. Total oxidation of toluene over calcined trimetallic hydrotalcites type catalysts. *J. Hazard. Mater.* **2010**, *177*, 407–413.

18. Tanasoi, S.; Mitran, G.; Tanchoux, N.; Cacciaguerra, T.; Fajula, F.; Săndulescu, I.; Tichit, D.; Marcu, I.-C. Transition metal-containing mixed oxides catalysts derived from LDH precursors for short-chain hydrocarbons oxidation. *Appl. Catal. A* **2011**, *395*, 78–86.

19. Gennequin, C.; Barakat, T.; Tidahy, H.L.; Cousin, R.; Lamonier, J.-F.; Aboukaïs, A.; Siffert, S. Use and observation of the hydrotalcite "memory effect" for VOC oxidation. *Catal. Today* **2010**, *157*, 191–197.

20. Genty, E.; Cousin, R.; Capelle, S.; Siffert, S. Influence of Gold on Hydrotalcite-like Compound Catalysts for Toluene and CO Total Oxidation. *Catalysts* **2013**, *3*, 966–977.

21. Carpentier, J.; Lamonier, J.F.; Siffert, S.; Zhilinskaya, E.A.; Aboukaïs, A. Characterisation of Mg/Al hydrotalcite with interlayer palladium complex for catalytic oxidation of toluene. *Appl. Catal. A* **2002**, *234*, 91–101.

22. Kovanda, F.; Rojka, T.; Dobešová, J.; Machovič, V.; Bezdička, P.; Obalová, L.; Jirátová, K.; Grygar, T. Mixed oxides obtained from Co and Mn containing layered double hydroxides: Preparation, characterization, and catalytic properties. *J. Solid State Chem.* **2006**, *179*, 812–823.

23. Kovanda, F.; Koloušek, D.; Cílová, Z.; Hulínský, V. Crystallization of synthetic hydrotalcite under hydrothermal conditions. *Appl. Clay Sci.* **2005**, *28*, 101–109.

24. Ran, R.; Weng, D.; Wu, X.; Fan, J.; Qing, L. Rapid synthesis of $La_{0.7}Sr_{0.3}MnO_{3+\lambda}$ catalysts by microwave irradiation process. *Catal. Today* **2007**, *126*, 394–399.

25. Kaddouri, A.; Ifrah, S. Microwave-assisted synthesis of $La_{1-x}B_xMnO_{3.15}$ (B = Sr, Ag; x = 0 or 0.2) via manganese oxides susceptors and their activity in methane combustion. *Catal. Commun.* **2006**, *7*, 109–113.

26. Zhou, X.; Chen, Q.; Tao, Y.; Weng, H. Influence of ultrasound impregnation on the performance of $Co/Zr/SiO_2$ catalyst during Fischer-Tropsch synthesis. *Chin. J. Catal.* **2011**, *32*, 1156–1165.

27. Climent, M.; Corma, A.; Iborra, S.; Epping, K.; Velty, A. Increasing the basicity and catalytic activity of hydrotalcites by different synthesis procedures. *J. Catal.* **2004**, *225*, 316–326.

28. Benito, P.; Labajos, F.M.; Rives, V. Microwaves and layered double hydroxides: A smooth understanding. *Pure Appl. Chem.* **2009**, *81*, 1459–1471.

29. Rivera, J.; Fetter, G.; Bosch, P. Microwave power effect on hydrotalcite synthesis. *Microporous Mesoporous Mater.* **2006**, *89*, 306–314.

30. Pérez, A.; Lamonier, J.-F.; Giraudon, J.-M.; Molina, R.; Moreno, S. Catalytic activity of Co–Mg mixed oxides in the VOC oxidation: Effects of ultrasonic assisted in the synthesis. *Catal. Today* **2011**, *176*, 286–291.

31. Neto, O.R.; Ribeiro, N.F.P.; Perez, C.A.C.; Schmal, M.; Souza, M.M.V.M. Incorporation of cerium ions by sonication in Ni–Mg–Al layered double hydroxides. *Appl. Clay Sci.* **2010**, *48*, 542–546.

32. Mokhtar, M.; Saleh, T.S.; Ahmed, N.S.; Al-Thabaiti, S.A.; Al-Shareef, R.A. An eco-friendly *N*-sulfonylation of amines using stable and reusable Zn-Al-hydrotalcite solid base catalyst under ultrasound irradiation. *Ultrason. Sonochem.* **2011**, *18*, 172–176.

33. Fetter, G.; Hernández, F.; Maubert, A. Microwave irradiation effect on hydrotalcite synthesis. *J. Porous Mater.* **1997**, *30*, 27–30.

34. Lars, S.; Andersson, T. Reaction Networks in the Catalytic Vapor-Phase Oxidation of Toluene and Xylenes. *J. Catal.* **1986**, *98*, 138–149.

35. Kannan, S.; Swamy, C.S. Catalytic decomposition of nitrous oxide over calcined cobalt aluminum hydrotalcites. *Catal. Today* **1999**, *53*, 725–737.

36. Gabrovska, M.; Edreva-Kardjieva, R.; Tenchev, K.; Tzvetkov, P.; Spojakina, A.; Petrov, L. Effect of Co-content on the structure and activity of Co–Al hydrotalcite-like materials as catalyst precursors for CO oxidation. *Appl. Catal. A Gen.* **2011**, *399*, 242–251.

37. Pérez-Ramírez, J.; Mul, G.; Kapteijn, F.; Moulijn, J.A. *In situ* investigation of the thermal decomposition of Co–Al hydrotalcite in different atmospheres. *J. Mater. Chem.* **2001**, *11*, 821–830.

38. Babay, S.; Bulou, A.; Mercier, A.M.; Toumi, M. The decomposition of the layered double hydroxides of Co and Al: Phase segregation of a new single phase spinel oxide. *Spectrochim. Acta Part A* **2015**, *141*, 80–87.

39. Alvarez, A.; Ivanova, S.; Centeno, M.A.; Odriozola, J.A. Sub-ambient CO oxidation over mesoporous Co_3O_4: Effect of morphology on its reduction behavior and catalytic performance. *Appl. Catal. A* **2012**, *431–432*, 9–17.

40. Xie, X.; Li, Y.; Liu, Z.-Q.; Haruta, M.; Shen, W. Low-temperature oxidation of CO catalysed by Co_3O_4 nanorods. *Nature* **2009**, *458*, 746–749.

41. Arnoldy, P.; Moulijn, J.A. Temperature-Programmed Reduction of CoO/Al_2O_3 Catalysts. *J. Catal.* **1985**, *93*, 38–54.

42. Sexton, B.A.; Hughes, A.E.; Turney, T.W. An XPS and TPR Study of the Reduction of Promoted Cobalt-Kieselguhr Fisher-Tropsch Catalysts. *J. Catal.* **1986**, *406*, 390–406.

43. Busca, G.; Finocchio, F.; Ramis, G.; Ricchiardi, G. On the role of acidity in catalytic oxidation. *Catal. Today* **1996**, *32*, 133–143.

44. Cook, K.M.; Poudyal, S.; Miller, J.T.; Bartholomew, C.H.; Hecker, W.C. Reducibility of alumina-supported cobalt Fischer-Tropsch catalysts: Effects of noble metal type, distribution, retention, chemical state, bonding, and influence on cobalt crystallite size. *Appl. Catal. A* **2012**, *449*, 69–80.

45. Dupin, J.-C.; Gonbeau, D.; Vinatier, P.; Levasseur, A. Systematic XPS studies of metal oxides, hydroxides and peroxides. *Phys. Chem. Chem. Phys.* **2000**, *2*, 1319–1324.

46. Delimaris, D.; Ioannides, T. VOC oxidation over MnO_x–CeO_2 catalysts prepared by a combustion method. *Appl. Catal. B Environ.* **2008**, *84*, 303–312.

47. Liu, Y.; Dai, H.; Deng, J.; Xie, S.; Yang, H.; Tan, W.; Han, W.; Jiang, Y.; Guo, G. Mesoporous Co_3O_4-supported gold nanocatalysts: Highly active for the oxidation of carbon monoxide, benzene, toluene, and *O*-xylene. *J. Catal.* **2014**, *309*, 408–418.

48. Sampanthar, J.T.; Zeng, H.C. Synthesis of $Co^{II}Co^{III}_{2-x}Al_xO_4$-$Al_2O_3$ Nanocomposites via Decomposition of $Co^{II}_{0.73}Co^{III}_{0.27}(OH)_{2.00}(NO_3)_{0.23}(CO_3)_{0.02} \cdot 0.5H_2O$ in a Sol-Gel-Derived γ-Al_2O_3 Matrix. *Chem. Mater.* **2001**, *13*, 4722–4730.

49. Garcia, T.; Agouram, S.; Sánchez-Royo, J.F.; Murillo, R.; Mastral, A.M.; Aranda, A.; Vázquez, I.; Dejoz, A.; Solsona, B. Deep oxidation of volatile organic compounds using ordered cobalt oxides prepared by a nanocasting route. *Appl. Catal. A* **2010**, *386*, 16–27.

50. Cheng, J.; Yu, J.; Wang, X.; Li, L.; Li, J.; Hao, Z. Novel CH_4 Combustion Catalysts Derived from Cu-Co/X-Al (X = Fe, Mn, La, Ce) Hydrotalcite-like Compounds. *Energy Fuels* **2008**, *22*, 65–67.

51. Solsona, B.; Davies, T.E.; Garcia, T.; Vázquez, I.; Dejoz, A.; Taylor, S.H. Total oxidation of propane using nanocrystalline cobalt oxide and supported cobalt oxide catalysts. *Appl. Catal. B* **2008**, *84*, 176–184.

52. Bahlawane, N. Kinetics of methane combustion over CVD-made cobalt oxide catalysts. *Appl. Catal. B* **2006**, *67*, 168–176.

Oxygen Storage Capacity and Oxygen Mobility of Co-Mn-Mg-Al Mixed Oxides and Their Relation in the VOC Oxidation Reaction

María Haidy Castaño, Rafael Molina and Sonia Moreno

Abstract: Co-Mn-Mg-Al oxides were synthesized using auto-combustion and co-precipitation techniques. Constant ratios were maintained with $(Co + Mn + Mg)/Al$ equal to 3.0, $(Co + Mn)/Mg$ equal to 1.0 and Co/Mn equal to 0.5. The chemical and structural composition, redox properties, oxygen storage capacity and oxygen mobility were analyzed using X-ray fluorescence (XRF), X-ray diffraction (XRD), Raman spectroscopy, scanning electron microscopy (SEM), temperature-programmed reduction of hydrogen (H_2-TPR), oxygen storage capacity (OSC), oxygen storage complete capacity (OSCC) and isotopic exchange, respectively. The catalytic behavior of the oxides was evaluated in the total oxidation of a mixture of 250 ppm toluene and 250 ppm 2-propanol. The synthesis methodology affected the crystallite size, redox properties, OSC and oxide oxygen mobility, which determined the catalytic behavior. The co precipitation method got the most active oxide in the oxidation of the volatile organic compound (VOC) mixture because of the improved mobility of oxygen and ability to favor redox processes in the material structure.

Reprinted from *Catalysts*. Cite as: Castaño, M.H.; Molina, R.; Moreno, S. Oxygen Storage Capacity and Oxygen Mobility of Co-Mn-Mg-Al Mixed Oxides and Their Relation in the VOC Oxidation Reaction. *Catalysts* **2015**, *5*, 905–925.

1. Introduction

A number of studies have investigated the individual oxidation of volatile organic compounds (VOCs), such as toluene, ethyl acetate, 2-propanol, among others [1–4]. However, in practical applications, emissions from industrial sources or mobile sources contain VOC mixtures with different compositions and concentrations, which increases the difficulty of predicting catalytic behavior; thus, new catalytic materials must be developed that can achieve a greater efficiency in removing mixtures of these contaminants.

According to certain authors, the oxidation of a mixture provokes the inhibition of the reaction, which is manifested as an increase in the temperature required for the conversion of a compound. This behavior could occur as a result of competition among VOCs for adsorption at the active sites [5,6] or competition by oxygen atoms chemisorbed [7].

According to the literature [8,9], the behavior of components in a mixture has been difficult to predict because such behavior is dependent on the composition of the mixture, number and proportion of components, and nature of the catalyst.

Mixed Mn and Co oxides are considered valuable materials in oxidation reactions because of their ability to occur in multiple oxidation states and optimum oxygen storage capacity [10]. In general, a redox mechanism occurs in the oxidation of volatile organic compounds in which the surface oxygen or lattice oxygen in the oxide may be involved. In Co and Mn oxides, the lattice oxygen has been found to participate in the oxidation of propene, CO, and VOCs [11–13] demonstrating an elevated oxygen mobility in these materials.

Numerous synthesis methods have been used to prepare metal oxides [14,15] and co-precipitation is one of the most frequently used for the generation of hydrotalcites as precursors of mixed oxides [16,17]. The thermal decomposition of hydrotalcites leads to the formation of oxides with interesting properties, such as elevated surface area (100–300 m^2/g), homogeneous element interdispersion, active phase dispersion, synergistic effects and basic characteristics [17,18]. Similarly, auto-combustion represents another important synthesis method in the preparation of mixed oxides in shorter synthesis times and without a precursor requirement. The characteristics that contribute to the unique properties of synthesized products can be summarized as follows: (i) initial reaction medium (aqueous solution) can mix reactants at a molecular level; (ii) high reaction temperatures ensure high purity and crystallinity products; and (iii) short process duration and gas formation inhibit increases of particle size and favor the synthesis of solids with large surface area [19–21].

Studies on mixed oxides obtained by the use of hydrotalcite-type precursors and prepared by the auto-combustion method that focus on their oxygen storage ability, reoxidation capacity, oxygen mobility and relationship between these properties and their catalytic performance are limited in number. In the present work, relationships between the catalytic activity in the oxidation of a VOC mixture and oxygen mobility and oxygen storage capacity in the Co-Mn mixed oxides are established.

2. Results and Discussion

2.1. Precursors Characterization

The diffraction patterns of the Mn and Co samples obtained by the auto-combustion and co-precipitation methodologies prior to the calcination process at 500 °C are presented in Figure 1. In the material prepared by co-precipitation, the typical formation of the hydrotalcite-type precursor (HTCoMn) is observed, with intense signals at 11.3, 22.6 and 34.4° 2θ, which indicate the generation of a well-crystallized layered structure with 3R symmetry [22]. In addition to the hydrotalcite phase, signals attributed to the $MnCO_3$ rhodochrosite phase and the

Mn_3O_4 hausmannite phase (JCPDS 44-1472 and JCPDS 24-0734) are observed. The formation of the $MnCO_3$ phase in the HTCoMn sample is consistent with that reported for Mn hydrotalcites; however, the presence of Co most likely contributes to the greater generation of this phase because Co^{2+} can be incorporated more efficiently into the layered structure as a result of similarities in the ionic radii of Co^{2+} and Mg^{2+} (0.65 Å and 0.72 Å, respectively) compared with that of Mn^{2+} (0.83 Å), increasing the presence of Mn^{2+} surface species capable of reacting with the carbonates and producing more $MnCO_3$ [23,24].

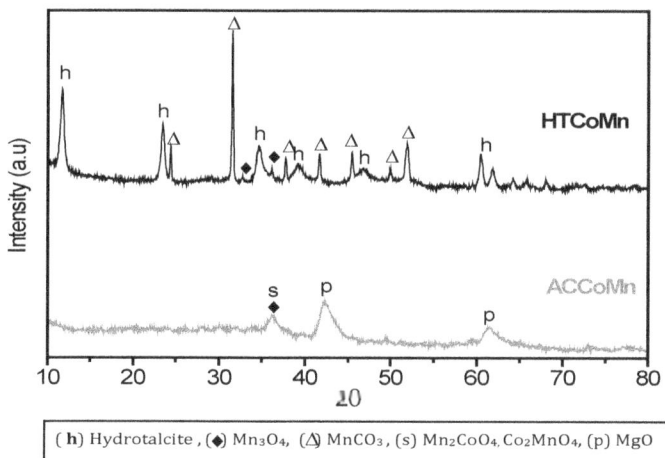

(**h**) Hydrotalcite , (◆) Mn_3O_4, (△) $MnCO_3$, (s) Mn_2CoO_4, Co_2MnO_4, (p) MgO

Figure 1. Diffraction patterns of the samples before the calcination process at 500 °C.

The lattice parameters a and c (distance between two neighboring cations and thickness of three brucite layers along with their interlayer space, respectively) calculated for HTCoMn correspond to 0.306 nm and 2.27 nm, respectively. The similarity between the values calculated for HTCoMn and values reported in the literature (a = 0.306 nm and c = 2.34 nm) confirm the formation of a layered structure [25]. However, the decrease in value of parameter c in HTCoMn is most likely because of an increment of the electrostatic forces between the layers and interlayer of the structure, an effect that is only possible if ions such as Mn^{3+} are present in the brucite layers.

On the other hand, the X-ray diffraction (XRD) profile of the sample obtained by auto-combustion exhibits direct formation of the mixed oxide. The peaks at 42.4 and 61.7° 2θ are attributed to the MgO periclase phase (JCPDS 45-0946), whereas the signal at 36.2° 2θ is attributed to the formation of Mn or Co spinels. The lattice parameter a calculated for the oxide using the $d_{(200)}$ reflection of the periclase phase corresponds to 0.427 nm. This value is higher than that found for the Mg-Al oxide synthesized by auto-combustion (0.419 nm) suggesting the partial incorporation of Co and/or Mn ions into the MgO-Al_2O_3 oxide matrix.

2.2. Characterization of the Mixed Oxides

2.2.1. Chemical and Structural Composition

In hydrotalcites to maintain the ratio $M^{2+}/M^{3+} = 3.0$ or $M^{3+}/(M^{2+} + M^{3+}) = 0.25$ is of great importance because it can form a layered structure that is characteristic of layered double hydroxides without the presence of undesired species, such as $Al(OH)_3$ or $Mg(OH)_2$ [18].

The elemental chemical analysis of the Co-Mn mixed oxides is indicated in Table 1. The results reveal that the molar ratios of the mixed oxides synthesized by both methodologies present values similar to the nominal values for $(Mn^{2+} + Co^{2+} + Mg^{2+})/Al^{3+} \cong 3.0$, $(Co^{2+} + Mn^{2+})/Mg^{2+} \cong 1.0$ and $Co^{2+}/Mn^{2+} \cong 0.5$.

Table 1. Chemical composition of the mixed oxides and particle size (Dp).

Sample	$(Mn^{2+} +Co^{2+} + Mg^{2+})/Al^{3+}$	$(Mn^{2+} + Co^{2+})/ Mg^{2+}$	Co^{2+}/Mn^{2+}	Dp (nm) $\pm$ 3nm
ACCoMn	3.3	1.6	0.5	7
CPCoMn	2.9	1.4	0.6	20

On the other hand, it is widely reported in the literature that the thermal decomposition of Mg-Al hydrotalcites starts with the loss of interlayer water molecules at 150–200 °C, followed by the collapse of the hydroxide layers in the temperature range of 300–400 °C and the total loss of the layered structure above 500 °C [26]. Figure 2 shows the diffraction patterns of the Co-Mn mixed oxides obtained by auto-combustion (ACCoMn) and generated after the thermal decomposition of the hydrotalcite-type precursor at 500 °C (CPCoMn). The CPCoMn profile does not show signals associated with the hydrotalcite phase, confirming the total destruction of the layered structure and formation of the corresponding mixed oxide.

After the calcination process, more spinel-type species are segregated in the ACCoMn because of the oxidation of Mn and Co ions after the thermal process. These species can be derived from the Co-Mn-Mg-Al mixed oxide matrix as indicated by the increased signal intensity at ~36° 2θ and decreased signal intensity corresponding to the MgO phase, which shifts towards higher 2θ values (~ 44°) with respect to the non-calcined ACCoMn oxide (Figure 1).

The diffraction profiles of ACCoMn and CPCoMn exhibit signals that may correspond to the Mn_3O_4, Mn_2MgO_4, Mg_2MnO_4, $MnAl_2O_4$, Co_3O_4, $CoAl_2O_4$ and Co_2AlO_4 phases (JCPDS 24-0734, JCPDS 23-0392, JCPDS 19-0773, JCPDS 29-0880, JCPDS 042-1467, JCPDS 044-0160 and JCPDS 030-0814, respectively) or mixed oxides of the type $Mn_{3-x}Co_xO_4$ ($x = 1, 2$), whose maxima are similar and their exact assignment is unknown. Furthermore, these oxides show the formation of an MgO phase with peaks that correspond to the signals at ~42.9 and ~63.7° 2θ.

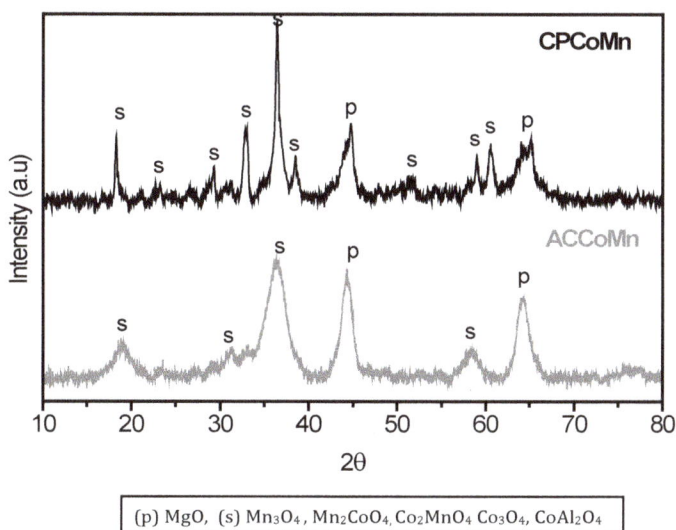

(p) MgO, (s) Mn_3O_4, Mn_2CoO_4, Co_2MnO_4, Co_3O_4, $CoAl_2O_4$

Figure 2. Diffraction patterns of the mixed oxides synthesized by co-precipitation and auto-combustion.

The presence of Mn_3O_4 and Co_3O_4 in the ACCoMn and CPCoMn oxides coincide with results reported in the literature, which indicate that when Co and Mn are present in solution, the simultaneous precipitation of Co_3O_4 and Mn_3O_4 can occur [27]. Furthermore, the formation of a solid solution of $Mn_{3-x}Co_xO_4$ (for $x = 0$–3) can be conducted because of the similarity of ionic radii between Co^{2+} (0.65 Å) and Co^{3+} (0.55 Å) ions and Mn^{2+} (0.83 Å) and Mn^{3+} (0.58 Å) ions; thus, a dissolution of Co ions can occur in the Mn oxide lattice to form solid solutions [23]. This result was reported by Sinha *et al.* [28] who indicated that Co^{2+} can randomly substitute Mn^{2+} or Mn^{3+} at tetrahedral or octahedral sites, respectively, because of its affinity towards both sites.

The average size of the crystalline aggregate of the ACCoMn and CPCoMn mixed oxides was calculated using the Scherrer equation and hausmannite (211) plane (Table 1), and this value correspond to the mixture of different spinel-type species.

The ACCoMn solid exhibits a much smaller particle size than that obtained by co-precipitation, which is attributed to the great amount of gaseous products formed during synthesis that avoid interparticle contact and improve dispersion in terms of the crystallite size. The greater crystallite size of CPCoMn is justified because of the aging processes that occur during synthesis by co-precipitation, which favors crystal growth.

The N_2 adsorption desorption isotherms of the Co-Mn mixed oxides are shown in Figure 3. CPCoMn reveals a type II isotherm characteristic of macroporous materials with an H1 hysteresis loop, associated with a uniform pore arrangement. On the other hand, ACCoMn exhibits a type IV isotherm characteristic of mesoporous materials with a type H3 hysteresis loop proper of pores in a slit form [29].

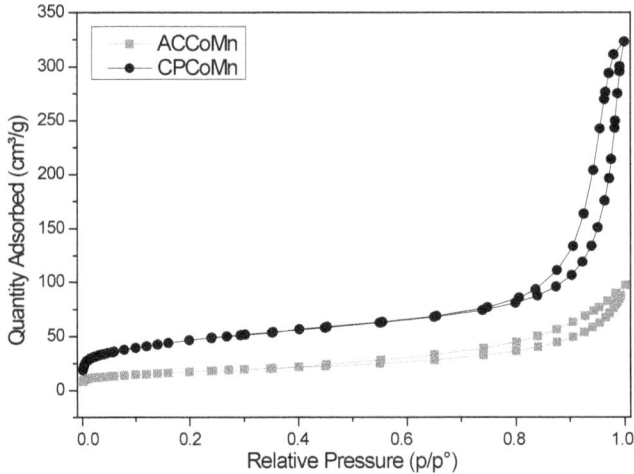

Figure 3. Nitrogen adsorption–desorption isotherms of the Co-Mn oxides.

The BET specific areas for CPCoMn and ACCoMn correspond to 161 m^2/g and 61 m^2/g, respectively. The area of the oxide synthesized by co-precipitation is attributed to the formation of pores resulting from the destruction of the laminar spaces of the hydrotalcite by calcination process (loss of water and carbonates) whereas in the oxide prepared by auto-combustion, the area is the result of the evolution of gases such as CO_2 and H_2O generated during the combustion process. However, the specific area determined in the ACCoMn oxide corresponds to an approximation of the real area of the solid due to the presence of pores of a large size (micrometer order) which limit the determination of the textural properties by means of the adsorption of N_2 technique [30].

Within Mn oxides, the hausmannite phase (Mn_3O_4) is the most sensitive in Raman spectroscopy, and it has a characteristic intense peak at approximately 654 cm^{-1}($A_{1\,g}$) that corresponds to the Mn-O vibration of Mn^{2+} in a tetrahedral configuration, a weak signal at 485 cm^{-1} and two signals with medium intensity at approximately 310 cm^{-1} (E_g) and 360 cm^{-1}($T_{2\,g}$). [31].

Oxide Co_3O_4 has a cubic structure, with the ions Co^{2+} and Co^{3+} located in the tetrahedral and octahedral positions, respectively, and it presents five Raman-active modes (A_{1g}, E_g and $3F_{2g}$) [32]. One band attributed to octahedral sites (CoO_6) at approximately 680 cm^{-1}, three bands with medium intensity at 480 cm^{-1}, 530 cm^{-1} and 580 cm^{-1}, and one band characteristic of tetrahedral sites (CoO_4) at 190 cm^{-1}.

Figure 4 shows the Raman spectra of the mixed oxides prepared by auto-combustion and co-precipitation. It is important to note that the vibrational modes are only associated with the Co-Mn oxides because the modes corresponding to MgO and Al_2O_3 are not present within the studied range [33,34].

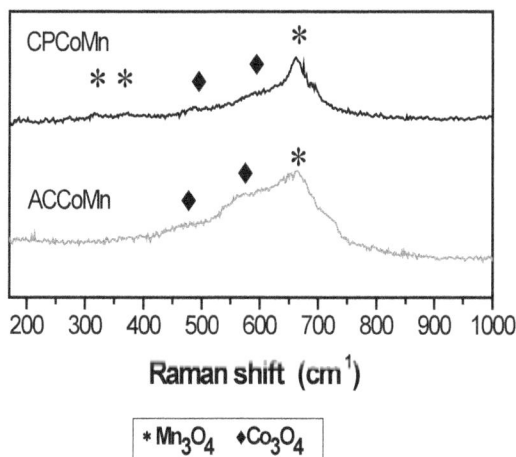

Figure 4. Raman spectra of the Co-Mn mixed oxides.

The CPCoMn oxide exhibits an intense signal at 660 cm^{-1} and signals at 315 cm^{-1} and 370 cm^{-1}, which are attributed to the presence of the Mn_3O_4 phase, along with Co_3O_4 signals. The ACCoMn oxide exhibits a signal at 665 cm^{-1}, which is characteristic of the hausmannite phase, along with Co_3O_4 signals. In the spectra of these two solids, a shift towards higher wavelengths is observed in the peak associated with the Mn_3O_4 phase, which can be attributed to the incorporation of Co ions into the structure of the Mn_3O_4 phase. However, according to the results found for these oxides, the co-existence of individual Mn_3O_4 and Co_3O_4 phases cannot be ignored. These results verify the predominant formation of the Mn_3O_4 phase in the mixed oxides obtained by the two methodologies, which is consistent with XRD results.

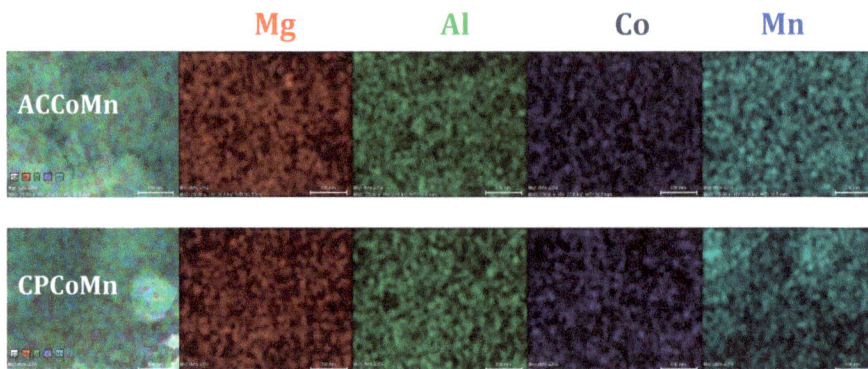

Figure 5. Chemical mapping performed by EDX for the mixed oxides.

The chemical mapping technique with energy-dispersive X-ray spectroscopy (EDX) was used to study the chemical composition distribution in the mixed oxides in specific regions of the materials, which is presented in Figure 5.

A homogeneous distribution of elements is observed in both solids, indicating that equivalent oxides are obtained by using the auto-combustion and co-precipitation methodologies.

2.2.2. Redox Properties and Oxygen Mobility

Taking into account the Mars Van Krevelen mechanism [35] which is commonly accepted in oxidation reactions with metal oxides, a catalyst with a redox pair and high oxygen mobility is required to ensure the reoxidation of the reduced catalyst.

The redox process of the mixed oxides can be described according to the following general mechanism:

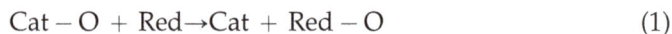

$$Cat - O + Red \rightarrow Cat + Red - O \tag{1}$$

$$Cat + Ox - O \rightarrow Cat - O + Ox \tag{2}$$

The catalyst (Cat-O) is reduced by the substrate (reducer) and then reoxidized by an oxidizer (Ox-O) to its initial state. The net results indicate an oxygen transfer from one species to another, which is verified in the specific reaction of VOCs by the oxidation of the organic compound, reduction of the oxide surface by the loss of surface oxygen atoms, and the subsequent reoxidation of the catalyst [36].

In the present work, the temperature-programmed reduction of hydrogen (H_2-TPR), oxygen storage capacity (OSC), and oxygen storage complete capacity (OSCC) techniques were used to evaluate the amount and mobility of available oxygen in the materials, and these methods provide complementary information on the redox properties of the oxides.

The H_2-TPR is appropriate to investigate the amount of active oxygen species and the steps involved in the reduction process of oxides [37]. For Co oxides derived from hydrotalcite precursors, the reduction processes occur at low temperatures (<500 °C) and high temperatures (>700 °C) [38]. The reduction at $T < 500$ °C is attributed to the reduction of Co oxide to Co metal, $Co_3O_4 \rightarrow CoO \rightarrow Co$, and reduction processes at high temperatures are associated with the presence of Co-Al-type or species with strong interactions within the solid solution [39,40].

The H_2-TPR profiles of Mn oxides are complex because of the multiple oxidation states of Mn and different chemical environments in which it can be found. However, these profiles can be classified into three main regions: low temperature <300 °C; intermediate temperature 300–550 °C; and high temperatures >550 °C. The reduction processes of MnOx species with diverse particle sizes as well as that of Mn_5O_8 species are assigned to the low temperature region; the transitions of MnO_2/Mn_2O_3 to Mn_3O_4 and Mn_3O_4 species to the reduced form of Mn (MnO) are assigned to the intermediate region [41,42] and reduction processes of the Mn spinel phases, such as $MnAl_2O_4$, and Mn species that present strong interactions in solid solution are assigned to the high temperature region [43].

Figure 6 shows the H_2-TPR profiles of the Co-Mn-Mg-Al mixed oxides. The ACCoMn and CPCoMn solids show reduction signals along all of the evaluated temperature ranges. In general, the signals observed for these oxides at temperatures <500 °C are assigned to the reduction of $Mn^{4+}/Mn^{3+}/Mn^{2+}$ species to MnO and Co^{3+}/Co^{2+} to CoO, whereas the signals observed at temperatures >500 °C are assigned to reduction events of the Co or Mn spinels with Al [27]. The presence of a signal at 895 °C is indicative of the formation of other species in addition to the Mn_3O_4, Co_3O_4 and $Mn_{3-x}Co_xO_4$ phases found by XRD for these solids (cobalt aluminates).

The signal at 680 °C in the ACCoMn solid indicates the presence of Mn species with significant interactions in the solid solution or $MnAl_2O_4$-type spinels that are generated because of the high temperatures that can be reached in the auto-combustion process (1400–1600 °C) [44], which decreases the oxygen mobility inside the material. This signal is not present in the CPCoMn oxide.

According to the XRD and Raman results, the predominant phase in the ACCoMn and CPCoMn oxides corresponds to the Mn_3O_4 phase, which is reduced to MnO at approximately 490 °C [24]. Figure 6 shows that this reduction process occurs at 480 and 450 °C for ACCoMn and CPCoMn, respectively, indicating that a lower temperature is required to carry out the Mn_3O_4 reduction in the solid synthesized by co-precipitation (CPCoMn). According to certain reports [4,30] this behavior reveals that in the Co-Mn mixed oxides obtained by co-precipitation, the reduction temperatures decrease because of a possible substitution of Co ions by Mn ions in the structure, which leads to a distortion of the lattice and generation of oxygen vacancies that destabilize the Mn-O bond.

In the case of the ACCoMn oxide, the elevated synthesis temperatures favor fusion and coalescence processes in the metals [44], provoking the formation of species with strong interactions that decrease their mobility. Thus, the presence of oxygen vacancies is the determining factor for oxygen mobility, and such mobility increases when the strength of the M-O bond decreases.

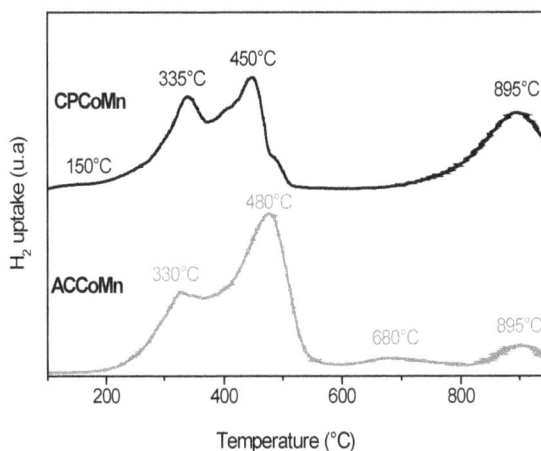

Figure 6. Reduction profiles of the Co-Mn mixed oxides.

The quantity of hydrogen consumed by means of the area below the curve permits the estimation of the quantity of species reducible with the temperature. For CPCoMn a H_2 consumption of 1257 μmol g^{-1} at $T < 350$ °C was observed which corresponds to the range where the catalytic oxidation of VOCs occurs. For ACCoMn the consumption was 965 μmol g^{-1} in this temperature region. These H_2 values correspond to 25% and 18% of the total quantity of H_2 consumed for CPCoMn and ACCoMn, respectively, which indicates that the oxide CPCoMn presents the greatest quantity of reducible species within the temperature range evaluated.

Additionally, the reduction of the CPCoMn oxide begins at lower temperatures, with a peak around 150 °C attributed to the presence of very active species on the surface of this oxide, capable of participating in the catalytic process. This peak is absent in the ACCoMn oxide.

On the other hand, the OSC and OSCC techniques are used to complement the study of surface and bulk oxygen reactivity, which is directly related to the oxygen mobility and redox properties of the material. The OSC was calculated from the amount of O_2 consumed after an H_2 pulse, and this value corresponds to the most active and available oxygen species for reaction. The OSCC corresponds to the total amount of reactive oxygen species (surface plus bulk), and it was evaluated based on the total amount of oxygen consumed after a series of H_2 pulses until complete reaction of the materials. These analyses also provide information on the reoxidation capacity of the material using oxygen from the gas phase, which is a determining property to avoid the loss of the catalyst activity [45].

To determine the oxygen storage capacity of the catalysts in the temperature range used in VOC oxidation (100–400 °C), the OSC of the materials was evaluated under dynamic conditions at different temperatures. This study collected data on the more labile or reactive oxygen species that can participate in oxidation reactions as well as the redox ability of the materials, which stimulates the electron transfer in the VOC oxidation mechanism.

Figure 7 summarizes the OSC results under dynamic conditions at different temperatures for the Co-Mn mixed oxides synthesized by the two methodologies. It should be noted that the OSC obtained for the Mg-Al oxide obtained by auto-combustion does not indicate important oxygen consumption in the studied temperature range; thus, the redox behavior and OSC are attributed exclusively to the presence of Mn and Co in the oxides.

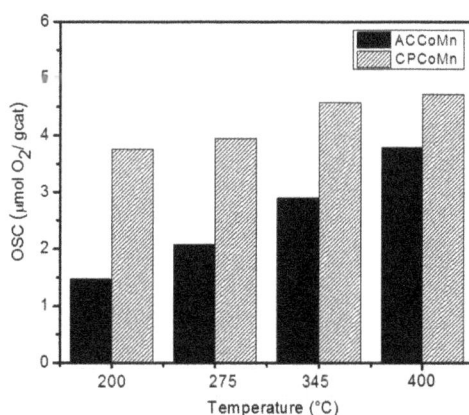

Figure 7. OSC of the mixed oxides at different temperatures.

The OSC increases with increases of temperature because of the participation of lattice oxygen that migrate and are available for the reaction [46]. At temperatures <300 °C, the oxygen at the surface and close to the surface contribute to the OSC, whereas at temperatures >300 °C, the migration of bulk oxygen plays an important role in the OSC performance [47].

The temperature dependence of the OSC is also verified in Table 2, where values for the mixed oxides are recorded after two H_2-O_2 pulse cycles at different temperatures, which indicated that the oxygen consumption was similar in the two cycles in both materials, suggesting reversibility in the redox process. These results show successful electron transfer by the Co and Mn mixed oxides because of the presence of phases such as Mn_3O_4 and Co_3O_4, (M^{2+} and M^{3+}), which present different oxidation states inside the structure and act as active components in the redox system.

Table 2. Oxygen storage capacity of the mixed oxides at different temperatures.

Sample	OSC ($\mu molO_2$/g cat) two H_2-O_2 pulse cycles							
	200 °C		275 °C		345 °C		400 °C	
ACCoMn	1.6	1.5	2.1	1.9	3.0	2.9	3.9	3.7
CPCoMn	3.9	3.7	4.0	4.2	4.7	4.5	4.7	4.8

Figure 7 shows that the redox properties and oxygen mobility are superior in the oxide obtained by co-precipitation; this result is most likely because of the incorporation of Co into the Mn oxide structure, which increases the migration of lattice oxygen and generation of vacancies in the structure and leads to the formation of weak bonds with oxygen species and increased OSC. The OSCC can be used to study the total amount of oxygen species available in the oxides (surface plus bulk) that act as a reserve bank and contribute to the reoxidation of the catalyst, continuing the redox cycle even when gaseous oxygen is absent in the reaction current [12].

Figure 8 shows the OSCC of the Co-Mn mixed oxides at 400 °C along with that of the Mg-Al oxide obtained by auto-combustion. Although the ACCoMn and CPCoMn oxide present the same chemical composition, the oxygen storage capacity is affected by the synthesis method, with the oxide obtained by co-precipitation (CPCoMn) showing the most important OSCC value.

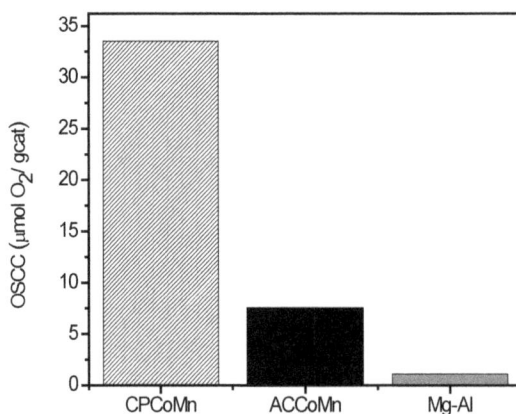

Figure 8. OSCC of the Co-Mn mixed oxides evaluated at 400 °C.

The maximum amount of both surface as well as lattice oxygen species that can participate in the catalytic oxidation of VOCs is determined at 400 °C, and the average OSC values of ACCoMn and CPCoMn are 3.8 and 4.8 $\mu mol O_2/g$ cat., respectively, whereas the OSCC at that temperature corresponds to 7.6 and 33.5 $\mu mol O_2/g$ cat., respectively. It is evident that there is a greater contribution of lattice oxygen in the CPCoMn solid (~86%) compared with that of the ACCoMn oxide (~50%). However, the results show the joint participation of the surface and lattice oxygen in both materials during the oxidation reactions.

The H_2-TPR, OSC and OSCC analyses revealed that the synthesis method influences the oxygen storage capacity, redox properties and oxygen species migration in the structure and demonstrated that the co-precipitation methodology promotes such properties. Although small crystallite sizes favor the formation of easily reducible species [48], crystallite size did not appear to determine the redox behavior of the evaluated materials because CPCoMn exhibits the most important values of OSC, OSCC and lowest reduction temperatures, but shows the largest particle size.

Considering that the materials with high specific areas show more active sites to adsorb O_2 from the gaseous phase and more active oxygen species [49], the study of the relation between the specific area and OSC was investigated by several authors [16,50,51]. Kamiuchi et al. [50] studied the OSC of Pt catalysts supported on Ce-Zr oxides calcined at 800 °C and 1000 °C whose areas correspond to 64 m^2/g and 25 m^2/g, respectively. However, the OSC values were comparable for materials with different areas (~400 $\mu mol\ g^{-1}$), indicating that the OSC of the Pt/CZ catalysts is independent of the specific area. Likewise, Mamontov et al. [51] found a poor relation between the specific area and the OSC in Ce-Zr oxides while they established a direct correlation between the oxygen vacancies, the presence of defects and the OSC. On the other hand, in Ce-Co oxides, Wang et al. [47] suggested that the incorporation of Co in the structure of CeO_2 leads a lattice distortion with the generation of oxygen vacancies, which lead to a promotion of OSC in the Co-Ce oxides with respect to the individual Ce oxide.

Taking into account the aforementioned research, the OSC values determined in the present work are attributed mainly to the presence of vacancies by the incorporation of Co in the structure of the manganese oxide, being most important in the oxide obtained by co-precipitation CPCoMn as demonstrated by the TPR-H_2 y OSC results. Then, the specific area does not appear to be a determining factor in the OSC of the Co-Mn oxides evaluated in the present work.

The isotopic exchange technique is one of the most frequently used methods of evaluating oxygen mobility.

There are three pathways through which O^{18}/O^{16} exchange can be conducted [30,52,53]:

1. Homogeneous exchange: this reaction does not require the participation of oxygen atoms in the solid, and the concentrations of ^{16}O and ^{18}O remain constant during the test:

$$^{16}O_2(g) + {}^{18}O_2(g) \Leftrightarrow 2^{18}O^{16}O(g) \tag{1}$$

2. Heterogeneous single exchange: this reaction involves an oxygen atom from the oxide and an oxygen atom from the gas phase:

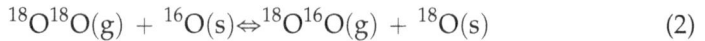

$$^{18}O^{18}O(g) + {}^{16}O(s) \Leftrightarrow {}^{18}O^{16}O(g) + {}^{18}O(s) \tag{2}$$

$$^{18}O^{16}O(g) + {}^{16}O(s) \Leftrightarrow {}^{16}O^{16}O(g) + {}^{18}O(s) \tag{3}$$

3. Heterogeneous multiple exchange: this reaction assumes the participation of two atoms of the solid in each step:

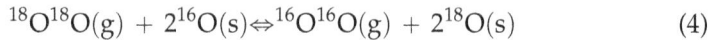

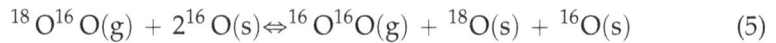

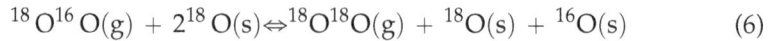

$$^{18}O^{18}O(g) + 2^{16}O(s) \Leftrightarrow {}^{16}O^{16}O(g) + 2^{18}O(s) \tag{4}$$

$$^{18}O^{16}O(g) + 2^{16}O(s) \Leftrightarrow {}^{16}O^{16}O(g) + {}^{18}O(s) + {}^{16}O(s) \tag{5}$$

$$^{18}O^{16}O(g) + 2^{18}O(s) \Leftrightarrow {}^{18}O^{18}O(g) + {}^{18}O(s) + {}^{16}O(s) \tag{6}$$

Isotopic exchange experiments with O^{18}/O^{16} were performed in a temperature range from 200 °C to 400 °C to evaluate oxygen mobility under reaction conditions (VOC oxidation). Figure 9 shows the isotopic exchange of Co-Mn mixed oxides obtained by the two synthesis methodologies as well as Mg-Al oxide. In general, the exchange increases with temperature with exchange representing a decrease in the concentration of signal 36 ($^{18}O_2$) and increase of the concentrations of signals 32 ($^{16}O_2$) and 34 ($^{18}O^{16}O$) in the reaction current. However, the isotopic exchange mainly occurs through the Mn and Co species because in the oxide consisting of Mg-Al, the concentrations of $^{18}O_2$ and $^{16}O_2$ species are approximately 85% and 14%, respectively, and a significant change does not occur with temperature. Furthermore, the maximum amount of $^{18}O^{16}O$ occurs at 400 °C and corresponds to a concentration of approximately 3%.

Because the concentration of $^{18}O^{16}O$ in the oxides is variable and depends on the nature of the solid, the gas phase reaction may be considered negligible (Equation 1). In addition, the difference in concentrations between signals 34 and 32 ($^{18}O^{16}O$ and $^{16}O_2$) suggests the presence of the two heterogeneous exchange mechanisms in the oxides: simple (Equations 2–3) and multiple (Equations 4–6).

By comparing the concentrations of $^{18}O_2$, $^{16}O_2$ and $^{18}O^{16}O$ at 400 °C in the mixed oxides, a beneficial effect was observed in the oxygen mobility in the CPCoMn oxide related to increased concentrations of $^{16}O_2$ (48%) and $^{18}O^{16}O$ (15%) and decreased concentration of $^{18}O_2$ (38%) concentration compared with that of ACCoMn ($^{16}O_2$ (22%), $^{18}O^{16}O$ (5%), and $^{18}O_2$ (73%).

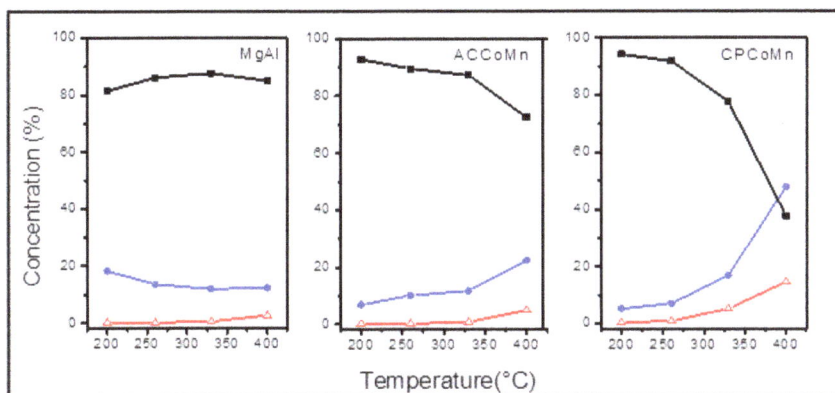

Figure 9. Isotopic exchange with the Co-Mn and Mg-Al mixed oxides. (■) $^{18}O_2$ (36); (●)$^{16}O_2$ (32); (△)$^{18}O^{16}O$ (34).

In addition, isotopic exchange mainly occurs after 320 °C and 270 °C in ACCoMn and CPCoMn, respectively, indicating that in this last solid, the mobility of the lattice oxygen is high because oxygen is capable of participating in the exchange process even at temperatures <300 °C, which usually indicates the participation of surface oxygen species.

The isotopic exchange results confirm that the lattice oxygen from the Co-Mn oxides synthesized by both methodologies participates in the oxidation reactions, and their participation is more important in the CPCoMn solid. These results are consistent with the analysis performed based on the H_2-TPR, OSC and OSCC results.

2.2.3. Oxidation Reaction

VOCs from mobile and stationary sources of emission are present as mixtures and not individually, which has increased interest in the evaluation of the oxidation of VOC mixtures. Figure 9 shows the conversion of mixtures of two VOCs (toluene and 2-propanol) towards CO_2 in the studied oxides.

The oxidation of toluene and 2-propanol mixture is not 100% selective towards CO_2 because acetone and propene appear as intermediate products of the reaction. These results are consistent with those reported for the individual oxidation of 2-propanol, where ketone and alkene are formed from the dehydrogenation and dehydration reactions of the alcohol [54].

181

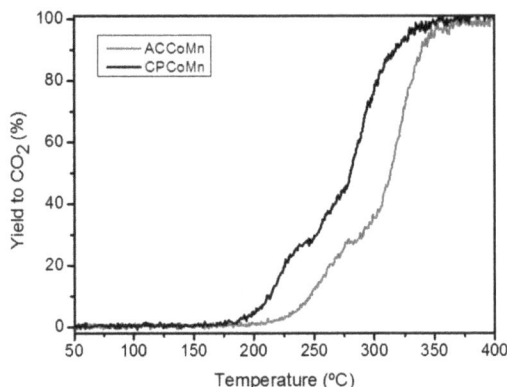

Figure 10. Catalytic performance of the mixed oxides in the oxidation of the VOC mixture.

The T_{50} and T_{90} values that correspond to the temperatures required to reach 50 and 90% conversions towards CO_2, respectively, are used to compare the catalytic performance of the oxides. Table 3 presents the T_{50} and T_{90} values of the ACCoMn and CPCoMn oxides, where lower conversion temperatures are observed with the solid synthesized by co-precipitation, which implies a better catalytic performance in the oxidation of the VOC mixture. Furthermore, the conversion temperatures are related to the OSC and the exchange capacity in terms of the $^{16}O_2$ concentration, which indicates that the catalytic properties of the materials are directly dependent on their redox properties and oxygen mobility. Thus, CPCoMn shows greater OSC values at the respective T_{90} and T_{50} because of the greater concentrations of $^{16}O_2$. In addition, CPCoMn exhibits the greatest catalytic activity towards the total combustion of the VOC mixture (Figure 10), with complete conversion towards CO_2 and H_2O occurring at 360 °C, which is lower than the temperature required for ACCoMn to achieve total oxidation (370 °C). The relation between oxygen mobility and catalytic behavior is likewise reported by Cimino et al. [55] regarding hexaaluminates of Mn and La synthesized by auto-combustion and co-precipitation, where the catalysts prepared by co-precipitation showed better oxygen mobility and therefore a better catalytic behavior in the combustion of methanol.

Based on the information provided by the OSC values on the most labile oxygen species and isotopic exchange analysis on the lattice oxygen, the correlation results demonstrate the joint participation of surface and lattice oxygen in the oxidation of the VOC mixture on the mixed oxides.

Additionally, in Table 3 the catalytic behavior of the Co-Mn oxides is compared with that of the reference 1%Pt/Al_2O_3 [9] where it is shown that the mixed oxides present higher conversion temperatures to the latter. The reason for this behavior is that the total oxidation of VOCs is determined by the compound that is the most difficult to oxidize which in this case is toluene—a compound easily oxidized by the Pt/Al$_2$O$_3$ that is the most active metal for the conversion of aromatic structures [56].

Table 3. Temperatures of conversion of the VOC mixture and their relationship with the OSC and isotopic exchange.

Sample	Yield to CO$_2$ temperatures ($^\circ$C)		OSC (μmolO$_2$/ g cat.)*		^{16}O$_2$ concentration (%) *	
	T_{90}	T_{50}	T_{90}	T_{50}	T_{90}	T_{50}
ACCoMn	340	312	2.9	2.4	12	10
CPCoMn	320	280	4.4	4.1	14	8
1%Pt/Al$_2$O$_3$	195	180	-	-	-	-

* The values are calculated at the respective temperatures to reach 50% and 90% conversion of the VOC mixture.

3. Experimental Section

3.1. Synthesis of Catalysts

3.1.1. Auto-Combustion

For the synthesis of the Co-Mn-Mg-Al mixed oxide by the auto-combustion method, solutions of Mg(NO$_3$)$_2$.6H$_2$O, Al(NO$_3$)3.9H$_2$O, Mn(NO$_3$)$_2$.6H$_2$O and/or Co(NO$_3$)$_2$.6H$_2$O nitrates were used as oxidizers, and a solution of glycine [CH$_2$NH$_2$COOH] was used as fuel, maintaining constant molar ratios of [(Mg^{2+} + Mn^{2+} + Co^{2+}) /Al^{3+} = 3], [(Mn^{2+} + Co^{2+})/Mg^{2+} = 1] and [Co^{2+} /Mn^{2+} = 0.5] and a fuel/oxidizer ratio of 0.56.

The ACCoMn mixed oxide was produced by the addition of the respective nitrates to glycine, maintaining constant agitation. The resulting solution was slowly evaporated until a gel was obtained, and it was heated at approximately 100 $^\circ$C to initiate the ignition process (~10 min). Once the ignition reaction is completed (~3 min), a powder is produced, and it is then calcined at 500 $^\circ$C for 4 h to remove the remaining carbonaceous residues and obtain the corresponding oxide.

3.1.2. Co-Precipitation

The CPCoMn mixed oxide was synthetized by thermal decomposition of the hydrotalcite-type precursor (HTCoMn) obtained by the simultaneous co-precipitation method, which used solutions of the respective nitrates. The molar ratios used in the mixed oxide synthesized by auto-combustion were maintained.

The mixture Mg^{2+}, Al^{3+}, Mn^{2+} and/or Co^{2+} nitrates was added drop by drop to a 0.2 M K$_2$CO$_3$ solution under constant agitation, with the pH maintained at 9.5 and 10.5 by addition of a 1 M NaOH solution. The obtained precipitate was aged for 18 h without agitation. Subsequently, the solid was washed and dried at 60 $^\circ$C in air for

18 h to obtain the hydrotalcite and calcined at 500 °C for 16 h to obtain the respective mixed oxide.

3.2. Characterization

The chemical analysis of the materials was performed using an X-ray fluorescence spectrometer MagixPro PW-2440 (Phillips), which has a maximum power of 4 KW.

The composition of the phases present in the oxides was determined by the XRD in a Panalytical X´Pert PRO MPD instrument equipped with a copper anode (λ = 1.5406 Å), using a scan speed of 1° θ min^{-1} and step size of 0.02° θ. The particle size was calculated with the Scherrer equation using the most intense signal corresponding to each oxide.

The Raman spectra were collected in a dispersion spectrometer (Horiva Jobin Yvon LabRam HR800) with a He-Ne green laser at 20 mW (532.1 nm) without a filter and with a 600 g mm^{-1} grid. The microscope uses an objective of 50 with a confocal aperture of 1000 μm [45].

The scanning electron micrographs (SEM) were collected using a Hitachi S2700 equipped with an energy-dispersive X-ray spectrometer (Bruker Quantax 400).

The H_2-TPR was performed in a ChemBET 3000 Quantachrome equipped with a thermal conductivity detector. The materials (0.100 g) were previously degassed at 400 °C for 1 h in the presence of Ar and reduced using a 10% (v/v) H_2/Ar mixture at a flow rate of 30 mL/min and a heating ramp of 10 °C/min. The hydrogen consumption was calculated from a CuO calibration curve, and OriginPro 8.0 software was used to quantify the areas associated with the H_2 consumption.

The OSC and $^{18}O/^{16}O$ isotopic exchange analyses were performed in a device designed and tuned by a group at Estado Sólido y Catálisis Ambiental (ESCA). For the OSC analyses, two measurements were performed: (i) OSCC, and (ii) OSC under dynamic conditions.

Prior to performing the two measurements, the samples (0.015 g in 0.100 g of SiC) were pre-oxidized under an air flow of 10 mL/min (21% O_2) at 400 °C for 40 min and subsequently purged with Ar until the $^{16}O_2$ was not detected.

In the OSC measurements, after the pretreatment, the samples were cooled to 200 °C under Ar flow (10 mL/min) and alternating pulses of H_2-O_2 at 50 μL were performed until two cycles were completed. To determine the amount of available oxygen in the solids within a temperature range, this same procedure was used at temperatures of approximately 275, 345 and 400 °C.

The concentration of the output gases H_2, O_2, H_2O, Ar, and He were monitored by mass spectroscopy (Omnistar mass spectrometer) at m/z = 2, 32, 18, 40 and 4, respectively.

For the OSCC analysis, the temperature was kept constant at 400 °C, Ar flow was maintained at 10 mL/min, successive H_2 (99.99%) pulses of 50 μL were injected until saturation of the samples, and two successive O_2 pulses (15.2% in He) of 50 μL were injected to reoxidize the solid until O_2 consumption was not observed.

The OSCC and OSC were calculated based on the amount of oxygen consumed during the reoxidation stage after the H_2 pulses, and the values were expressed in terms of $\mu molO_2$ g^{-1} of catalyst. OriginPro 8.0 software was used to quantify the areas of each pulse associated with the O_2 signal.

For the isotopic exchange tests, the samples (0.030 g with 0.100 g of SiC) were treated under an air current (10 mL/min) at 400 °C for 1 h. After the pretreatment, the sample was left to cool down to 200 °C under Ar flow. Pulses of $^{18}O_2$ (20 μL) were performed on the samples at 200, 260, 330 and 400 °C in the presence of an Ar current (10 mL/min). The output gas composition was monitored by mass spectroscopy (Omnistar mass spectrometer) at m/z = 36, 34 and 32, which corresponded to $^{18}O_2$, $^{18}O\,^{16}O$ and $^{16}O_2$, respectively. OriginPro 8.0 software was used to quantify the areas under the signals associated with the concentration of each of the species.

3.3. Catalytic Evaluation

To evaluate the catalytic performance of the Co-Mn mixed oxides obtained by the auto-combustion and co-precipitation methodologies, a mixture of 250 ppm toluene and 250 ppm 2-propanol in air was employed.

The catalysts were pretreated in air at 400 °C for 1 h before starting the reaction. The complete oxidation of the VOCs was conducted in a fixed-bed tubular reactor at atmospheric pressure with a heating ramp of 1.5 °C/min. The VOC liquid mixture was fed with a high-performance liquid chromatography (HPLC) pump (Jasco PU 1585) and diluted with air until the desired composition was obtained. A total flow of 500 mL/min and 0.200 g catalyst (sieved at <250 μm) were used. The reaction products were detected with a mass spectrometer (Balzers Instruments, Omnistar) and CO_2 detector (Sensotrans IR). The performance towards CO_2 was calculated as the ratio between the CO_2 concentration at any temperature divided by its value when complete conversion is reached at high temperatures and no other oxidation products are observed.

4. Conclusions

Co-Mn-Mg-Al oxides were successfully obtained by the auto-combustion and co-precipitation methodologies. The formation of a hydrotalcite-type precursor was verified in the production of the CPCoMn oxide, whereas the mixed oxide is directly generated through the synthesis by auto-combustion.

A Co-Mn mixed oxide with greater crystallite size was obtained through co-precipitation. However, a cooperative effect may occur between Co and Mn that increases the redox properties and oxygen mobility and improves the catalytic activity, demonstrating that crystal size is not a determinant in the catalytic performance of the material. The catalytic behavior in the total oxidation of the binary mixture of toluene and 2-propanol is dependent on the redox properties of the solid; thus, the catalyst that presents greater OSC values exhibits the lowest conversion temperatures.

Acknowledgments: The authors would like to thank Professor Mario Montes of the Universidad del País Vasco (Spain), for his support during the catalytic tests and the Instituto de Ciencia de Materiales de Sevilla (ICMS), from the Universidad de Sevilla for conducting the Raman and SEM analyses of the solids studied in the present work.

Author Contributions: M. Haidy Castaño carried out the experimental work and analyzed the data; Rafael Molina and Sonia Moreno supervised all of the study.

Conflicts of Interest: The authors declare no conflict of interest

References

1. Bastos, S.S.T.; Carabineiro, S.A.C.; Órfão, J.J.M.; Pereira, M.F.R.; Delgado, J.J.; Figueiredo, J.L. Total oxidation of ethyl acetate, ethanol and toluene catalyzed by exotemplated manganese and cerium oxides loaded with gold. *Catal. Today* **2012**, *180*, 148–154.

2. Domínguez, M.I.; Sánchez, M.; Centeno, M.A.; Montes, M.; Odriozola, J.A. 2-Propanol oxidation over gold supported catalysts coated ceramic foams prepared from stainless steel wastes. *J. Mol. Catal. A* **2007**, *277*, 145–154.

3. Liu, S.Y.; Yang, S.M. Complete oxidation of 2-propanol over gold-based catalysts supported on metal oxides. *Appl. Catal. A* **2008**, *334*, 92–99.

4. Aguilera, D.A.; Perez, A.; Molina, R.; Moreno, S. Cu–Mn and Co–Mn catalysts synthesized from hydrotalcites and their use in the oxidation of VOCs. *Appl. Catal. B* **2011**, *104*, 144–150.

5. Tsou, J.; Magnoux, P.; Guisnet, M.; Órfão, J.J.M.; Figueiredo, J.L. Catalytic oxidation of volatile organic compounds: Oxidation of methyl-isobutyl-ketone over Pt/zeolite catalysts. *Appl. Catal. B* **2005**, *57*, 117–123.

6. He, C.; Li, P.; Cheng, J.; Hao, Z.-P.; Xu, Z.-P. A Comprehensive Study of Deep Catalytic Oxidation of Benzene, Toluene, Ethyl Acetate; their Mixtures over Pd/ZSM-5 Catalyst: Mutual Effects and Kinetics. *Water Air Soil Pollut.* **2010**, *209*, 365–376.

7. Burgos, N.; Paulis, M.a.; Mirari Antxustegi, M.; Montes, M. Deep oxidation of VOC mixtures with platinum supported on Al_2O_3/Al monoliths. *Appl. Catal. B* **2002**, *38*, 251–258.

8. Beauchet, R.; Mijoin, J.; Batonneau-Gener, I.; Magnoux, P. Catalytic oxidation of VOCs on NaX zeolite: Mixture effect with isopropanol and o-xylene. *Appl. Catal. B* **2010**, *100*, 91–96.

9. Sanz, O.; Delgado, J.J.; Navarro, P.; Arzamendi, G.; Gandía, L.M.; Montes, M. VOCs combustion catalysed by platinum supported on manganese octahedral molecular sieves. *Appl. Catal. B* **2011**, *110*, 231–237.

10. Fierro, J.L.G. Applications of Metal Oxides for Volatile Organic Compound Combustion. In *Metal Oxides, Chemistry and Applications*; CRC Press: New York, NY, USA, 2006; pp. 543–555.

11. Liotta, L.F.; Wu, H.; Pantaleo, G.; Venezia, A.M. Co_3O_4 nanocrystals and Co_3O_4-MO_x binary oxides for CO, CH_4 and VOC oxidation at low temperatures: A review. *Catal. Sci. Tech.* **2013**, *3*, 3085–3102.

12. Santos, V.P.; Pereira, M.F.R.; Órfão, J.J.M.; Figueiredo, J.L. The role of lattice oxygen on the activity of manganese oxides towards the oxidation of volatile organic compounds. *Appl. Catal. B* **2010**, *99*, 353–363.

13. Liotta, L.F.; Ousmane, M.; di Carlo, G.; Pantaleo, G.; Deganello, G.; Marcì, G.; Retailleau, L.; Giroir-Fendler, A. Total oxidation of propene at low temperature over Co_3O_4–CeO_2 mixed oxides: Role of surface oxygen vacancies and bulk oxygen mobility in the catalytic activity. *Appl. Catal. A* **2008**, *347*, 81–88.

14. Zhu, L.; Lu, G.; Wang, Y.; Guo, Y.; Guo, Y. Effects of Preparation Methods on the Catalytic Performance of $LaMn_{0.8}Mg_{0.2}O_3$ Perovskite for Methane Combustion. *Chin. J. Catal.* **2010**, *31*, 1006–1012.

15. Avgouropoulos, G.; Ioannides, T.; Matralis, H. Influence of the preparation method on the performance of CuO-CeO_2 catalysts for the selective oxidation of CO. *Appl. Catal. B* **2005**, *56*, 87–93.

16. Kovanda, F.; Jirátová, K. Supported layered double hydroxide-related mixed oxides and their application in the total oxidation of volatile organic compounds. *Appl. Clay. Sci.* **2011**, *53*, 305–316.

17. Vaccari, A. Preparation and catalytic properties of cationic and anionic clays. *Catal. Today* **1998**, *41*, 53–71.

18. Cavani, F.; Trifirò, F.; Vaccari, A. Hydrotalcite-type anionic clays: Preparation, properties and applications. *Catal. Today* **1991**, *11*, 173–301.

19. Mukasyan, A.; Dinka, P. Novel approaches to solution-combustion synthesis of nanomaterials. *Int. J. Self-Propag. High-Temp Synth.* **2007**, *16*, 23–35.

20. Tahmasebi, K.; Paydar, M.H. The effect of starch addition on solution combustion synthesis of Al2O3-ZrO2 nanocomposite powder using urea as fuel. *Mater. Chem. Phys* **2008**, *109*, 156–163.

21. Castaño, M.H.; Molina, R.; Moreno, S. Catalytic oxidation of VOCs on $MnMgAlO_x$ mixed oxides obtained by auto-combustion. *J. Mol. Catal. A* **2015**, *398*, 358–367.

22. Tsyganok, A.; Sayari, A. Incorporation of transition metals into Mg–Al layered double hydroxides: Coprecipitation of cations *vs.* their pre-complexation with an anionic chelator. *J. Solid State Chem.* **2006**, *179*, 1830–1841.

23. Xu, Z.P.; Zhang, J.; Adebajo, M.O.; Zhang, H.; Zhou, C. Catalytic applications of layered double hydroxides and derivatives. *Appl. Clay. Sci.* **2011**, *53*, 139–150.

24. Lamonier, J.-F.; Boutoundou, A.-B.; Gennequin, C.; Pérez-Zurita, M.; Siffert, S.; Aboukais, A. Catalytic Removal of Toluene in Air over Co–Mn–Al Nano-oxides Synthesized by Hydrotalcite Route. *Catal. Lett.* **2007**, *118*, 165–172.

25. Evans, D.G.; Duan, X. *Structural Aspects of Layered Double Hydroxides, in Structure and Bonding*; Springer-Verlag: Berlin & Heidelberg: Germany, 2006; p. 15.

26. Tsai, Y.-T.; Mo, X.; Campos, A.; Goodwin, J.G., Jr.; Spivey, J.J. Hydrotalcite supported Co catalysts for CO hydrogenation. *Appl. Catal. A* **2011**, *396*, 91–100.

27. Hem John, D. Redox Coprecipitation Mechanisms of Manganese Oxides. In *Particulates in Water*; American Chemical Society: Menlo Park, CA, USA, 1980; pp. 45–72.

28. Sinha, A.P.B.; Sanjana, N.R.; Biswas, A.B. On the structure of some manganites. *Acta Crystallogr.* **1957**, *10*, 439–440.

29. Leofanti, G.; Padovan, M.; Tozzola, G.; Venturelli, B. Surface area and pore texture of catalysts. *Catal. Today* **1998**, *41*, 207–219.

30. Castaño, M.H.; Molina, R.; Moreno, S. Cooperative effect of the Co–Mn mixed oxides for the catalytic oxidation of VOCs: Influence of the synthesis method. *Appl. Catal. A* **2015**, *492*, 48–59.

31. Julien, C.M.; Massot, M.; Poinsignon, C. Lattice vibrations of manganese oxides: Part I. Periodic structures. *Spectrochim. Acta Part A* **2004**, *60*, 689–700.

32. Jiang, J.; Li, L. Synthesis of sphere-like Co_3O_4 nanocrystals via a simple polyol route. *Matter. Lett.* **2007**, *61*, 4894–4896.

33. Ulla, M.A.; Spretz, R.; Lombardo, E.; Daniell, W.; Knözinger, H. Catalytic combustion of methane on Co/MgO: Characterisation of active cobalt sites. *Appl. Catal. B* **2001**, *29*, 217–229.

34. Rezaei, E.; Soltan, J. Low temperature oxidation of toluene by ozone over MnO_x/γ-alumina and MnO_x/MCM-41 catalysts. *Chem. Eng. J.* **2012**, *198–199*, 482–490.

35. Mars, P.; van Krevelen, D.W. Oxidations carried out by means of vanadium oxide catalysts. *Chem. Eng. Sci.* **1954**, *3*, 41–59.

36. Fierro, J.L.G. *Metal Oxides, Chemistry and Applications*; CRC Press: New York, NY, USA, 2006.

37. Ran, R.; Weng, D.; Wu, X.; Fan, J.; Wang, L.; Wu, X. Structure and oxygen storage capacity of Pr-doped $Ce_{0.26}Zr_{0.74}O_2$ mixed oxides. *J. Rare Earth* **2011**, *29*, 1053–1059.

38. Muñoz, M.; Moreno, S.; Molina, R. Promoting effect of Ce and Pr in Co catalysts for hydrogen production via oxidative steam reforming of ethanol. *Catal. Today* **2013**, *213*, 33–41.

39. Coq, B.; Tichit, D.; Ribet, S. Co/Ni/Mg/Al Layered Double Hydroxides as Precursors of Catalysts for the Hydrogenation of Nitriles: Hydrogenation of Acetonitrile. *J. Catal.* **2000**, *189*, 117–128.

40. Pérez, A.; Lamonier, J.-F.; Giraudon, J.-M.; Molina, R.; Moreno, S. Catalytic activity of Co-Mg mixed oxides in the VOC oxidation: Effects of ultrasonic assisted in the synthesis. *Catal. Today* **2011**, *176*, 286–291.

41. Döbber, D.; Kießling, D.; Schmitz, W.; Wendt, G. MnO_x/ZrO_2 catalysts for the total oxidation of methane and chloromethane. *Appl. Catal. B* **2004**, *52*, 135–143.

42. Stobbe, E.R.; de Boer, B.A.; Geus, J.W. The reduction and oxidation behaviour of manganese oxides. *Catal. Today* **1999**, *47*, 161–167.

43. Li, Q.; Meng, M.; Xian, H.; Tsubaki, N.; Li, X.; Xie, Y.; Hu, T.; Zhang, J. Hydrotalcite-Derived MnxMg3−xAlO Catalysts Used for Soot Combustion, NOx Storage and Simultaneous Soot-NOx Removal. *Environ. Sci. Technol.* **2010**, *44*, 4747–4752.

44. Hwang, C.-C.; Wu, T.-Y.; Wan, J.; Tsai, J.-S. Development of a novel combustion synthesis method for synthesizing of ceramic oxide powders. *Mater. Sci. Eng. B* **2004**, *111*, 49–56.

45. Gonzalez Castaño, M.; Reina, T.R.; Ivanova, S.; Centeno, M.A.; Odriozola, J.A. Pt *vs.* Au in water–gas shift reaction. *J. Catal.* **2014**, *314*, 1–9.

46. Zhao, M.; Shen, M.; Wang, J. Effect of surface area and bulk structure on oxygen storage capacity of $Ce_{0.67}Zr_{0.33}O_2$. *J.Catal.* **2007**, *248*, 258–267.

47. Wang, J.; Shen, M.; Wang, J.; Gao, J.; Ma, J.; Liu, S. CeO_2–$CoOx$ mixed oxides: Structural characteristics and dynamic storage/release capacity. *Catal. Today* **2011**, *175*, 65–71.

48. de Rivas, B.; López-Fonseca, R.; Jiménez-González, C.; Gutiérrez-Ortiz, J.I. Synthesis, characterisation and catalytic performance of nanocrystalline Co_3O_4 for gas-phase chlorinated VOC abatement. *J. Catal.* **2011**, *281*, 88–97.

49. Cui, M.; Li, Y.; Wang, X.; Wang, J.; Shen, M. Effect of preparation method on MnO_x-CeO_2 catalysts for NO oxidation. *J. Rare Earth* **2013**, *31*, 572–576.

50. Kamiuchi, N.; Haneda, M.; Ozawa, M. Enhancement of OSC property of Zr rich ceria–zirconia by loading a small amount of platinum. *Catal. Today* **2014**, *232*, 179–184.

51. Mamontov, E.; Egami, T.; Brezny, R.; Koranne, M.; Tyagi, S. Lattice Defects and Oxygen Storage Capacity of Nanocrystalline Ceria and Ceria-Zirconia. *J. Phys. Chem. B* **2000**, *104*, 11110–11116.

52. Nováková, J. Isotopic exchange of oxygen ^{18}O between the gaseous phase and oxide catalysts. *Catal. Rev.* **1971**, *4*, 77–113.

53. Royer, S.; Duprez, D.; Kaliaguine, S. Role of bulk and grain boundary oxygen mobility in the catalytic oxidation activity of $LaCo_{1-x}Fe_xO_3$. *J. Catal.* **2005**, *234*, 364–375.

54. Díez, V.K.; Apesteguía, C.R.; Di Cosimo, J.I. Acid–base properties and active site requirements for elimination reactions on alkali-promoted MgO catalysts. *Catal. Today* **2000**, *63*, 53–62.

55. Cimino, S.; Nigro, R.; Weidmann, U.; Holzner, R. Catalytic combustion of methanol over La, Mn-hexaaluminate catalysts. *Fuel Process. Technol.* **2015**, *133*, 1–7.

56. Golodets, G.I. *Heterogeneous Catalytic Reactions Involving Molecular Oxygen*; Elsevier: Amsterdam, The Netherlands, 1983.

Utilization of Volatile Organic Compounds as an Alternative for Destructive Abatement

Satu Ojala, Niina Koivikko, Tiina Laitinen, Anass Mouammine, Prem K. Seelam, Said Laassiri, Kaisu Ainassaari, Rachid Brahmi and Riitta L. Keiski

Abstract: The treatment of volatile organic compounds (VOC) emissions is a necessity of today. The catalytic treatment has already proven to be environmentally and economically sound technology for the total oxidation of the VOCs. However, in certain cases, it may also become economical to utilize these emissions in some profitable way. Currently, the most common way to utilize the VOC emissions is their use in energy production. However, interesting possibilities are arising from the usage of VOCs in hydrogen and syngas production. Production of chemicals from VOC emissions is still mainly at the research stage. However, few commercial examples exist. This review will summarize the commercially existing VOC utilization possibilities, present the utilization applications that are in the research stage and introduce some novel ideas related to the catalytic utilization possibilities of the VOC emissions. In general, there exist a vast number of possibilities for VOC utilization via different catalytic processes, which creates also a good research potential for the future.

Reprinted from *Catalysts*. Cite as: Ojala, S.; Koivikko, N.; Laitinen, T.; Mouammine, A.; Seelam, P.K.; Laassiri, S.; Ainassaari, K.; Brahmi, R.; Keiski, R.L. Utilization of Volatile Organic Compounds as an Alternative for Destructive Abatement. *Catalysts* **2015**, *5*, 1092–1151.

1. Introduction

Volatile organic compounds (VOCs) are emitted to our environment from various industrial and natural sources. Since one can count several hundreds of compounds that belong in the VOCs category, one can also easily imagine that the environmental and health effects caused by these compounds are diverse. They range from local annoying odor emissions to global effects such as greenhouse phenomena. Furthermore, certain VOCs are even more dangerous as carcinogens and their utilization is limited, for example carbon tetrachlorine and benzene. Due to these harmful effects, different countries have set growingly stringent emission levels for the industrial activities [1]. One example of the legislation is the Industrial Emissions Directive set by the European Commission [2] that replaced the old VOC-directive from 1999 [3].

The VOC emissions can be reduced in several different ways. The primary way should be avoiding the formation of the emissions by process changes and replacing the organic raw materials used. However, this is not always possible or practical. In those cases, the VOCs treatment can be done either by destruction or by recovery-based technologies [4]. In both categories, catalysis has taken its role. Instead of only treatment, this review will concentrate on the recovery of the VOC emissions and further valorization of them. Utilization of gaseous emissions is a rather recent idea, since typically the circular economy principle is connected to solid or liquid waste. There exists research related to the utilization of CO_2, but the utilization of organic gaseous emissions is a somewhat less studied area. In this review, we will take into account also methane, while it is not strictly speaking a VOC compound, but it offers a good potential for utilization. Based on our surveys it seems that the utilization of VOCs as an energy source, either additional or in co-firing purposes [5–10] is the most currently used solution. However, we exclude this alternative from this review and we concentrate on either the direct use, synthesis gas production or hydrogen production from VOC emissions. In this review, different types of utilization of organic compounds in the production of chemicals are presented including existing solutions and those under research. Most of the processes presented here are catalytic, apart from the actual recovery and purification technologies presented here as well. When considering the total concept of the VOC emission utilization, an important part related to the recovery, purification of emission gas to be utilized and the purification of the final product. Since the catalytic processes are considered, the presence of catalyst poisons and the deactivating nature of the VOC compound itself need to be considered. At the end of the review, certain economic issues and the sustainability of the VOC utilization are discussed.

2. Pre- and Post-Processing of the Emissions to Be Utilized

2.1. Recovery of the Raw Emissions

Even in certain rare cases where the direct utilization of VOCs would be possible, typically the concentrations of VOC emissions are so low that one or two pre-treatment methods are needed to recover the compounds from the waste gas stream in order to valorize them. The recovery method therefore needs to be efficient and economically feasible. The technical feasibility can be evaluated by the efficiency of the recovery, the quality of the product, the physical and chemical characterization of the product and the characteristics of the emission stream [11]. In general, the recovery of VOCs is economically feasible from the solvents vapors of low-boiling products and when VOCs are used as energy carriers in incinerators or oxidizers or boilers. The techniques that can be used for the recovery of VOCs are adsorption, membrane separation, wet scrubbing, and condensation and cryogenic condensation [12]. Table 1 presents these most common techniques for the recovery of VOC emissions and they are discussed in more detail in this chapter.

Table 1. Characteristics of the volatile organic compounds (VOC) recovery techniques [13].

Recovery technique	Efficiency (%)	Indication of applied flow ($m^3\ h^{-1}$)	Critical parameters
Adsorption:			
Activated carbon	80–98	100–100,000	Fluid percentage, VOC concentration
Zeolite	80–99	<100,000	Fluid percentage, dust in inlet gas
Polymeric	95–98	-	Dust in inlet gas
Membrane separation	99.9	<3000	-
Wet scrubbing	30–99	50–500,000	Temperature
Condensation	60–90	100–100,000	Saturation of inlet gas
Cryocondensation	>99	<5000	Fluid percentage in inlet gas

2.1.1. Adsorption

In VOC adsorption the gaseous compounds (adsorbate) are adhered to the surface of the solid sorbent (adsorbent). The interaction between adsorbent and adsorbate may be physical or chemical in nature. In the physical adsorption the gaseous compounds adhere to the adsorbent by van der Waal forces and are therefore more easily desorbed from the sorbent. Desorption of the adsorbed molecule can be carried out either by lowering the pressure or rising the temperature or by a combination of these two [13]. After desorption the concentrations of the molecules are higher than in the original emission gas, and the molecules are therefore more easily used as raw material for new products [12].

Adsorption by activated carbon is the most commonly used method for recovering VOC emissions since it is the most economic process for different kind of gas streams [7]. Activated carbons can be prepared from basically any carbon source, such as coconut, wood, peat, coal, tar, sawdust and cellulose residues [14]. Recent research has been conducted on preparation of activated carbons from many agricultural waste materials [15–18]. Also activated carbon fiber cloth with regeneration by electrothermal desorption has been applied for the recovery of methylene chloride in an industrial scale [19]. The main reason for the activated carbons' very good ability to adsorb molecules is the high surface area and vast structure of pores and micropores they have [20].

Zeolites are an alternative for activated carbons when thermal stability and hydrophobicity are needed. They are inorganic materials, made of Si and Al, and have a crystalline structure and fixed pore sizes. The precise pore size allows the selective adsorption of certain compounds [20]. Zeolites appear naturally, but they can also be produced synthetically. The main applications of zeolite adsorption in VOC recovery are, e.g., in spraying cabins and varnish production. Since zeolites are more expensive than activated carbons, it is rational to combine them together, using zeolites in the final recovery phase [13].

In the polymer adsorption process small porous plastic balls are used as adsorbents. However, the small size of micropores in activated carbon cannot be achieved in polymers even if polymers have rather high adsorption capacity. Polymers have low selectivity in VOC adsorption, but different types of polymers adsorb different types of VOCs. Polymers can also be combined with other adsorbents, activated carbon or zeolites, zeolites being typically the last purification phase [13].

2.1.2. Membrane Separation

Membrane separation of VOCs is based on selective permeability of the membrane for different species. Several organic compounds can be recovered, including alkenes, olefins, chlorinated hydrocarbons, aromatics, ketones, ethers, alcohols and monomers, like vinyl chloride, ethylene or propylene, from gas streams containing nitrogen, oxygen, hydrogen or carbon dioxide. Membrane recovery is applied in the chemical industry, petrochemical industry and pharmaceutical industry [21]. The membrane technologies used in the VOC gas recovery are mainly gas permeation and reverse osmosis [21]. The concentration of VOC gases is rising in the membrane unit as the separation proceeds, and the limits from the lower to the upper explosive limit have to be taken into consideration to prevent the formation of an explosive mixture [12]. However, membrane separation is only a concentration technique and if complete purification of the emission stream is wanted, membrane technology needs to be combined with another technique, e.g., adsorption [13]. If we consider membranes to recover selected VOC compounds, it seems to be a rather feasible method.

2.1.3. Wet Gas Scrubbing

In wet gas scrubbing a gaseous compound is absorbed from a gas phase to a liquid phase. For the recovery of chemical compounds physical absorption is the only possibility since after that process the compounds can be desorbed. Chemical scrubbing is used only for the abatement of compounds from gaseous emissions. The suitable chemicals that can be scrubbed with water are those that are soluble in water, such as alcohols, acetone and formaldehyde. However, absorption is mainly used for the recovery of inorganic compounds rather than VOCs [12].

2.1.4. Condensation and Cryocondensation

In condensation the temperature of the gas is decreased or the pressure is increased to reach its dew point. The use of condensers is favorable when the concentration of VOCs is high in the gas stream. Low-volatility VOCs are easier to condense, since their condensing temperatures are higher. Therefore, volatility is a crucial property of the compound when considering condensation as a recovery method. Condensers can be used as a primary recovery method or together with another recovery technique [22].

Cryocondensation uses very low temperature refrigerants, e.g., liquid nitrogen, as coolants to reduce the temperature for condensation. Cryocondensation is very effective with all VOCs and can be used as the final method for the VOC emission control. The method is very flexible for the changes in the VOC flow rate and solvent loading [12].

2.2. Purification of the Raw Emissions and the Products

In some cases, the raw emission needs to be pretreated for the removal of particulates or water before the recovery of valuable VOC compounds [23]. Often, these compounds are separated from VOCs in the recovery phase and no separate pretreatment is needed. The recovered emission is typically a mixture of several organic compounds and for the chemical utilization it is usually necessary to purify the desired compound from other organic compounds. In this case, a selective technique is needed and also the sequence of the separation of compounds has to be considered carefully.

The recovered gas may contain several impurities, such as catalyst poistons, that are harmful in the utilization of the compounds. The source of the emission gas and the application of the recovered gas determine the need for purifying the recovered gas. Particulates and non-organic gases are typical impurities found in VOC gases. In catalytic utilization of the recovered gas, it is vital that the gas does not affect the activity of the catalyst nor contain catalyst poisons, such as siloxanes and compounds of sulphur, phosphorus and halogens at intolerable amounts. It may also be necessary to concentrate the gas with one organic compound and remove other VOCs that complicate the utilization of the major compounds [24].

Particulates present in the gas mixture to be utilized may cause erosion, compression damages and explosion risks [7]. Methods that remove particles, but not VOCs are gravitational separations and several filtration techniques, namely dry and wet electrostatic precipitator, fabric filter, ceramic and metal filter, two-stage dust filter, absolute filter (HEPA) and high-efficiency air filter (HEAF) [12]. Nitrogen removal is the most critical and expensive process. However, it is possible to remove nitrogen and also CO_2 by pressure swing adsorption with certain carbon molecular sieves. Water can be removed, e.g., by adsorbing it to a zeolite. Oxygen removal is also essential, since it may cause combustion or explosion of VOCs. Oxygen removal can be carried out by catalytic process where some of the organic compounds are oxidized to CO_2. However, this reduces the amount of valuable VOCs to be recovered [7].

One special case of impurities in VOCs is the presence of siloxanes in landfill gases. Siloxanes are added to many domestic products, which is the reason that their concentrations are increasing in landfill gases. Siloxans are non-toxic silicon containing organic molecules. They are volatile and thus evaporate to the recovered landfill gas. The problem of siloxanes is, that they for example burn together with the other organic gases, but instead of gaseous products they form crystalline solid silica. This product is then fouling the equipment surfaces, deactivating the catalysts, *etc.* Siloxanes can be removed from the gas by condensation (refrigeration) and adsorption (such as carbon adsorption) [25].

To summarize, several mature techniques are available for the recovery of VOCs. Some techniques include the regeneration phase, which increase the costs compared to the abatement of VOCs from emission gases. Membrane techniques, condensation and cryocondensation have the benefit that no regeneration is needed. Selectivity of the recovery technique plays also an important role when the recovered compound will be utilized as raw material to form new products. A thorough planning needs to be made when the suitable recovery and purification technology is considered. The selection is very case-dependent and not only dependent on the recovered VOC, but also on the potential products concerned.

3. Catalytic Utilization of Methane Emissions

3.1. From Methane to Syngas and Hydrogen

Methane is the simplest form of hydrocarbon and organic gas. Methane is also the major compound in natural gas (NG) and it is the second largest energy source after crude oil. Around the world, abundant NG reserves exist and it is ubiquitous material distributed more evenly than oil and coal [26]. Methane is mainly utilized as a fuel and energy source for many decades, for example in electricity production and as a transportation fuel in the form of liquefied natural gas (LNG) or compressed natural gas (CNG). In this present section, the main focus is on the utilization of methane gas (or NG) emissions from different possible sources. Globally, enormous amounts of NG is extracted, produced, and transported. Methane is one of the major greenhouse gases (GHGs), even though it is present in the atmosphere in low concentrations. It is noteworthy that methane has ~20–23 times higher global warming potential (GWP) than CO_2 [27,28]. Methane emissions originate from a wide variety of natural (such as oceans and wetlands) and anthropogenic sources, typically from oil drilling (NG is vented and flared), shale gas, automobiles' exhaust gas, landfill gas (LFG), coal bed methane (CBM), hydrocarbon processes, agro-farming, mining and metal industries [29,30]. In many places, methane is a waste hydrocarbon source from different process industries and exploration sites, which is not utilized efficiently. There are many ambitious projects around the world on reduction of NG flaring and utilization [31,32]. Moreover, in many geological deposits, in coal mining and process industries, methane exists with CO_2. Henceforth, both, *i.e.*, CH_4 and CO_2 GHGs can be utilized in synthesis gas (*syngas* as short form) production via the dry reforming process. Further, syngas can be converted to valuable fuels, intermediates, and chemicals via Fischer-Tropsch synthesis [26].

Utilization of methane emissions depends on many factors, for example, the amount and concentration or purity levels of CH_4. Further, gas stream originated from different sources contain different concentration levels of methane (e.g., ~50% by volume exist in LFG) [29]. Even the low-concentrations of methane from CBM can be utilized to produce useful energy, for example, by catalytic combustion over a heterogeneous catalyst CuO/γ-Al_2O_3 [33]. The economic viability and thermal efficiency are the topics that need to be considered in the case of methane emission utilization.

Methane is a symmetrical molecule, with no dipole moment and has strong C–H bond energy (493 kJ mol^{-1}). In order to convert CH_4, a lot of energy is required to activate the C–H bond. This activation is one of the biggest challenges in the catalysis field and has been an important topic of research for academic and industrial communities [34–37]. The kinetic and thermodynamic constraints are critical for developing energy efficient processes in methane conversion technologies [34,37]. As mentioned in Tang *et al.* [38], the C–H bond activation can be done by mainly four ways by reducing the oxidation state of carbon atom, such as (i) complete C–H bond removal by partial oxidation, for example to syngas (CO + H_2), oxygenates (CH_3OH) and hydrocarbons compounds (olefins and aromatics) production; (ii) complete C–H bond removal by decomposition to carbon and H_2; (iii) complete combustion in mobile and stationary applications (such as fuel and energy) and finally; (iv) partial C–H bond removal by substituting heteroatoms (e.g., CH_3Cl). In the following sections, catalysts development in these four strategies of methane C-H bond activation will be discussed.

3.1.1. Methane Dry Reforming and Partial Oxidation to Syngas

The first and foremost route was the direct methane conversion to syngas production. Currently, there are three main routes to convert CH_4 to syngas, *i.e.*, partial oxidation, and dry and steam reforming (Figure 1). Methane dry reforming (MDR) is a promising route for syngas production, and moreover, it is beneficial in utilizing CO_2 emissions [39]. In the MDR process, the CO/H_2 ratio of one can be achieved, which is ideal for the F-T synthesis. Whereas, methane steam reforming

(MSR) is more appropriate for H_2 production than syngas, due to a higher H_2/CO ratio that can be achieved. This topic is discussed more in the next section. Few processes are successful, but still to reach the industrial scale, many challenges exists in catalysis and reactor design. In this perspective, the biggest challenge occurs in developing catalytic materials for the methane conversion routes (Figure 1).

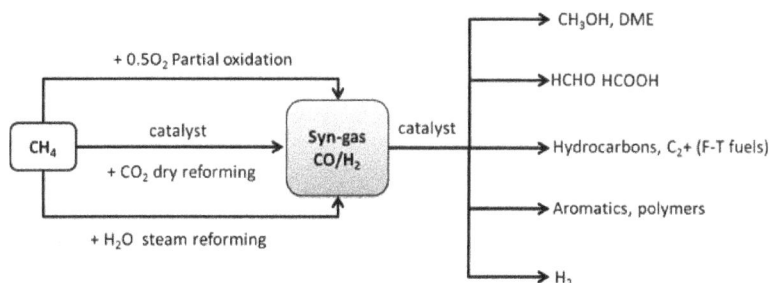

Figure 1. Indirect methane conversion to produce fuels and value-added chemicals via intermediate syngas production.

A list of chemical reactions involved in methane conversion to syngas, hydrogen and direct decomposition is as follows [35].

(E1) *Methane dry reforming (MDR)*

$$CH_4 + CO_2 \leftrightarrow 2H_2 + 2CO \quad \Delta H^\circ{}_{298} = 247 \, kJ \, mol^{-1} \tag{1}$$

(E2) *Methane partial oxidation (POM)*

$$CH_4 + 0.5O_2 \leftrightarrow 2H_2 + CO \quad \Delta H^\circ{}_{298} = -36 \, kJ \, mol^{-1} \tag{2}$$

(E3) *Methane steam reforming (MSR)*

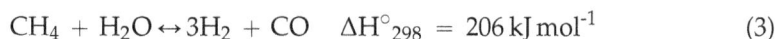

$$CH_4 + H_2O \leftrightarrow 3H_2 + CO \quad \Delta H^\circ{}_{298} = 206 \, kJ \, mol^{-1} \tag{3}$$

(Insufficient steam)(E4) *Methane steam reforming*

$$CH_4 + 2H_2O \leftrightarrow 4H_2 + CO_2 \quad \Delta H^\circ{}_{298} = 165 \, kJ \, mol^{-1} \tag{4}$$

(Sufficient steam)(E5) *Methane decomposition*

$$CH_4 \leftrightarrow C + 2H_2 \quad \Delta H^\circ{}_{298} = 75 \, kJ \, mol^{-1} \tag{5}$$

Several catalysts are developed and tested in syngas production via the MDR and POM process. In the literature, one can find quite a good number of studies and excellent reviews on thermodynamics and heterogeneous catalysts for methane dry reforming reaction (e.g., [40–45]) and also for the methane partial oxidation [46–48]. The recently studied most active catalysts published on MDR and POM process are summarized in Table 2 and Table 3. Most of the studies in MDR are related to high-temperature catalysis operated at above 600 °C [41].

Recently, micro-plasma technology coupled with catalysis has gained interest in research communities and promising results at room temperature reaction have been obtained, e.g., using dielectric barrier discharge (DBD) plasma assisted reactor with a packed bed catalyst [49]. This technology is also very promising for the preparation of catalytic materials since it produces a uniform heat distribution [50].

In general, platinum group metals (noble) [41] and non-noble [42–51] metal catalysts were studied in MDR, but non-noble metal catalysts are the preferred ones due to the cost and availability. The noble metal based catalysts have also shown comparable activities and higher durability than non-noble metal catalysts. As presented in Table 2, Ni based catalysts were found to be most popular and widely studied in MDR due to their superior activity and low cost. Most of the studies employed stoichiometric molar ratio $CH_4/CO_2 = 1$ and the syngas production was close to the ideal, *i.e.*, $H_2/CO = 1$. The influence of the preparation method on the catalytic properties was a crucial parameter. Moreover, the operating conditions had a greater influence on catalytic activity in MDR and POM, such as reaction temperature, which had a significant effect on the products formation at ambient pressure. The catalyst development for MDR was mostly investigated on Ni based catalysts on different carrier materials such as oxides (CeO_2, MnO_2, SiO_2, Al_2O_3, MgO_2, ZrO_2, *etc.*), mixed oxides (CeO_2-Al_2O_3, CeO_2-ZrO_2), tri and ternary oxides, and moreover bimetallic materials such as NiPt, NiCo, NiFe based catalysts [40,41]. Preparation methods for the carrier and active metal catalysts were found to be the most important parameter in preparing active and selective catalysts. In this perspective, calcination temperature had greater influence on the active particle size. As reported by Gould *et al.* [52], an atomic layer deposition method is effective in preparing very small active Ni particles with controlled growth and distribution compared to impregnation. In methane activation, the metal-support interactions are important and they affect strongly the overall catalytic activity [53,54]. In MDR, carbon formation is a serious problem and this results in catalyst deactivation. There are different strategies to reduce and avoid carbon formation [55]. For example, addition of small amounts of noble metals such as Pt, Rh can improve the catalyst performance and coking resistance [56]. Another effective way is to prepare a catalyst with high oxygen storage capacity (OSC) materials such as CeO_2, for effective carbon removal via gasification by lattice oxygen [57]. In many catalytic applications, OSC materials renders the surface by oxidizing the carbon species. Novel methods were developed by modifying the conventional methods to prepare catalyst with desired physicochemical properties for example refluxed co-precipitation and sequential impregnation [51,58]. The effect of calcination method had a pronounced effect on the activity and selectivity of MDR, for example, freeze drying and NO calcination of Ni based catalysts were superior compared to conventional and air drying and calcination [53]. Nevertheless, the nature of the support material also plays vital role in catalytic activity and deactivation [59,60]. Nano based catalytic materials are most active and robust with optimal size, shape and uniform particle distribution [52,61]. New class of materials are needed and also modification of conventional mixed metal oxides with lower amount of active metal can lead to reduction of the overall costs.

Table 2. Most recently studied Ni based catalysts in methane dry reforming (MDR) to syngas production.

Catalyst	Prepa-ration method	Reaction conditions	Conversion (%)	H_2/CO	Comments	Ref.
$LaNiO_3@SiO_2$	Stöber	RT $Q_{tot} = 50$ mL min^{-1} $CH_4/CO_2 = 1$	$X_{CH4} = 67$ $X_{CO2} = 57$	0.9	NT-plasma, DBD reactor, core shell	[49]
$Ni_{4.7}Pt_1$	ALD	$T = 600$ °C $Q_{tot} = 40$ mL min^{-1} $CH_4/CO_2 = 1$	$X_{CH4} = 30$	na	small Ni particles ~8 nm with ALD	[52]
$Ni_{10}Pt_{0.5}/Al_2O_3$	WI	$T = 750$ °C, $Q_{tot} = 20$ mL min^{-1} $CH_4/CO_2 = 1$	$X_{CH4} = 78$ $X_{CO2} = 95$	0.63	Pt addition reduce the carbon deposition and improve the geometric effects	[56]
$Ni_5Mn_{95}O$	CP	$T = 750$ °C, $Q_{tot} = 30$ mL min^{-1} $CH_4/CO_2 = 1$	$X_{CH4} = 75,$ $X_{CO2} = 75$	na	rapid deactivation due to Mn phase change leads to metal sintering	[59]
$Ni_5/diatomites$	two solvents and DI	$T = 650$ °C, GHSV = 21,120 mL g^{-1} h^{-1} $CH_4/CO_2 = 1$	$X_{CH4} = 39$ $X_{CO2} = 49$	0.88	deactivation of Ni^0 by re-oxidation, only C-β (reactive C-nanotubes)	[62]
Ni_5/SiO_2	two solvents and DI	$T = 650$ °C, GHSV = 21,120 mL.g^{-1} h^{-1} $CH_4/CO_2 = 1$	$X_{CH4} = 61,$ $X_{CO2} = 70$	0.9	deactivation of Ni^0 by re-oxidation, two types of carbon forms: C-β and C-γ (less reactive sp^3 C-graphite)	[62]
$(NiCo)_{2.5}/Ce_{77}Zr_{20}$	Glycothermal synthesis + DP + FD calcination	$T = 750$ °C $p = 1.2$ bar	$X_{CH4} = 90$ $X_{CO2} = 90$	0.85	TOS 20 h, calcination effect	[53]
$Ni_5-Ce_6/SBA-15$	two solvents and IWI	$T = 600$ °C, GHSV = 264 Lg^{-1}h^{-1}	$X_{CH4} = 95,$ $X_{CO2} = 90$	0.96	high dispersion inside pores, smaller NiO particles, mode of NiO addition after Ce	[51]
$Ce_{2.5}-Co_5-Ni_{10}/Al_2O_3$	IWI	$T = 750$ °C $CH_4:CO_2 = 1$	$X_{CH4} = 60$		TOS 5 h, effect of forced periodic cycling	[63]
$Ni_{15}CeMgAl$	refluxed co-precipitation	$T = 750$ °C GHSV = 48,000 h^{-1} $CO_2/CH_4 = 1.04$	$X_{CH4} = 96.5$ $X_{CO2} = 92$	0.8	bimodal pore, Ni sites in $NiAl_2O_4$ spinel structure had longer stability	[58]

Table 2. *Cont.*

Catalyst	Prepa-ration method	Reaction conditions	Conversion (%)	H_2/CO	Comments	Ref.
Ni_5/La_2O_3-ZrO_2	IWI	$T = 400\ ^\circ C$, 12 mL min^{-1}	$X_{CH4} = 7.5$ $X_{CO2} = 12$	na	effect of calcination temperature relation with Ni particle size	[64]
$NiFe_2O_4/SiO_2$	sol–gel	$T = 800\ ^\circ C$ $Q_{tot} = 30$ mL min^{-1} $CH_4/CO_2 = 2$	$X_{CH4} = 66$ $X_{CO2} = 93$	1.1	surface acidity is crucial in limiting RWGSR	[65]
$Rh_{0.2}$-Ni/SBA-15	DI	$T = 650\ ^\circ C$ GHSV = 48,000 mL $h^{-1}\ g^{-1}$	-	0.82	Rh promoted the Ni reducibility and decrease the Ni particle size and enhances the stability	[54]
Ni/CNT	IWI	$T = 750\ ^\circ C$ $W/F = 1$ g.h.mol^{-1}	$X_{CH4} = 66$ $X_{CO2} = 79$	0.9	TOS 8 h, Ni particles inside the CNTs are more active than outside, tube confinement effect	[66]
Ni/Al_2O_3	IWI	$T = 850\ ^\circ C$ GHSV = 24 $L\cdot g^{-1}\cdot h^{-1}$	$X_{CH4} = 90$ $X_{CO2} = 94$	1	preparation of nano catalyst with high SA, good dispersion,	[61]
Ni_{15}/TiO_2	sonication	$T = 700\ ^\circ C$, GHSV=95,500 h^{-1} $H_2O/CH_4 = 1.2$	$X_{CH4} = 86$ $X_{CO2} = 84$	0.9	sonication is more efficient than impregnation, fine dispersion of Ni NPs,	[67]
Rh_1/NiO_{20}-Al_2O_3	sol-gel/IWI	$T = 700\ ^\circ C$ $Q_{tot} = 37$ L min^{-1} $CH_4:CO_2 = 1:1$	$X_{CH4} = 89$ $X_{CO2} = 91$	1.22	210 min TOS, deactivation of Rh avoided by $NiAl_2O_4$ spinel phase, Ni enhance the Rh dispersion on the surface	[68]

T = reaction temperature ($^\circ C$); p = reaction pressure; RT = room temperature; Q_{tot} = CO_2 + CH_4 volumetric feed gas flow rate; GHSV = gas-hourly-space-velocity; TOS = time-on-stream; X_{CH4} = methane conversion (%); X_{CO2} = carbondioxide conversion (%); IWI = Incipient wetness impregnation; NT = non-thermal; DBD = dielectric barrier discharge; CP = co-precipitation; ALD = atomic layered deposition; DI = dry impregnation; FD = freeze drying.

In POM, stoichiometric $CH_4:CO_2$ ratio is critical in determining the syngas composition (E2). Most widely studied catalytic materials in POM are the transition metal catalysts (*i.e.*, PGM) due to their high activity in oxidation [47]. Among them, Rh based catalysts were found to be most active and stable and widely reported in scientific literature (Table 3). As reported earlier, noble metals are highly selective for oxidation reactions due to high O_2 adsorption and resistance against carbon formation [69]. POM is thermodynamically an exothermic reaction, but most of the catalysts are active and selective at higher temperatures 700–1000 $^\circ C$ (shown in Table 3). At higher temperatures, deactivation of the catalysts takes place due to active metal sintering or agglomeration [55,70]. Developing new catalytic materials, which are active at low temperatures can be a possible solution to avoid metal sintering and also make the process less energy intensive. Most of the catalysts used in the literature achieved the conversion and selectivity close to equilibrium values (Table 3). Durability of the catalyst is still a challenge, since to the best of our knowledge, only a few catalysts performed excellently when the time-on-stream was longer than 150 h [70].

Table 3. Recently studied catalysts in partial oxidation of methane to syngas.

Catalyst	Prepa-ration Method	Reaction conditions	Conversion and selectivity (%)	H_2/CO	comments	ref.
$Rh_{0.005}/\gamma\text{-}Al_2O_3$	IWI	$T = 900\ ^\circ C$, GHSV = 24,640 h^{-1}, $CH_4/O_2 = 2$	$X_{CH4} = {\sim}88$, $S_{CO} = 90$, $S_{H2} = 90$	2	TOS 160 h, CH$_4$ conversion and CO selectivity decrease with Rh NP size	[70]
Rh_5-doped CeO_2	solution-based hydrothermal	$T = 700\ ^\circ C$, $Q_{tot} = 100\ mL\cdot min^{-1}$, $CH_4/O_2 = 2$	$X_{CH4} = 95$, $S_{H2} = 93$, $S_{CO} = 86$	2.2	surface chemistries with *in situ* studies are important to correlate the catalysts performance	[69]
Rh-honeycombs in microreactor	micro-structuring Rh foils	$T = 1100\ ^\circ C$, $p = 0.15\ Mpa$, GHSV = 195·10^3 h^{-1}, $CH_4/O_2 = 2$	$X_{CH4} = 90$, $S_{H2} = 88$, $S_{CO} = 87$	2	$t > 10$ ms; high pressure used in Rh honeycomb catalyst; minimizing pressure drop	[71]
Ni-Th-O	controlled oxidation	$T = 750\ ^\circ C$, $CH_4/N_2O = 1$, GHSV = 8500 mL·$g_{cat}^{-1}\cdot h^{-1}$	$X_{CH4} = 50$, $S_{CO} = 90$, $S_{H2} = 90$	2	using N$_2$O as oxidant, bimetallic Ni thorium used, accessibility and acidity	[72]
$Ni\text{-}Pt/La_{0.2}Zr_{0.4}Ce_{0.4}O_x$	wash coating and IWI	$T = 800\ ^\circ C$, GHSV = 820,000 h^{-1}, $Q_{tot} = 4.0\ mL\cdot s^{-1}$, $\tau = 4.4$ ms, $CH_4/O_2 = 2$	$X_{CH4} = 82$, $S_{CO} = 60$	1.5	microchannel reactor, very short residence time, high flow rates	[73]
$Ce_7\text{-}Fe_3\text{-}O$ mixed oxide	CP	$T = 900\ ^\circ C$	$X_{CH4} = 95$, $S_{H2} = 99$, $S_{CO} = 98.8$	2	oxygen carrier showed good, lattice oxygen is crucial and high O mobility	[62]
Ni_{30}/MgO	IWI	$T = 800\ ^\circ C$, $P = 0.1\ MPa$, GHSV = 1200 h^{-1}, $CH_4/O_2 = 2$	$X_{CH4} = 94$, $S_{CO} = 94$, $S_{H2} = 94$	2	higher conversion with increasing calcination temperature, highly Ni dispersed	[74]
$Rh_{1.5}/CeO_2/$ monolith cordierite	solution combustion synthesis	$T = 800\ ^\circ C$, GHSV = 400,000 mL $g_{cat}^{-1}\ h^{-1}$, $S/C = 1.2$, $O/C = 0.55$	$X_{CH4} = 100$	3	Structured catalysts cordierite honeycombs, high performance at high WSV, process intensification	[75]
$(CeO_2)_5/Pt_2\text{-}SiC$	micro-emulsion	$T = 805\ ^\circ C$, $CH_4/O_2 = 2$, GHSV = 1500 h^{-1}	$X_{CH4} = 93$	1	composites with optimal Ce and Pt content to achieve high activity,	[76]
$(NiO)_{48}/CeO_2$	one-step template	$T = 850\ ^\circ C$, GHSV = 9.5×10^6 h^{-1}, $CH_4/O_2 = 2$	$X_{CH4} = 90$, $S_{CO} = 84$, $S_{H2} = 90$	2	Fibrous nanocatalyst, highly active at very short contact time of 98 µs,	[77]
$Ni_{10}/CeO_2\text{-}SiO_2$	sol-gel/DI	$T = 750\ ^\circ C$, $CH_4/O_2 = 2$, $Q_{tot} = 30\ mL\cdot min^{-1}$	$X_{CH4} = {\sim}84\%$	2	calcined at 700 $^\circ C$ had a weak surface acidity with high surface area and low carbon formation, atomic Ce/Si = 1, 15 h TOS	[78]
$Pt_{10}Ce_{0.5}Zr_{0.5}O_2/Al_2O_3$	IWI	$T = 900\ ^\circ C$, $Q_{tot} = 100\ mL\cdot min^{-1}$, $CH_4/O_2 = 2$	$X_{CH4} = 70$, $X_{CO} = 80$, $S_{H2} = 96$		high reducibility and O$_2$ mobility reduce the carbon formation	[79]

T = reaction temperature ($^\circ C$); p = reaction pressure; τ = residence time; Q_{tot} = volumetric feed gas flow rate; IWI = Incipient wetness impregnation; CP = co-precipitation; ALD = atomic layered deposition; GHSV = gas-hourly-space-velocity; TOS = time-on-stream; DI = dry impregnation; X_{CH4} = methane conversion (%); X_{CO2} = carbondioxide conversion (%); S_{CO} = CO selectivity (%); S_{H2} = CO selectivity (%).

3.1.2. Methane Steam Reforming and Decomposition to H_2

As reported by many scientists, environmental panelists and governmental organizations, possible solution for the current environmental problems and clean energy security can be the H_2 economy. Hydrogen is known as a clean energy carrier and it is used as a fuel in fuel cell systems. Today, ~50% of H_2 is produced by natural gas steam reforming over industrially applied Ni-based catalyst [80]. Still challenges exist in catalysts development for MSR. MSR is widely studied over several catalytic systems with different configurations, intensified reactors, and process conditions [42,81]. There is a need to improve the efficiency of the existing MSR process using new robust catalytic materials (Table 4) and also by intensifying reformer system (such as membrane assisted reactors (MRs)). Steam reforming of hydrocarbons is studied extensively in MR related to CO_x-free H_2 production [82,83]. MSR at low temperature operation will be one of the main research objectives, but thermodynamically temperatures less than 600 °C are challenging for the reforming reaction [42]. There is a need for active and selective catalysts which can activate the C–H bond and the water molecule at moderate temperatures. Another route in methane to H_2 is the catalytic decomposition. For low temperature fuel cell applications, catalytic decomposition of methane (CDM) is more suitable than MSR since then CO_x free H_2 can be produced [84]. In MSR, carbon formation is relatively lower than in CDM due to steam environment. The challenge in CDM is to avoid deactivation of catalyst due to carbon formation during the H_2 production. The type of carbon formed is very much dependent on the type of the feed and operating conditions. Interestingly, carbon nanotubes can be grown along with H_2, which does not have any influence on catalyst deactivation [84]. Still, MSR is highly matured technology and cost-effective in H_2 production, but process modifications and new catalysts are needed to improve the existing efficiency and to reduce the emissions.

3.2. From Methane to Chemicals

Transportable liquids are more cost-effective than NG due to its low volumetric density. Gas-to-liquid (GTL) technology is a dominant approach to convert gases to liquid fuels and chemicals. Direct methane conversion in a single-step can reduce operating costs, energy consumption, and it is more efficient than producing first syngas and then the chemicals. Transformation of methane to value added intermediates and chemicals is still at the research stage compared to methane to syngas production (Figure 2). Hydrogen cyanide production from methane in the presence of oxygen was the only industrially established process for many decades [94]. One of the strategies in the C–H bond activation is the catalytic oxidation process. Oxidation process is one of the most viable, easy and efficient methods in VOC abatement and utilization technologies [95].

Table 4. Recently studied catalysts in methane steam reforming to hydrogen.

Catalyst	Prepa-ration method	Reaction conditions [#]	Conversion and H_2 selectivity (%), yield (% or mole)	Comments	Ref.
Ni_{10}/ZrO_2-CeO_2-La_2O_3	WI	T = 500 °C GHSV = 70,000 h^{-1} S/C = 3	X_{CH4} = 25 Y_{H2} = 30	Low temperature steam reforming	[85]
$Ni/K_2Ti_xO_y$-Al_2O_3	WI	T = 750 °C GHSV = 15,000 h^{-1} S/C = 2.5 TOS = 10 h	X_{CH4} = 97 Y_{H2} = 3 mol	high activity due to weak interactions between Ni and support, addition of secondary support K_2TiO_2 improved the resistance to deactivation	[86]
NiO/SiO_2	sol-gel	T = 700 °C ST = 11.31 kg_{cat} h.$kmol^{-1}$ S/C = 3.5	X_{CH4} = 96 Y_{H2} = 3.8 mol	crystallite size of NiO catalysts can be controlled by calcination	[81]
$Ni/Ce_{0.65}Hf_{0.25}Pr_{0.1}O_2$	EDTA-citrate	T = 700 °C S/C = 2	X_{CH4} = 85	Dopants redox property crucial in enhancing the OSC in rare earth metals-doped	[87]
Cu_5/Co_6Al_2	precipitation and IWI	T = 650 °C S/C = 3 Q_{tot} = 50 mL·min^{-1}	X_{CH4} = 96 Y_{H2} = 2.60 mol	Cu favors the WGS reaction	[88]
$Ni_{0.5}Mg_{2.5}AlO_9$	co-precipitation	T = 900 °C S/C = 3.1, Q_{tot} = 50 mL·min^{-1}	X_{CH4} = 100	150 h, 20 ms, high WHSV, higher dispersion of Ni particles, small particles	[89]
Ru/Co_6Al_2	IWI	T = 600 °C, S/C = 3, Q_{tot} = 20 mL·min^{-1},	X_{CH4} = 95 Y_{H2} = 24	100 h, well dispersed at the surface and higher Co loading	[90]
$Ni_{0.15}Al_{0.85}$	Solution-combustion/ IWI	T = 850 °C Q_{tot} = 100 mL·min^{-1} S/C = 4	X_{CH4} = 97.8 Y_{H2} = 2.9 mol	high surface area and strong interaction between Ni and Al	[91]
$Ir_5/MgAl_2O_4$	IWI	T = 850°C GHSV = 284,000 h^{-1} S/C = 3	X_{CH4} = 55	Ir particles bind strongly $MgAl_2O_4$ surface via redox process leading to a strong metal–support interaction and activate facile water dissociation	[92]
Au/Ni_5-LaAl	anionic exchange /IWI	T = 700 °C S/C = 1.24 GHSV = 135,000 mL·g^{-1}.h^{-1}	X_{CH4} = 34	Au addition enhance the stability and decreases the carbon growth rate	[93]

[#] atmospheric pressure; T = reaction temperature (°C); p = reaction pressure; Q_{tot} = volumetric feed gas flow rate; S/C = steam-to carbon ratio; ST = space time; IWI = Incipient wetness impregnation; CP = co-precipitation; ALD = atomic layered deposition; DI = dry impregnation; GHSV = gas-hourly-space-velocity; TOS = time-on-stream; OSC = oxygen storage capacity; X_{CH4} = methane conversion (%); Y_{H2} = hydrogen yield (%).

One of the commercially applied conversion technology is the catalytic methane combustion (CMC) (in the form of LNG and/or CNG) in automotive applications. A significant amount of research has been done on catalyst materials for methane combustion in internal combustion engine (ICE) applications. Uncombusted methane is lost through exhaust pipe to the environment. Huge amount of research is devoted to develop an efficient catalytic converter to improve the CH_4 fuel economy [96]. In CMC, PGM metals are good candidates for complete combustion and Pd is more selective in oxidizing the methane. In catalytic combustion of methane, Pd was found to be most active and reported in several studies [96,97].

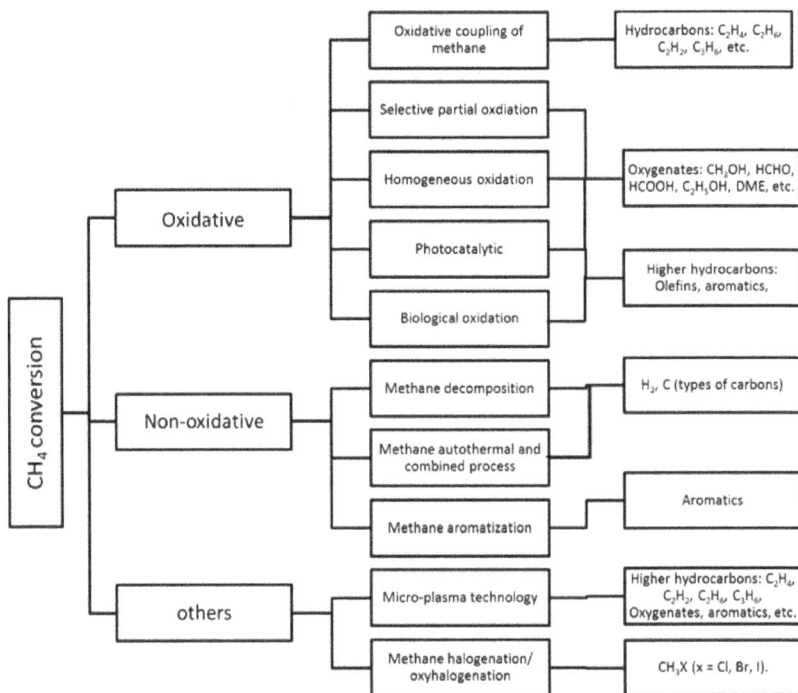

Figure 2. Direct methane conversion technologies to various valuable chemicals, intermediates and fuel compounds.

List of reactions involved in methane to chemical synthesis [35]:

(E6) *Methane partial oxidation*

$$CH_4 + 0.5O_2 \leftrightarrow CH_3OH \quad \Delta H^\circ_{298} = 127 \, kJ.mol^{-1} \tag{6}$$

$$CH_4 + O_2 \leftrightarrow HCHO + H_2O \quad \Delta H^\circ_{298} = -276 \, kJ.mol^{-1} \tag{7}$$

(E7) *Methane combustion or total oxidation*

$$CH_4 + 2O_2 \leftrightarrow 2H_2O + CO_2 \quad \Delta H^\circ_{298} = -802 \, kJ.mol^{-1} \tag{8}$$

(E8) *Oxidative coupling of Methane (OCM)*

$$2CH_4 + 0.5O_2 \leftrightarrow C_2H_6 + H_2O \quad \Delta H^\circ_{298} = -174.2 \, kJ.mol^{-1} \tag{9}$$

$$C_2H_6 + 0.5O_2 \leftrightarrow C_2H_4 + H_2O \quad \Delta H^\circ_{298} = -104 \, kJ.mol^{-1} \tag{10}$$

(E9) *Methane dehydroaromatization (MDA)*

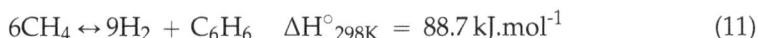

$$6CH_4 \leftrightarrow 9H_2 + C_6H_6 \quad \Delta H^\circ_{298K} = 88.7\,kJ.mol^{-1} \tag{11}$$

(E10) *Methane halogenation and oxyhalogenation*

$$CH_4 + HCl + 2O_2 \leftrightarrow CH_3Cl + HCl \tag{12}$$

$$CH_4 + HCl + 0.5O_2 \leftrightarrow CH_3Cl + H_2O \tag{13}$$

Methane can be selectively converted to a wide variety of products such as oxygenates, olefins, aromatics and methyl halides by oxidation and non-oxidation processes (Figure 2).

3.2.1. Catalysts for Selective Partial Oxidation of Methane to Oxygenates

Thermodynamically the direct conversion of methane to C_1 oxygenates via partial oxidation takes place at temperatures above 300 °C and the process is restricted by the products stability [98,99]. Large amounts of O_2 are needed for methane to chemical synthesis. Instead of using expensive pure O_2, several authors have also utilized oxidants such as GHGs (N_2O, CO_2, CO), H_2O_2 and air [72,100]. Thus, the methane oxidation in the N_2O stream makes the process more efficient and greener. In Table 5, results obtained over few studied catalysts with different operating conditions are summarized (thermal and non-thermal catalysis). Methane to methanol (MTM) synthesis over a Cu_4/MOR catalyst was studied at low temperatures by Alayon in 2012 [101], and he reported that the reaction conditions affect the Cu structure. During the reaction, O_2 is activating the Cu^{2+} active sites; thereafter react with CH_4 to form Cu^+. The nature of active sites plays an important role in the product selectivity [102]. Wei *et al.* investigated low temperature (at 60 °C) conversion of CH_4 to HCOOH over a $VOSO_4$ catalyst and they were able to achieve 7% conversion with 70% HCOOH selectivity due to the formation of V^{+5} species, which activate methane in the H_2O_2 media. In CH_4 to C_1 oxygenates synthesis, photo-, bio-, plasma, and homogeneous liquid phase oxidation (in H_2SO_4) are widely studied at relatively low temperatures to improve the overall yield and selectivity especially for CH_3OH and HCHO [103–107]. Recently, much attention has been paid to CH_4 photocatalytic conversion and some promising results are shown at relatively low temperatures (below 60 °C) operating in a simplified reactor apparatus [103,108]. Another interesting route is the selective oxidation of methane in homogeneous liquid phase (HLP). HLP carried out over an $OsCl_3$ catalyst produced highest turnover frequency for the formation of oxygenates compared to other transition metal chlorides reported in Ref. [109].

Table 5. A few selected catalysts studied in methane to C_1 oxygenates (mostly CH_3OH and HCHO) under different reaction conditions.

Catalyst	Preparation method	Reaction conditions	Conversion, selectivity and yield (%)	Comments	Ref.
nano-Au/SiO$_2$	HAuCl$_4$ + IL in SiO$_2$ sol	$T = 90\ ^\circ C$ $p = 20$ atm	$X_{CH4} = 25$ $S_{CH3OH} = 72$ $Y_{CH3OH} = 18$	[Bmim]Cl ionic liquid as dissolution solvent, Liquid phase oxidation, 97% Au is recovered	[104]
CuFe$_2$(P$_2$O$_7$)$_2$	IWI	$T = 630\ ^\circ C$ $p_{CH4} = 65$ kPa $p_{N2O} = 26$ kPa $Q_{tot} = 3.6$ L/h, CH$_4$/N$_2$O = 1.2	$X_{CH4} = 6.4$ $S_{HCHO} = 26.6$ $S_{CH3OH} = 1.8$ $Y^*_{CH3OH+HCHO} = 1.8$	interactions and synergistic effects of Fe and Cu leads to enhanced activity, crystalline phase	[110]
Fe$_2$O$_3$-CuO/γ-Al$_2$O$_3$	IWI	$T = 300\ ^\circ C$	$X_{CH4} = 43$ $Y_{CH3OH} = 1.5$	Plasma-catalysis, inside packed catalyst configuration is more effective than post catalysis	[105]
Y$_3$/Cu$_6$-Zn$_{2.5}$-Al$_{1.5}$	co-precipitation	CH$_4$/O$_2$ = 4	$X_{CH4} = 25$ $S_{CH3OH} = 27$	Plasma-Catalysis, dielectric barrier discharge addition of Yttrium, also Pt and Fe, found to enhance performance	[106]
(V$_2$O$_5$)$_{0.03}$/SiO$_2$	IWI	$T = 650\ ^\circ C$ GHSV = 6.6×10^4 h^{-1} $Q_{tot} = 71$ mL· min^{-1} CH$_4$/O$_2$ = 2	$X_{CH4} = 34$ $Y_{CH3OH+HCHO} = 16$	1% NO in feed enhanced the conversion and selectivity	[111]
PPFe^{+3}OH/AlSiMg Fe/protoporphyrin	Activation	$T = 180\ ^\circ C$ H$_2$O$_2$/CH$_4$ = 1.4	$X_{CH4} = 50$ $Y_{CH3OH} = 60$ $S_{CH3OH} = 97$	Biomimetic enzyme catalysis, using H$_2$O$_2$ oxidant, highest hydroxylating activity due to active O$_2$ carriers, deactivates in 5 h	[112]
MoO$_3$/SiO$_2$	sol-gel	$T = 500$	$X_{CH4} = 2.9$ $S_{HCHO} = 52$ $S_{CH3OH} = 8$ $Y_{CH3OH+HCHO} = 0.86$	catalyst preparation method influences the products selectivity, formation of silicomolybdic acid on the surface reduce the successive oxidation	[113]

T = reaction temperature ($^\circ$C); p = reaction pressure; Q_{tot} = volumetric feed gas flow rate; S/C = steam-to-carbon ratio; ST = space time; IWI = incipient wetness impregnation; CP = co-precipitation; ALD = atomic layered deposition; DI = dry impregnation; GHSV = gas-hourly-space-velocity; TOS = time-on-stream; OSC = oxygen storage capacity; X_{CH4} = methane conversion (%); S_{CH3OH} = CH$_3$OH selectivity (%); S_{HCHO} = HCHO selectivity (%); Y^* = yield of CH$_3$OH and HCHO (%).

3.2.2. Oxidative Coupling of Methane (OCM) to Ethylene

OCM is extensively studied for C_2 olefin/alkane synthesis over heterogeneous catalysts at higher temperatures (above 700 $^\circ$C) and recently reported catalysts are Mn-Na$_2$WO$_4$/SiO$_2$ based catalysts [114], Li/MgO [115], Mg- and Al-doped SrTiO$_3$ and Sr$_2$TiO$_4$ [116], Ce-doped La$_2$O$_3$ [117], BaCl$_2$-TiO$_2$-SnO$_2$ [118] and La$_{0.6}$Sr$_{0.4}$Co$_{0.8}$Fe$_{0.2}$O$_{3-\delta}$ [119]. The most commonly studied OCM catalysts are Mn-Na$_2$WO$_4$/SiO$_2$ based catalysts, which are found to be very active and selective when prepared by a sol-gel method. Recently, Godini *et al.* reported promising results on OCM with 78% C_2-selectivity, 64% C_2H_4-selectivity and 24.2% C_2-yield at a low CH$_4$/O$_2$ ratio and N$_2$ dilution. A comparative analysis on the performance was made between different Mn-Na$_2$WO$_4$/SiO$_2$ based catalysts, *i.e.*, taking into account catalyst composition (different Mn (1.9%–2%) and Na$_2$WO$_4$ (4.5%–5%)), reaction conditions, reactor type, catalyst preparation methods, and the following results are summarized: methane conversion 7%–40%, C_2-selectivity 58%–80%, yield 5%–27% and the C_2H_4/C_2H_6 ratio of 0.45–2.1 [120]. The synergistic effect of Mn_2O_3 and Na_2WO_4 phases are very important in the OCM activity [114]. Wang *et al.* reported that the optimal conditions for OCM are 800 °C, flow rate of CH_4/O_2 of 2, and a space velocity (SV) 5000 mL.h^{-1}g^{-1}. In these conditions, 44% conversion with 20% C_2H_4 yield, and C_2 selectivity of 53.3% were achieved over the $BaCl_2$-TiO_2-SnO_2 catalyst [118]. Still many challenges are involved in improving the conversion and yield of ethylene in OCM. Developing Mn-Na_2WO_4/SiO_2 based catalysts can be one approach. However, there is a need for alternative catalysts for OCM to improve the conversion and yield such as binary, ternary and quaternary mixed oxides (e.g., alkali modified MCl-Mn-Na_2WO_4/SiO_2 (M = Na, Li, K, *etc.*) [114]. One approach to enhance the performance was to intensifying the OCM reaction by using novel reactors such as membrane assisted reactors (MR) and/or microreactor technology. OCM is performed in oxygen permeable MRs to shift the equilibrium towards the forward direction to improve the overall selectivity and yield. In Oshima *et al.* [100], plasma catalysis in OCM showed good results using a La_5-ZrO_2 catalyst and CO_2 as an oxidant. The electric field and synergistic phase of the La-cation and stabilized tetragonal ZrO_2 were the three main reasons for high activity at 200 °C [100].

3.2.3. Non-oxidative or Pyrolysis of Methane to Aromatics

Methane aromatization process to produce value added H_2 and aromatic compounds such as benzene offers many potential benefits. Non-oxidative conversion of methane (NOM) is a highly endothermic reaction that takes place at higher temperatures (E9). An excellent review on the studied catalysts on methane dehydroaromatization (MDA) or NOM was reported by Spivey and Hutchings [41] and Ma *et al.* [66]. Thermodynamically MDA is an extremely difficult reaction to proceed and involves a highly complex reaction mechanism over the studied catalysts [41]. In MDH, most of the studies dealt with zeolite based catalysts such as Mo/ZSM-5 and Mo/MCM-22 (Ma *et al.* 2013). Zeolite based catalysts exhibit a bifunctional mechanism in the MDA reaction. The additives in the methane feed such as olefins, alkanes and oxygenates had a significant effect on the methane conversion and benzene selectivity. A higher co-conversion of methane/methanol was achieved with high hydrocarbon selectivity (mainly aromatics) over a Zn_2-Mo_5/HZSM-5 catalyst prepared by co-impregnation [121]. Recently, the effect of Fe and Zn promoters on Mo/ZSM-5 were studied in MDA [122] and promoters were observed to change the surface acidity of the ZSM-5. The Fe promoted catalyst had created a higher number of Brönsted acid sites and Fe was bound more strongly to the zeolite structure than Zn, thus, Fe enhances the benzene selectivity [122]. Still more research needs to be done to improve the net aromatic yield, conversion and selectivity before making this process commercially viable.

Membrane assisted reactor (MR) technology is widely studied in gas separation and purification processes [123]. In MR, *in situ* reaction and selective separation takes place simultaneously, thus shifting the equilibrium towards the forward direction. In methane conversion technologies (MCT), oxidant O_2 is a reagent in processing CH_4, and H_2 is one of the products from non-oxidative methane processing, e.g., from OCM, MDH, and steam reforming. Most of the MCT processes are equilibrium limited and kinetically controlled reactions. One approach to enhance the MCT performance is by applying membrane-assisted reactors for example in MDH, wherein H_2 is selectively removed through the membrane and reaction proceeds more towards the forward direction. According to Cao *et al.* [124], MDH in an O_2 permeable MR resulted in a selective removal of product H_2 that leads in superior activity in methane conversions and also avoids the catalyst deactivation.

In methane conversion, predominantly the products are more reactive than methane itself. There is a need for effective product separation and kinetic control over the process. The most favorable option to utilize methane emissions will be syngas and/or H_2 production than chemicals synthesis due to the maturity of the technology and significant research and development that has already been done. Methane to chemicals holds a promising route, but many challenges exist in converting methane efficiently to chemicals with high yield and conversion. In the long run, methane transformation to fuels and chemicals via bio- or photochemical conversion will be potential technologies. Furthermore, biological and photochemical renewable processes are very promising due to enormous amounts of living biological sources and surplus amount of solar energy. Nevertheless, thermal catalytic conversion holds potential route for the short term implementation to convert methane to chemicals due to the existing industrial expertise and infrastructure. The future work will be more focused on developing new catalytic materials (e.g., for low temperature methane activation), process intensification and reactor design (e.g., microstructured reactors), integration of two or more processes such as tri-reforming (MDR+POM+MSR), O_2 and H_2 permeable membrane reactors for yield, conversion and selectivity enhancement.

4. Catalytic Utilization of Volatile Organic Compounds

As an alternative for VOC abatement, VOCs can be used as important sources of energy if collected and valorized effectively. Methanol and ethanol are interesting examples of the use of VOCs in energy conversion systems. However, scientific data are scarcer in the case of other molecules especially those encountered from industrial emanations. The previous chapter of the review was concentrating on the utilization of methane. Now we will proceed further to the other organic emissions.

4.1. From Volatile Organic Compounds to Syngas and Hydrogen

Reforming reaction of alkanes is an important energy conversion process for synthesis gas (mainly H_2 and CO) generation. Methane, as the main component of natural gas, is one of the most studied molecules for reforming reactions. However, several studies have been centered on processing and valorizing a number of higher alkanes such as ethane [125,126], propane [127–130], *n*-butane [131–133] which are present in natural gas, but also generated in engine burners. Among reforming reaction, the steam reforming (SR) of higher alkanes is probably the most common and cost-effective industrial process for H_2 generation. Rhodium based materials have been reported as highly active catalysts for steam reforming of alkane molecules at low temperatures [131]. Schädel *et al.* [125] studied the steam reforming of methane, ethane, propane, butane, separately, over rhodium-based monolithic honeycomb catalyst, and found that ethane, propane, and butane are converted at much lower temperatures than methane. For industrial application, Ni based materials are more studied due to their reasonable activity and significantly lower costs [133–135]. As in the case of methane discussed before, the coke deposition and high temperature of reforming when conventional $Ni/\gamma-Al_2O_4$ is used, lead to rapid catalytic deactivation and reduced H_2 yield. Support modification by catalytic promoters and/or atomic dispersion of Ni in mixed oxide structure has been reported to stabilize the catalytic activity of Ni [133,135]. For instance, CeO_2-doped Ni/Al_2O_3 displayed enhanced resistance toward carbon formation [135].

The generation of aromatic hydrocarbons as by-products of biomass [136] and plastic gasification to synthesis gas [137,138] have driven more interest into investigating in details their catalytic conversion to hydrogen-rich gas streams. To assess the efficiency of a classical reforming process, model molecules such as benzene [139–141], toluene [142,143] are usually studied and can provide insight for the VOCs utilization and processing. The existing studies are more oriented on steam reforming reaction using conventional catalysts such as Rh based catalyst [144–147], supported Ni catalyst [148–154], Ni-mixed oxide: perovskite [155–159] and hexaaluminate [148]. Rh based catalyst are reported to be highly active for toluene and benzene reforming within the temperature range of 700–850 °C [147,156]. The catalytic activity of Rh is reported to be dependent on the nature of the support used, where benzene/toluene is activated on Rh particles and water activated on the site associated with the support [144]. In concordance with a binfunctional mechanism, Colby *et al.* [146] have reported an enhanced catalytic activity of Rh/Al_2O_3 in the presence of CeO_2 in benzene steam reforming due to probable activation of H_2O on the surface of reduced ceria and by consequence increasing the rate of hydroxyl transfer between the oxide support and metal [146]. Nickel based catalysts were also investigated as a viable alternative to Rh based materials. Yoon *et al.* have evaluated the efficiency of two commercial nickel catalyst, *i.e.*, Katalco 46-3Q and

Katalco 46-6Q (from Jonson Matthey) for toluene steam reforming and found that Katalco 46-6Q (zirconia-promoted nickel-based catalyst) displays high efficiency for toluene reforming reaching 100% conversion at relatively low temperature of 600 °C. Although, nickel supported catalysts have been widely proposed as potential substitutes for Rh based catalysts in steam reforming reaction, nickel is more vulnerable to coking leading to severe deactivation [157]. The general scheme adopted to enhance Ni stability towards coking is based on the modification of Ni active site by support modification. The addition of alkali, alkali earth oxides and rare earth metal oxides to the Al_2O_3 support or the use of basic metal oxides as the support improves Ni resistance to coking [152]. In fact, the presence of alkaline additives on the catalyst surface is believed to accelerate the reaction of steam with adsorbed species decreasing the accumulation of carbon deposits on the catalysts. Nevertheless, due to the high temperature of the catalytic process, Ni oxides interacting weakly with the support tend to aggregate and sinter. The presence of large ensemble of Ni atoms stimulates coke formation leading to strong deactivation of the catalyst [158]. In this perspective perovskite materials with the general formula ABO_3, where A is an alkali, an alkali earth oxides or a rare earth metal oxide presenting a basic character, and B is a transition metal principally Ni, Fe and Co, inserted into defined structures, are interesting materials [150,155]. The role of the cation A in the catalytic activity of Ni based perovskite was reported in several studies. For instance, the catalytic activity of $La_{1-x}A_xNi_{0.8}Fe_{0.2}O_3$ (A = Sr, Ca and Mg) perovskite in toluene steam reforming was found to vary with the nature of the cation A. The catalytic activity was found to vary following the order $La_{0.8}Sr_{0.2}Ni_{0.8}Fe_{0.2}O_3$ > $La_{0.8}Ca_{0.2}Ni_{0.8}Fe_{0.2}O_3$ > $La_{0.8}Mg_{0.2}Ni_{0.8}Fe_{0.2}O_3$. Oemar et al. [156] reported also a lower carbon formation on $La_{0.8}Sr_{0.2}Ni_{0.8}Fe_{0.2}O_3$ which was correlating according to the authors to the highest amount of lattice oxygen in this latter.

In the literature, only very few studies are available on the reforming of carbonyl-containing compounds and most of these studies have been conducted with acetone. Thermodynamic prediction for hydrogen production via steam reforming of acetone investigated in a broad range of conditions can be found in the work of Vagia et al. [159] and that maximum hydrogen content is obtained at the temperature of 627 °C. At low temperatures, i.e., ⩽500 °C, CO_2 and CH_4 are the main products expected. However, as the temperature increases H_2 and CO fractions are expected to be the main products. Steam reforming of acetone is often conducted on classical active reforming catalysts such as $Ni/\gamma-Al_2O_4$ [160,161], modified Ni-catalyst [162,163], PtNi alloys [164], Pt/ZrO_2 and Pt/CeO_2 [164]. However, acetone is known to be a coke precursor causing rapid deactivation of the supported Ni catalyst as observed in many reforming materials. PtNi alloys supported on modified alumina by La, Ce and Mg have been reported to display improved catalytic activity as well as enhanced water adsorption and OH-surface mobility of the Al_2O_3

support. Moreover, PtNi catalysts presented a lower carbon deposition and a higher thermal stability with respect to the monometallic Ni reference catalyst [163].

Generation of H_2 from an alcohol substrate, such as methanol and ethanol, is a well-documented subject and several reviews are available in the literature. Special interest is accorded to the 1st generation bioethanol and the 2nd generation bioethanol produced in a fermentation process [165]. Even if generation of H_2 from bioethanol is behind the scope of this review, some investigation related to the impact of by-products of fermentation on the H_2 generation can provide useful information. Devianto *et al.* [166] investigated the effect of higher alcohols as impurities on thedirect bioethanol internal reforming for molten carbonate fuel cells (MCFC) [166]. The activity test revealed that the presence of 1-propanol affected negatively the catalytic activity of Ni/MgO catalysts. 1-propanol was found to be partially reformed and negatively influenced the selectivity. In the contrary, methanol is reported to enhance the yield of H_2. The empoisoning effect of higher alcohols (1-propanol, 2-propanol) in the H_2 generation process was also reported by Rass-Hansen *et al.* [167] on the 10 wt % $Ni/MgAl_2O_4$ catalyst [167]. The presence of higher alcohols was found to favor carbon deposition on the Ni based catalyst. In the same manner steam reforming of butanol was found to show a much greater tendency towards coke deposition than ethanol [168]. However, using Gibbs free-energy-minimisation method Nahar *et al.* [169] predicted that carbon formation can be inhibited when steam reforming of butanol is operated between 600 and 800 °C at 1 bar and at water-to-butanol molar feed ratios of 9–12 [169].

From the existing literature, the current technologies for hydrogen production are absolutely transposable to VOCs valorization into to syngas and hydrogen. Steam reforming reaction is the most common and cost-effective industrial process to be envisaged for the H_2 generation from VOCs. The actual challenges facing VOC reforming reaction for H_2 production are in fact common to all reforming processes, which are mainly related to the high operating temperature and catalyst deactivation due to carbon deposition. The nature and origin of VOCs for reforming may arise additional difficulties. The low concentration of VOCs in industrial discharge may require an additional step to enhance the concentration of VOCs before the reforming reaction. Furthermore, the heterogeneous composition of VOCs encountered in an industrial site will complicate the optimization process of operating conditions.

4.2. From Volatile Organic Compounds to Chemicals

An interesting approach and not so intensively studied one, is the utilization of VOCs as reactants in the production of valuable chemicals. There exist actually only a few examples of VOC emission utilization, of which one is related to the utilization of contaminated methanol from pulp manufacturing industry. This section of the review concentrates on these few examples, but also presents some of the current technologies and production processes where the organic compounds are used as raw materials. In certain cases, especially in the production of high-value products, there could be room for the VOC utilization.

Organic reactions in general can be categorized based on the type of functional groups involved in the reaction as a reactant and the functional group that is formed as a result of this reaction. The basic organic chemistry reaction types are addition, elimination, substitution, pericyclic, rearrangement, photochemical and redox reactions. These may all have a potential role in the VOC utilization applications. To simplify the approach, we have grouped here the VOC compounds based on their functional groups, which classification can be found very often also in emission inventories. For example Piccot *et al.* [170] have made a Global inventory of VOC compounds in 1992 where they classified VOCs into paraffins, olefins, aromatic compounds, BTX compounds, formaldehyde, other aldehyde, other aromatics and marginally reactive organic compounds. Later, Wei *et al.* [171] have classified the VOC emissions in China as alkanes, aromatics, alkenes, alkynes and carbonyls. We have decided to group the different types of VOC in this case as *alkanes, alkenes, alkynes, aromatics, oxygenated and substituted* compounds meaning the compounds having any other atom than oxygen to replace one or more hydrogen atoms in the organic structure. Table 6 summarizes the typical sources of different types of VOCs.

Alkanes are introduced into the environment due to their extensive use as fuels and chemicals. The production, storage and transportation of oil and natural gas are one of the major global sources of hydrocarbons [172]. Due to the presence of single covalent C–C bonds, alkanes are less reactive compared to the other volatile organic compounds. However, several alkanes are natural metabolites and part of the lipophilic fraction in microorganisms, plants and animals [173,174].

Alkenes are hydrocarbons having one double bond between at least one of the carbon atoms. They belong to the group of unsaturated hydrocarbons and they are also called as olefins since they form oil-type liquids when reacting with gaseous chlorine. Alkenes are the raw materials for several different types of plastics, such as polyethylene, polypropylene and PVC. Alkenes are more reactive than alkanes and they are very often used in addition reactions. Alkene compounds exist also in the nature and they are responsible for the characteristic odors of trees and plants (limonene, mycrene) [175,176]. Some of the alkenes (ethane and propene) are efficient atmospheric precursors for the formation of formaldehyde that is typically the compound found at the highest concentrations in the boundary layer atmosphere. Interestingly, the direct emissions of formaldehyde to the atmosphere are small, and therefore formaldehyde can be used as a good indicator for atmospheric VOC oxidation. Alkene emissions followed by formaldehyde formation are typically emitted from oil refining industries [177,178].

An *alkyne* is a hydrocarbon that contains a carbon-carbon triple bond. Alkynes are less reactive and also less stable than alkenes. Acetylene (ethyne), C_2H_2, is the simplest alkyne. Combustion of acetylene produces a high-temperature flame, because of its triple bond and high enthalpy of formation [179–181]. Acetylene is one of the main alkyne emission compounds. Typically, acetylene emissions are formed from motor vehicle exhaust (mainly gasoline), industrial processes, as well as wood, biofuel and coal combustion [182,183]. Acetylene is an important source material for the production of several chemicals and carbon materials [184].

The simplest and most industrially important *aromatic compound* is benzene (C_6H_6). Benzene is the precursor for many chemicals, which could further be used as end products or intermediates. Benzene is used to produce for example styrene, phenol, cyclohexane, and maleic anhydride. Toluene (C_7H_8) and xylene (C_8H_{10}) are substituted benzenes. Most of the toluene extracted for the chemical use is actually converted to benzene. Toluene is also used to produce important synthetic intermediates such as nitrotoluene and dinitrotoluenes. Xylenes are an aromatic mixture composed of three isomers, *o*-, *m*-, and *p*-xylene. Xylenes are an important chemical raw material and for example *p*-xylene is an important isomer for producing terephthalic acid to manufacture polyesters [185,186].

Aromatic organic emissions can also be classified as the BTEX (Benzene, toluene, ethyl benzene, *p*-xylene, *m*-xylene, *o*-xylene) and BTX (Benzene, toluene, *p*-xylene, *m*-xylene, *o*-xylene) compounds [185,187]. Furthermore, another significant aromatic emission compound not mentioned before is styrene. Some of the aromatic organic compounds such as the Polycyclic Aromatic Hydrocarbon (PAH) compounds (e.g., Naphthalene, anthracene, pheanthrene) and phenols are classified as the semivolatile organic compounds [188]. BTEX compound emissions cause many environmental problems for groundwater, soil and air [187]. BTEX compounds take part in reactions promoting photochemical smog [189], and they have also high reactivity and ozone production potential [186]. BTEX compounds can cause variety of health problems such as respiratory irritation, and central nervous system damage. Benzene is also a carcinogenic compound [190].

Aromatic VOCs are formed in many industrial processes and are also present in vehicle exhausts. In some industrial processes a significant amount of aromatic compounds emissions are formed. For example in iron and steel industry (included coke making, sintering, hot forming, and cold forming) major emissions are aromatic compounds with 46%–64% of total emissions [191]. Aromatic VOCs are also an origin for the problems caused by urban city air. In Shanghai, the share of aromatic compounds from total VOC concentration is 30% [192] and in Los Angeles 20%–25% [193].

Traditionally aromatic VOCs are treated by different methods, such as catalytic oxidation, ozone-catalytic oxidation, and thermo-photocatalytic degradation. However, the reaction products of these treatment technologies are also an environmental concern (CO, CO_2) [187,194–199].

Oxygenated hydrocarbons are the compounds of hydrogen and carbon having at least one oxygen in the structure including alcohols (–OH), ethers (–O–), ketones (–CO–), aldehydes (–CO–H), carboxylic acids (–COOH), esters (–COO–), acid anhydrides (–CO–O–CO–) and amides (–C(O)–NR$_2$). The most commonly used oxygenates are alcohols and ethers, since they are used as fuel additives. Acetone and phenol are raw materials for many different products. Oxygenated VOCs in general are emitted from the traffic, solvent usage, industry and some natural sources [200–202]. It can be forecasted that the concentrations of oxygenated VOCs in urban atmosphere will increase due to the use of oxygenates to replace the gasoline.

In this paper, when the *substituted compounds* are discussed, we mean the compounds, where one or several hydrogens of an organic compound have been replaced by another element. For example, total reduced sulphur (TRS) compounds is a group of compounds that typically includes hydrogen sulphide, mercaptans, dimethyl sulphide, dimethyl disulphide and other sulphur compounds that are released to the environment by the steel industry, pulp and paper mills, refineries and sewage treatment facilities [203]. Alkyl halides are a class of compounds where a halogen atom or atoms are bound to an sp^3 orbital of an alkyl group with a general formula of RX where R is an alkyl or substituted alkyl group and X is a halogen (F, Cl, Br, I) [204]. Alkyl halides are used as flame retardants, fire extinguishers, refrigerants, propellants, solvents, and pharmaceuticals [205]. Dichloromethane (DCM, or methylene chloride) is an organic compound with the formula CH_2Cl_2. This colorless, volatile liquid with a moderately sweet aroma is widely used as a solvent. Alkyl halides are emissions sources are different industrial processes, e.g., paint removals, plastic industries, and the pharmaceutical sector [206]. The correct catalyst selection for utilization of substituted compounds is especially important, since the compounds themselces are potential deactivating agents for certain catalytic materials.

Table 6. Sources of the VOC emissions.

Sources of the Emission	Major Aromatic Compounds Emissions	Ref.
Vehicles exhaust (gasoline, diesel)	BTEX, trimethylbenzenes, ethylene, propene, 1-butene, ethane, acetylene, oxygenates	[182,183,200–202]
Transportation	1,3-butadiene, benzene, toluene, xylene, formaldehyde, acraldehyde	[207]
Fuel evaporation (gasoline, diesel)	Benzene, toluene, iso-pentane, pentenes, n-heptane	[182]
Biomass burning	BTEX, Benzene (major aromatic emission), acetylene, ethylene, propene, ethane, methylchloride, methanol, formaldehyde	[182,183,207]
Coal burning	BTEX, naphthalene, acetylene, ethylene, propylene, propane, ethane	[182,183]
Petrochemical industry	Styrene, benzene, hexane, methylcyclohexane, trichloroethylene, TRS compounds	[172,182,203,206]
Oil-refinery	Benzene, ethylene, hexane, cyclopentane, cyclohexane, methylcyclohexane, TRS compounds	[172,177,178,182,203]
Painting (Coating for building)	BTEX, styrene, n-butane	[182]
Electronics manufacturing (printed circuie board)	Toluene, 2-ethyl-1,3-dimethylbenzene, 1,2,4,5-tetramethylbenzene, ethanol, acetic acid, iso-propyl alcohol	[186]
Vehicle manufacturing	Toluene, ethylbenzene, p, o-xylene, trimethylbenzenes, ethyl acetate, 2-butanone, acetic acid	[186]
Flexographic Printing	o-xylene, acetic acid, ethyl acetate	[186]
Metal and plastic surface spraying	Ethylbenzene, p, o-xylene, ethyl acetate, 2-butanone, butyl acetate, ethanol	[186]
Furniture manufacturing	Toluene, ethylbenzene, p, o-xylene, trimethylbenzenes, trimethylbenzenes, acetic acid	[186]
Plastic waste recycling plants	BTEX, styrene, 1-butene, 2-hexene, pinenes, hexane, octane, 3-methylnonane	[194]
Iron and steel industry (cokemaking, sintering, hot forming, cold forming)	BTEX, trimethylbenzenes, isopentane, n-pentane, n-butane, methylhexanes, n-heptane, butenes, trichloroethylene, TRS compunds	[191,203]
Industrial solvent use (e.g., solvent production, paint and adhesive use)	Toluene, xylenes, n-hexane, oxygenates, substituted compounds	[200–202,205,207]
Pulp and paper production	oxygenates (mainly methanol), TRS compounds	204
Pharmaceutical industry	substituted compounds (alkyl halides)	207

TRS is Total reduced Sulfur, mixture of organic (e.g., methyl mercaptan, dimethyl sulphide, dimethyl disulphide) and inorganic (e.g., hydrogen sulphide) sulphur compounds.

4.2.1. Potential Existing Technologies and New Considerations for VOC utilization in Chemicals' Production

Burning the alkanes or using them as fuels stays the most common methods of valorization [6,208] due to their higher flammability. Another possibility for the utilization of the hydrocarbons that is under development is the metabolization of them following the aerobic or the anaerobic conversion to valuable chemicals [209–212]. Microorganisms have established effective strategies involving specialized enzyme systems and metabolic pathways to access n-alkanes as a carbon and energy source. The reactivity of the n-alkanes depends on the chain length and some of the short chain alkanes (<C9) may be toxic to micro-organisms. The aerobic alkane degraders use O_2 as a reactant for the activation of the alkane molecule. Oxidation of methane renders methanol that is subsequently transformed to formaldehyde and then to formic acid. In the case of longer chain alkanes, the oxidation starts usually in the terminal methyl group to give the primary alcohol, which is further oxidized to the corresponding aldehyde, and finally converted into fatty acid [210,213]. The alkane-activating enzymes are the mono-oxygenases, which are widely used as biocatalysts in several processes [214–216]. Temperature is an important factor in these processes, since it affects the solubility of the alkanes and the stability of the enzymes. For example stability of mono-oxygenase enzyme CYP P450 in Aspergillus terreus MTCC6324 ranges between 25 and 40 °C. The presence of molecular oxygen is affecting the acitivity of the enzymes as well as pH that needs to be between 7 and 8 to support the required reaction [210].

If the conversion of alkanes into more useful and reactive products, such as alkenes, alcohols, aldehydes, and carboxylic acids can occur under milder and better controlled catalytic conditions, then this can offer large economic benefits [217]. The oxidative functionalization following the catalytic route by functionalizing the C–H bond had been studied widely by many researchers [218–220]. Sivaramakrishna et al. [217] have reviewed the progress in the oxidation of n-alkanes by heterogeneous catalysis, from the partial oxidation to the oxidation, ammoxidation and the oxidative dehydrogenation, that are the three main processes in the heterogeneous catalysis to functionalize hydrocarbons. A mixed metal oxide catalyst containing Mo, V, and Nb was used in the oxidative dehydrogenation of ethane to ethylene. Supported iron (III) and manganese (III) catalysts in the presence of an oxidant were tested in the hydroxylation of alkanes to alcohols. Vanadium oxide supported metal oxides (e.g., MgO, TiO_2, Al_2O_3, ZrO_2, SiO_2) are an important group of catalysts for a variety of reactions, including selective oxidation of hydrocarbons [217].

Maybe the most important reactions of the alkenes are the addition reactions, especially electrophilic addition. For example ethene can be hydrated to ethanol with using phosphoric acid as a catalyst. However, more interesting direction in practice is the producing ethene from ethanol over Fe-ZSM-5 or carbon-based catalyst, since ethene is considered as more valuable than ethanol [221,222]. Ethene can be converted also to methyl propanoate via methoxycarbonylation over a palladium catalyst [223]. Methyl propionate is used as a solvent for cellulose nitrate and as a raw material in the production of paints and varnishes [224]. Other applications for ethene can be found in large amounts, since it is widely used in polymerization, oxidation, halogenation, alkylation, *etc*. About 90% of ethene is used for production of ethylene oxide, ethylene dichloride, ethylbezene and polyethylene in Europe and US [225]. Since the production of ethene is growingly increasing and it is produced in several countries, it is unlikely that the ethene from the waste gas would be economically interesting to collect and to be used in the near future. We have used ethene here as an example of alkene reactions due to its wide applications, even if the first alkenes that can be considered as VOCs are the pentenes, which have a rather low boiling point of 30 °C [226]. Pentene itself is not a very interesting chemical, since it does not have very many utilization areas. 1-hexene is used in polymer production as a co-monomer. Heptane, in turn, is used as an additive in lubricants, but also as a catalyst [227]. Alkenes, in general, are mostly emitted from oil refining, where the organic emissions in general are high, and being the potential industrial sector for the VOC utilization applications.

One possibly interesting reaction of the alkenes is the epoxidation. The epoxides are the main raw materials in the production of several chemicals, and therefore significant amount of research have been published related to the development of catalytic systems for alkene epoxidation. For example, Davarpanah and Kiasat [228] have proposed a novel solid polyperoxoacid catalyst for alkene epoxidation. The polyperoxoacid, that showed good activity, was produced by anchoring and oxidation of polyacrylic acid onto the surface of SBA-Im-Allyl and it showed to be stable and recyclable with the help of H_2O_2 solution. The yields of the epoxides were varying from 52% to 98% depending on the alkene used as a raw material [228]. Ionic liquids have also been used as co-solvents in epoxidation of alkenes. The aim of using aqueous/ionic liquid systems is in enhancing the recyclability of the catalysts.

Crosthwaite *et al.* [229] studied alkene epoxidation by using 2-alkyl-3,4-dihydroisoquinolinium salts as catalysts and Oxone™ as an ionic oxidant. Good results of epoxidation were found with the system they used, however, using the bi-phasic system to recycle the catalyst failed [229]. Pescarmona and Jacobs [230] applied high-throughput experimentation to develop transition metal free catalysts for epoxidation of alkenes. They tested epoxidation of 1-hexene, 1-octene, 1-decene, *R*-(+)-limonene, cyclohexene and cis-cyclooctene over gallium oxide, alumino- and gallosilicates (Al-MCM-41, Ga-MCM-41) and USY zeolites. Gallium catalysts showed high activities and the epoxide yields were improved when selecting the correct solvent for the system [230]. An interesting possibility for epoxide production is coming from the possibility to use microreactors in the production. Salmi *et al.* [231] have studied microreactors in the production of ethylene oxide over silver catalysts. They found out that ethylene oxide can be produced with high selectivity in microreactors [231,232]. The advantage of using microreactors in ethylene oxide production is the safety of the small-scale process that is easier to control and the reaction could be potentially run within the explosive regime and conditions where surplus oxygen is present. The usage of microreactors could be considered also in the case of VOC utilization. They allow flexible production from small to larger scale by adding several microreactors together. In addition, multiple products could be produced from the VOCs parallel and simultaneously. Another important industrial application is the production of aldehydes from alkenes. For example the Wacker-type of reaction is important in synthetic chemistry and pharmaceutical industry [233,234]. More recently, several researchers have reported the development of new types of catalysts for aldehyde production with better selectivities and without the C=C bond cleavage. These materials include palladium, copper, ruthenium, ruthenium(IV)-phorphyrin catalysts. With these catalysts the aldehyde yields range between 60%–99%, and especially ruthenium(IV)-phorphyrin seems to be very potential homogeneous catalyst for the aldehyde production reaction [235,236]. To conclude the findings related to the alkenes, it seems, that one potential route for VOC utilization could be via transforming the VOCs to different types of alkenes that could be further used in the production of variety of chemicals needed in chemical and pharmaceutical industry.

In the 1950s, acetylene was the one of the most important raw material for the production of the synthetic organic chemicals. Thermodynamically unstable acetylene can be used in many different kind of reaction such as carbonylation, dimerization, vinylation, polymerization, ethynylation, and oligomerization. Today, in many cases ethylene and propylene have replaced acetylene as a raw material, because price of the acetylene is high. Still, acetylene-based routes to produce chemicals have been operated on an industrial scale during the past century. For that reason there exists a lot of knowhow related to acetylene-based processes and therefore utilization of acetylene emissions for chemical production is possible. Typically, acetylene is manufactured from coal. Crude acetylene includes small quantities of impurities, such as phosphine, arsine, ammonia, and hydrogen sulfide. For chemicals manufacturing, the purity of acetylene have to be 99 vol % because impurities may poison the catalysts [180,184]. The demand for the high purity creates challenges for the utilization of acetylene emissions in manufacturing of chemicals. However, development of the new selective catalytic materials can be give opportunities for utilization of acetylene containing emissions also in this case.

In the commercial scale acetylene is used in the chemical industry to produce, e.g., acrylic acid, vinyl chloride, ethylene, acetaldehyde, acetic acid, and 1,4-butanediol [180,184,236,237]. Here we have reviewing shortly some chemical production possibilities starting from acetylene that could be interesting in the future. Worldwide annual production of acrylic acid that was traditionally produced from acetylene, is over 1 million metric tons. Acrylic acid has widely used in the manufacturing of acrylate, monomers for polymer synthesis, and plastics. Today, almost all acrylic acid is produced from propylene, because it is more economical. However, the raw material to produce acrylic acid is dependent on the country. Carbonylation of acetylene with a $Ni(CO)_4$ catalyst was firstly found by Reppe in 1939. Acrylic acid can be produced by hydrocarboxylation of acetylene with carbon monoxide and water over a catalyst [180,184,238,239]. Recently, many researchers have suggested new catalytic material to hydrocarbonylation of acetylene [238]. Ni_2O_3 and cupric bromide are effectively catalysing hydrocarboxylation of acetylene to acrylic acid. Yield of acrylic acid was found to be 72.1% and selectivity 85.2% with a Ni_2O_3 catalyst at 235 °C [180]. Good catalytic conversion of acetylene (85%) and selectivity of acrylic acid (99%) under mild conditions ($T = 40$–50 °C) is achieved with a catalyst system consisting of a homogeneous catalyst $Pd(OAc)_2$, with diphenyl-2-pyridylphospine as the phosphine ligands and trifluoromethane sulfonic acid as the acidic promoter [239]. Hydrocarboxylation of other alkynes is an important reaction for production of α,β-unsaturated carboxylic acids [240,241]. The most widely used catalyst in hydrocarboxylation of alkynes are based on Ni, Co, and Pd [242]. New interesting research area is the use of CO_2 as a renewable and environmentally friendly source material of carbon in the hydrocarboxylation process. The promising results for CO_2 process have been obtained with nickel and copper catalysts [241,243,244].

Worldwide annual production of vinyl chloride is over 30 million metric tons. Most of the vinyl chloride is produced from ethylene although the conversion of acetylene to vinyl chloride is inexpensive process when considering the capital and operating costs [180]. For that reason the utilization of acetylene emissions to produce vinyl chloride could be economically feasible. Vinyl chloride can be manufactured by hydrochlorination of acetylene on commercial scale. Vinyl chloride monomer (VCM) is the main raw material to produce polyvinylchloride (PVC). The traditional processes use mercuric chloride supported on activated carbon, which causes lot of environmental pollution problems. For this reason, new catalysts are developed [237,238,245]. N-doped carbon is used as a catalyst in converting acetylene to vinyl chloride with acetylene conversion 77% and vinyl chloride selectivity >98% [246]. With Au-La/SAC (SAC, spherical activated carbon) catalysts, the conversion of acetylene 98% and selectivity to VCM 99.8% can be obtained [245].

Hydrogenation of acetylene is often used to remove small amount of acetylene in the ethylene feedstock for polyethylene production. Hydrogenation can be used in connection with low concentrations (0.1–1 vol %) of acetylene to ethylene [247,248]. Several types of catalysts are suitable for the hydrogenation of acetylene [249], but the most widely studied hydrogenation catalyst is based on palladium and gold [248–250].

High conversion of acetylene (>99%) and selectivity to ethylene (>70%–80%) are achieved over Cu/Al_2O_3 catalysts modified with different Cu:Pd ratios at the temperature range of 90–100 °C [251]. With gold-based catalyst close to 100% selectivity for hydrogenation of acetylene to ethylene at temperature up to 300 °C is achievable [248]. Good results are obtained for example with a Au/CeO_2 catalyst [249] and a Au/C-TiO_2 catalyst with different Au loadings [248]. BTEX compounds are very important petrochemical intermediates and solvents for the synthesis of organic compounds [185]. A mixture of BTEX emissions compounds or individual compound could be utilized to produce many important intermediates and products. One of the most important processes is the production of phenol from benzene. Phenol is a very important intermediate for synthesis of petrochemicals, plastics, agrochemicals, and pharmaceuticals. Phenols are used in the production of bisphenol A, thermosetting phenolic resins, alkyl phenols, aniline, acetophenone, caprolactam, and other useful chemicals [252,253]. Currently, phenol is produced nearly 8 million tons/year worldwide. More than 90% of the world's phenol production technology is based on the three-step Hock process, which is also called the cumene process. Alternative new methods to the production of phenol are the toluene oxidation and direct hydroxylation of benzene [252–255].

The industrial Hock process is based on a three-step cumene synthesis and oxidation processes. First step of the Hock process is alkylation of benzene with propylene over an acid catalyst to form cumene. Alkylation reaction is catalyzed by, e.g., phosphoric acid, aluminium chloride, boron trifluoride, hydrogen fluoride or MCM22 zeolite. Today, almost all cumene is produced by using zeolite-based processes. The cumene can alkylate with propylene to form diisopropyl benzene (DIPB) by side reaction. The reactions of alkylation are highly exothermic [254–256]:

Main reaction:

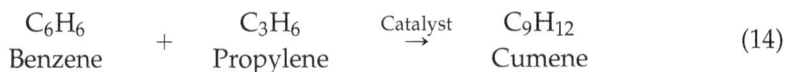

$$\underset{\text{Benzene}}{C_6H_6} + \underset{\text{Propylene}}{C_3H_6} \overset{\text{Catalyst}}{\rightarrow} \underset{\text{Cumene}}{C_9H_{12}} \tag{14}$$

Side reaction:

$$\underset{\text{Cumene}}{C_9H_{12}} + \underset{\text{Propylene}}{C_3H_6} \rightarrow \underset{\text{DIPB}}{C_{12}H_{18}} \tag{15}$$

The second step of the Hock process is oxidation of cumene to cumenehydroperoxide (CHP) with air proceeding via a free-radical mechanism that is essentially auto-catalyzed by CHP. (Busca 2008, Schmidt 2005)

Reaction:

$$C_9H_{12} \quad + \quad O_2 \quad \rightarrow \quad C_9H_{12}O_2 \tag{16}$$
$$\text{Cumene} \qquad \quad \text{Air} \qquad \qquad \text{CHP}$$

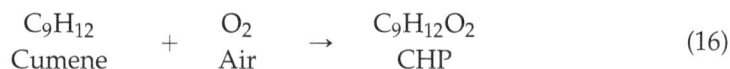

Last step of the phenol production is the decomposition of CHP to phenol and acetone. Decomposition reaction is catalyzed by strong mineral acid such as sulphuric acid at 60–100 °C [254,255,257].

Decomposition:

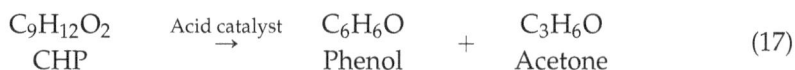

$$C_9H_{12}O_2 \xrightarrow{\text{Acid catalyst}} C_6H_6O \quad + \quad C_3H_6O \tag{17}$$
$$\text{CHP} \qquad\qquad\qquad \text{Phenol} \qquad\quad \text{Acetone}$$

The advantages of the Hock process is the inexpensive raw material (benzene and propylene), and high value useful products (phenol and acetone) [257]. Typical phenol production plant is producing phenol over 500,000 t/year. For example INEOS Phenol is producing almost 2 million tons phenol per year and 1.2 million tons acetone per year [258]. Utilization of the benzene emissions in economic way to produce phenol or other valuable products depends, e.g., price of the products and capital and operating cost of the small-scale plant.

Hydroxylation of benzene is an alternative method for producing phenol (Joseph 2009). The direct hydroxylation of benzene to phenol has potential advantages compared to the cumene process. Hydroxylation process is economic one-step method and no explosive intermediate (cumene hydroperoxide) is formed in the process. Challenge of the process is the thermodynamic limitation, because phenol is oxidized easier than benzene [253,259–261]. Commercially available AlphOx™ process by Solutia Inc. uses oxidation of benzene to phenol with N_2O oxidant. Selectivity to phenol is high 95%–100%, but it is dependent on the used catalyst [255,261]. Benzene's hydroxylation to phenol could be achieved using, e.g., zeolite catalysts containing rhodium, palladium, platinum, or iridium [185], as well as by using Cu-Ca-phosphate catalyst [262]. Multi-step Hock process could be lead to higher capital investment cost than direct benzene's hydroxylation. The simpler hydroxylation process could be more useful for the utilization method. Also, the efficiency of the Hock process strongly depends on price of the acetone. In additional, industrially valuable products can be produced by direct hydroxylation of other aromatic compounds in the same conditions. Therefore different kinds of aromatic emissions mixture could be utilized with direct hydroxylation [257,263].

In addition to N_2O, in hydroxylation of benzene to phenol, also three other oxidants can be used: O_2, H_2O_2, and mixture of H_2 and O_2 [253,259–261,264]. The same oxidants could be used also in hydroxylation of other aromatic compounds. Direct hydroxylation of toluene, ethylbenzene, xylenes and anisole to produce hydroxyaromatic compounds are a significantly less studied topic than benzene hydroxylation. Hydroxylation of toluene is producing o, m, and p-cresol that are valuable compounds to produce for example antioxidants, wood preservers, disinfectants in soap, ore flotation, fiber treatment, and precursors to synthetic tanning agents [264,265]. Hydroxylation of anisole is used to produce methoxy phenols such as p-hydroxyanisole [264]. Hydroxylation of toluene and anisole could be carried out over zeolite-based catalyst [266,267], vanadium [264], and iron [268–270]. Hydroxylation of ethylbenzene and xylenes has received limited attention in literature. Generally, ethylbenzene hydroxylation activity is low and often higher temperature and H_2O_2 concentration is needed to increase the reaction rate. Reaction mechanism is also significantly different for ethylbenzene hydroxylation than, e.g., toluene. Major products from ethylbenzene are acetophenone, benzaldehyde and 1-phenylethanol. Also small amounts of the styrene, cumene and phenol are formed [265]. Hydroxylation of m-xylene over zeolites produces 2,4- and 2,6-xylenols [271]. Development of new catalyst for hydroxylation of ethylbenzene and xylenes could be achieved higher selectivity and conversion. For example multi-walled carbon nanotube catalysts have been developed for hydroxylation [272]. Studies on the hydroxylation of a mixture of aromatic compounds could give new knowledge for utilization of aromatic emission compounds as well. At present, aromatic compounds' hydroxylation research is focused on the studies with separate compounds. Table 7 shows hydroxylation process catalysts and operation conditions for selected aromatic compounds.

The bromination and chlorination of BTEX compounds requires a Lewis acid catalyst such as ferric bromide ($FeBr_3$), ferric chloride ($FeCl_3$) or chlorides of aluminum or zinc [179,273]. The major disadvantages of the toxic and harmful Lewis acid catalysts are the catalyst disposal difficulties and corrosion problems. Also Lewis catalyst needs to be neutralized after the reaction. Bromination and chlorination of various aromatic compounds over zeolite catalysts are used in many studies [273–278]. Use of the zeolites in the processes will help in avoiding corrosion and disposal problems [273]. Bromination is catalyzed also with chlorides of aluminum and zinc [274]. Nitration of aromatic compounds could be used to produce organic intermediates. Nitration with nitric acid (HNO_3) requires sulphuric acid as a catalyst [179,279–281]. The Figure 3 summarizes possible intermediates and products that are achievable by bromination, chlorination, nitration and hydroxylation of BTEX compounds.

Table 7. Used catalyst and operation conditions hydroxylation of aromatic compounds.

Catalyst	Source compound	Oxidant	Conditions (°C)	Conversion	Product selectivity	Ref.
Zeolites	Benzene	N_2O	400–450	N_2O: 100%	Phenol: 80%–100%	[255,259]
Palladium membrane	Benzene	O_2	<250	Benzene: 2%–16%	Phenol: 80%–97%	[259]
VO(acac)2	Benzene	H_2O_2	65	Benzene: 11%	Phenol: 100%	[264]
VO(acac)2	Toluene	H_2O_2	65	Toluene: 5%	o-cresol 37% m,p-cresol: 56%	[264]
VO(acac)2	Anisole	H_2O_2	65	Anisole:2%	p-hydroxyanisole: 26%	[264]
VOPc [a]	Benzene	H_2O_2	65	Benzene: 22.4%	Phenol: 100%	[264]
VOPc [a]	Toluene	H_2O_2	65	Toluene: 18.74%	o-cresol: 38% m,p-cresol: 57%	[264]
VOPc [a]	Anisole	H_2O_2	65	Anisole: 6%	p-hydroxyanisole: 47%	[264]
Multi-walled carbon nanotubes	Benzene	-	50–70	Benzene: 2%–6%	Phenol: ~98%	[273]
Multi-walled carbon nanotubes	Toluene	-	50–70	-	o,m,p-cresol: <80%	[274]
H-[Al]ZSM-5 zeolites [b]	Benzene	N_2O	350	Benzene: 22%	Phenol: ~98%	[267]
H-[Al]ZSM-5 zeolites [b]	Toluene	N_2O	350	Toluene: 24%	-	[267]
H-[Al]ZSM-5 zeolites [b]	Anisole	N_2O	350	Anisole:53%	-	[267]

[a] Vanadyl tetraphenoxyphthalocyanine (VOPc); [b] Zeocat PZ-2/54 Uetikon.

Oxygenated VOCs form a great potential for chemicals production. For example methanol is very important raw material for variety of production. It is commercially used in the production of formaldehyde, esters of different acids that are further processed to the final products. Approximately 30% of methanol is used in formaldehyde production [282,283]. Ethanol, methanol and butanol can also be used as fuels to replace the fossil fuels currently in use [283–287]. There exist also so-called direct-methanol fuel cells that have the potential to be miniaturized and applied in consumer electronics [288–290]. Alcohols are directly used as solvents in production of medicines, perfumes and essential oils [282,285]. Methanol is emitted from pulping industry in rather large amounts [290], and therefore it is a suitable compound for VOC utilization purposes. There exist studies where methanol is used in production of formaldehyde over vanadium-based catalysts [291], but other potential products could be considered as well. An interesting approach to use alcohol is its possible usage in waste water treatment plants as a carbon source for denitrifying bacteria. [292] This could be potential and rather direct VOC emission utilization possibility.

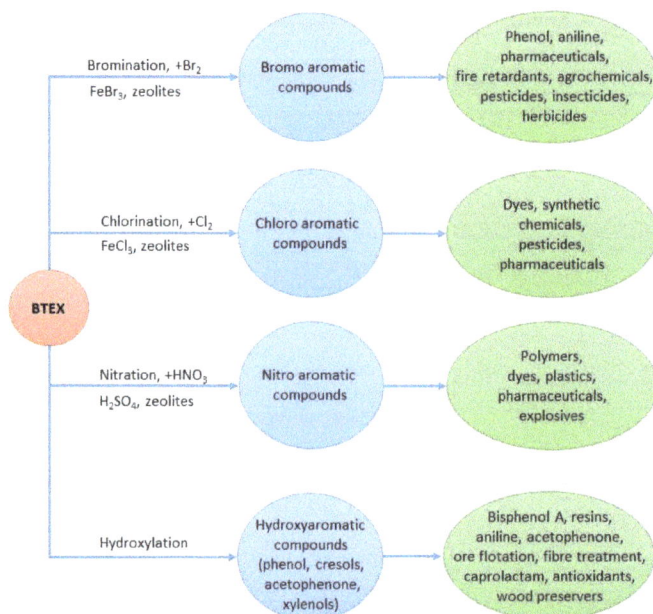

Figure 3. Products and intermediates from BTEX (benzene, toluene, ethylbenzene and xylenes) compounds [179,185,252,253,264,265,274–281].

Ethers are relatively unreactive. They have the uses as solvents for fats, waxes, resins and dyes, for example [293]. Certain ethers find applications as insecticides and they find also important applications in pharmacology. Ethyl ether is used as anesthetic, and codeine is the methyl ether of morphine [293,294]. Codeine can be produced also from diols via chemoenzymatic synthesis [295]. Dimethyl ether is used as a propellant and methyl t-butyl ether (MTBE) is used as octane number increasing additive in gasoline [296]. Dimethyl ether (DME) is considered as a potential alternative fuel for diesel engines, but it can also be used as a source for other hydrocarbons [297–299]. Yu *et al.* have studied SnO_2/MgO and SnO_2/CaO catalysts in the production of methyl formate, dimethoxyethane and ethanol from DME through catalytic oxidation. The DME conversion of 21.8% was obtained over SnO_2/CaO at 300 °C with the selectivity of 19.1% to methyl formate and 59% to dimethoxyethane [298]. DME can be also converted to propylene over MCM-68 zeolite [300].

Also ketones find applications in pharmacology, since they have important physiological properties. They are present in sugars and different medicines, including steroid hormones and cortisone [301]. Furthermore, hey are important chemical intermediates. The most important ketone is acetone, and it is used as a solvent in several applications. Only few ketones are manufactured in a large scale. The synthesis of them can be done via several different ways [301,302]. Ketones are used in for example asymmetric aldol reactions in production of chiral organic structures. Suzuki *et al.* have developed Ca complex structures that can be used as a catalyst in direct asymmetric aldol reaction of acetophenone and aliphatic aldehydes. These catalysts can be used to produce aldol products even with higher than 70% of yield and ee almost 90% [303]. Vetere *et al.* [304] studied hydrogenation of ketones over Pt-based heterogeneous catalysts. The aim was to obtain industrially interesting alcohols that could be used as chemical intermediates of fine chemicals, such as pharmaceuticals and flavors. The main challenge in this reaction is in the optimizing the selectivity of the reaction due to possibility to form several different products [304]. In this purpose, also Ni, Rh, Pd, Ag, Ru-based monometallic catalysts have been studied [305–308]. Unsaturated ketones can be hydrogenated in aqueous medium over anchored Ru(II)-phenanthroline complexes. Desmukh *et al.* [309] applied amino functionalized MCM-41 materials to immobilize the Ru-complexes and tested their activities in 3-methylpent-3-en-2-one hydrogenation to corresponding alcohol. They achieved complete conversion and up to 99% chemoselectivity for the reaction [309]. Chiral tertiary alcohols are produced via alkylation of ketones. They are very important in organic chemistry due to their wide use in pharmaceuticals and appearance in natural products. The main way to produce them is the asymmetric synthesis from ketones by the addition of organometallic complexes derived from Zn, Al, Mg and Li [310].

Maybe the most common use of carboxylic acids is their use in fatty acids in making different types of soaps and detergents [311,312]. They are also used, e.g., in polymer production (maleic acid, terephthalic acid, adipic acid), food preservatives (propionic acid) and beverages (citric acid). Carboxylic acids can be produced for example from aldehydes using Co and Mn catalysts or from alcohols via hydrogenation and using base-catalysts [312]. Carboxylic acid reacts with alcohols to produce esters of which the most important is the polyester [312]. The reduction of carboxylic acids leads to either alcohols or aldehydes, to do this selectively, remains still a challenge. Hydrosilylation of carboxylic acids can be made selectively with using iron-catalyst [313,314]. Recently, it was found that amide bond formation from formamides and carboxylic acids can be achieved over copper catalysts without producing considerable amounts of hazardous waste due to coupling reagents. Amide bonds are present for example in several drug molecules and agrochemicals. Liu *et al.* tested both aromatic and aliphatic carboxylic acids successfully with high yields with Cu-catalysts [315].

Esters are used as artificial flavors and aroma. They can be used in different consumer products and food production and especially they are extensively used in fragrance and flavor industry. They are also used as solvents in paints, for example, and as plasticizers. Nitroglycerin and polyester are well-known examples of esters [316]. Esterification reaction is reversible, and they can undergo hydrolysis under basic or acidic conditions. Hydrolysis under basic conditions leads to soap formation via saponification [316]. Esters can be also hydrogenated to fatty alcohols over copper chromite catalyst [317] or by iron complexes [313]. Direct reduction of esters to ethers is difficult, but it is possible to carry it out over triethylsilane or variety of Lewis acids [317]. The direct conversion of esters to nitriles is also possible [318]. There exist one interesting potential use for the esters, if we consider utilization of VOC emissions. The methyl esters of fatty acids can be used as bitumen fluxes and therefore in reducing the emissions of VOCs during the road construction [319]. Methyl acetate has also been used as a model for hydrogen production from biodiesel. For example, Rh and Rh/CeO_2-catalysts have been used in this purpose [320].

Converting methyl mercaptan in a gaseous feed to formaldehyde and sulfur dioxide is one rather recent example of the valorization of the TRS [291,320,321]. The process consists of contacting the gaseous stream containing methyl mercaptan with a catalysts under oxidizing conditions for sufficient time. The catalytic materials used during this process can be supported metal oxides or certain bulk metal oxides. Isreal E Wachs studied the activity of different catalysts with different compositions. Supported metal oxides catalysts 1% MO_3/TiO_2, 1% CrO_3/TiO_2 and 1% Re_2O_7/TiO_2 show good activity (around 80% of methyl mercaptan conversion) and important selectivity (around 80%) [320]. The production of formaldehyde from methyl mercaptan is also possible in presence of other compounds (*i.e.*, methanol). Titania-silica-supported vanadium catalysts have been studied in this case [291]. Later, cerium-titanium mixed oxide catalysts where studied in the partial oxidation of the methanol and methyl mercaptan to produce formaldehyde, interesting results were obtained in terms of activity and selectivity [321].

In the beginning of the last century, Krause and Rroka studied production of formaldehyde from methylene chloride. The process consisted of mixing DCM with water in a closed vessel, providing a heating from 140 °C to 170 °C. They found that the production of formaldehyde from DCM is possible [322,323]. From the same family, methyl chloride can be oxidized with air over a catalyst to give formaldehyde and hydrogen chloride. After separating the hydrogen chloride from the reaction products, formaldehyde is obtained. This reaction occur following a catalytic route, where the catalyst consists of particles or pellets of uniform compositions, or it can be an intimate mixture of particles or pellets of two distinct compositions. The first reaction step in which methyl chloride is hydrolyzed to methyl alcohol may be catalyzed by copper chloride, zinc chloride, bismuth chloride or by alumina gel, at a temperature in the range of 280 °C to 350 °C. The second reaction step whereby methyl alcohol is oxidized to formaldehyde is catalyzed by an iron-molybdenum oxide catalyst, which may be enhanced with chromium oxide. Other catalysts that have been reported to oxidize methyl alcohol include vanadium pentoxide and copper. The oxidation reaction is carried out at a temperature in the range of 250 °C to 400 °C [323].

These examples show that there exist indeed several possibilities for the utilization of VOC. It is more likely that economically feasible utilization will be based on the needs of the chemicals in the proximity of where the utilized VOC is produced. This calls for improvements in industrial ecology, Furthermore, the composition of emission mixture is important: Does the mixture contain catalyst poisons? Does it include enough the major compound to make the recovery and separation easier? Will the legislation set requirements for the utilization of gases in the future? What are the most potential industrial sectors for the utilization of VOCs? These questions are still open.

5. Considerations on Catalyst Durability in Utilization of Organic Emissions

As already highlighted before, the catalyst stability is one important issue related to the utilization of the VOC, since VOC emissions are mixtures containing several gaseous compounds and in some cases also solids. The recovery, concentration and purification possibilities are already discussed in detail in Chapter 2 and thus we will not consider the catalyst stability directly related to those aspects here. In general, the selection of the correct catalytic materials is central in VOC utilization. In any case, detailed measurements of the exact composition of the emission would be important to facilitate the selection. It is not very common that the industry knows the exact composition of their emissions, since the emission follow-up programs are related to those compounds that are limited or measurements are required by authorities. Also, one industrial site contains typically several different sources of emissions that compositions are different, which complicates making the general guidelines on the catalyst selection for example. This is why we approach here the catalyst durability through few case examples and by this way aim to demonstrate the importance of the issue.

In methane conversion, catalyst deactivation is a serious problem, which can occur due to various factors such as gas impurities and process conditions [324]. Firstly, the natural gas (NG) or landfill gas (LFG) quality is critical in processing methane further to syngas and hydrogen. Methane from LFG (40%–49% CH_4, 50% CO_2, traces of sulfides, chlorinated compounds, heavy metals, siloxanes, *etc.*) and coal bed methane (CBM, typically 30%–98% CH_4, other hydrocarbons, CO_2, H_2O) are the industrially relevant feedstocks for utilization [325–327]. Fossil based NG consists of many trace impurities as well, such as heavier hydrocarbons, sulfides and nitrous oxides besides CO_2. The contaminants such as H_2O, CO_2 and H_2S exist also in the extracted coal bed methane (CBM) [325]. The main causes for the catalyst deactivation in processing CBM are mainly fouling of the surfaces by ash (Na, K, Ca, and sulphates), chemical poisoning by sulfur and As compounds, hydrothermal sintering and fly ash erosion and abrasion [55]. Developing a catalyst which is sulfur resistant can be achieved by using high oxygen storage materials (such as CeO_2, ZrO_2), which reduce also the coke formation [57,328].

Large amounts of water is released during CBM production, thus there exist a big challenge in water management in a cost-effective and environmentally acceptable way [325,329]. One option in utilizing the methane and CO_2 along with huge quantity of water can be the combination of steam and dry reforming of methane to syngas. The addition of water has both positive effect, *i.e.*, by reducing coke formation and negative effect by energy consumption and re-oxidation of Ni particles to NiO. Nickel supported on alumina based catalysts is widely used industrially for steam reforming of methane and also studied extensively in methane dry reforming. Most common poisons, which are found in NG or coal mine streams are H_2S and As, which deactivate Ni catalysts. The traces of impurities and contaminants can poison the Ni particles and also enhance the coking during the reaction [55,56]. In syngas production via methane dry reforming, the carbon formation is inevitable during the reaction. Carbon diffuses to material and forms filamentous carbon around Ni particles at high temperatures. In order to avoid carbon formation during methane reforming, high steam to carbon ratio is used. Another common deactivation phenomena is Ni sintering at high temperatures, which can be solved by adding promoters (e.g., K, Ca, Mg), stabilizers (e.g., La and Al), and co-catalysts (e.g., noble metals Pt, Pd, metal oxides) and thus, enhance the catalyst durability [55].

The pulp mill chip bin emissions contain in addition to methanol and sulfur-containing volatile organic compounds also hydrogen sulfide, terpenic compounds, amines, higher alcohols, amines, water vapor and solid particulates [290,291,330,331]. Sulfur is a well-known catalyst poison, as discussed above, which is studied very widely in different applications [332–334]. There exist different solutions to avoid sulfur poisoning such as addition of sulfide-forming metal or an oxide sorbent to the catalyst [55]. It is also a good strategy to avoid the materials known to be sensitive to sulfur poisoning, if possible. Examples of sulfur-tolerant materials are the noble metals (except Ru) and high oxygen mobility/storage supports such as ceria, zirconia or their mixed oxides. Gd doping has been observed to improve sulfur tolerance [335].

Emissions of pharmaceutical industry are very complicated due to the batch-type of production. It means that the composition of the emissions may change in connection of different products made. Pharmaceutical industry uses substituted organic compounds, e.g., chlorinated compounds, for example as solvents [336]. In addition to chlorinated compounds, the emissions typically contain oxygenated organic compounds, such as, acetone and ethyl acetate. In these cases, for example Pt/Al_2O_3-CeO_2 catalyst seems to be stable at least to some extent [337]. However, it is always possible that platinum is gradually volatilized by the formation of $PtCl_4$ [338]. As catalyst support material, Ce-Zr mixed oxides show interesting behavior against Cl-poisoning [339].

To conclude, in the cases of VOC utilization to produce chemicals, the most significant deactivation mechanism is poisoning that is more pronounced in connection of utilizing the substituted compounds. When it comes to syngas production, several deactivating phenomena exist depending on the source of the gas. If proper purification of emission before reforming is done, the major cause of deactivation at high temperature is sintering and at low temperature, the coking. Regarding the potential catalytic materials, noble metals seem to be a good choice as well as high oxygen mobility support materials.

6. Economic Issues Related to Utilization of Organic Emissions

To get an idea about economic issues related to utilization of volatile organic compounds emissions, certain subjects need to be summarized. As a starting point, the emission amounts and sources give a view what is the maximum potential of the utilization of VOCs. Assessment of the sources of the emissions can be used to evaluate the industrial sectors that could be potential users of the utilization technologies. The replacement of the emission treatment system by a process that is able to utilize the emissions is still possible in many cases. The feasibility of different options depends on the qualities of emitted compounds and the need of the products made via utilization. It has been estimated by the U.S. Department of energy that in short term, the market for the utilization of the industrial VOC emissions as the fuel is limited to industrial plants that produce and collect the volatile compounds [9].

The NMVOC emissions, reported by EEA (European Environment Agency), have decreased by 57% compared with 1990 levels in European countries. In 2011, the most significant sources of NMVOC emissions are solvent and product use (43%), commercial, institutional and households (17%), road transport (15%), and industrial processes (8%), energy production and distribution (10%) (see Figure 4). The decline in emissions since 1990 has primarily been due to reductions achieved in the road transport sector due to the introduction of catalytic converters [324].

Different emission data registers and research inventories are using different kind of classification to the industrial sectors that emit NMVOC. The European Pollutant Release and Transfer Register (E-PRTR) divides the NMVOC emissions into nine sectors:

(1)   Energy sector

(2)   Production and processing of metals

(3)   Mineral industry

(4)   Chemical industry

(5)   Waste and waste water management

(6)   Paper and wood production processing

(7)   Intensive livestock production and aquaculture

(8)   Animal and vegetable products from the food and beverage sector

(9)   Other activities

E-PRTR contains annual data reported from 28,000 industrial facilities in Europe. In year 2012 the total NMVOC amount from these facilities was 457,862 t. The biggest emitters were mineral oil and gas refineries (27.9%), surface treatment of substances, objects or products (15.6%), and industrial scale production of basic organic chemicals (14.2%) as presented in Figure 5. United Kingdom, Spain, France, Germany and Norway are the top five countries when checking the releases per country [340].

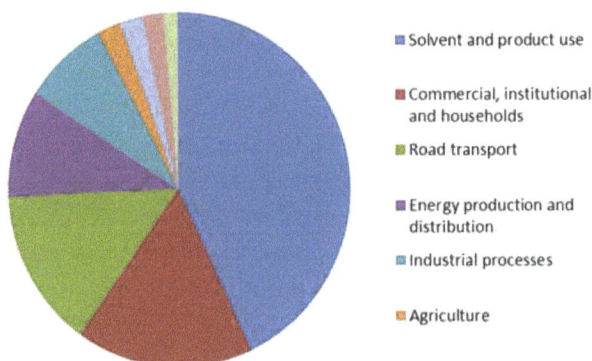

Figure 4. NMVOC emissions shares from different source categories in European countries in 2011 (Data collected from [324]).

When we consider different types of VOCs, as an example, Theloke *et al.* [342] have reported the share of the total VOC emissions (in Germany 1998) from four different categories; transport, solvent use, production and storage processes and combustion processes. From these sources, 33% of the VOC emissions were alkanes, 23% aromatics, 14% alcohols, 8% alkenes, 5% esters, aldehydes 4%, glycol derivates 4%, and 3% ketones. It is clear that different emissions sources have different profiles of compounds (see Table 8); for the solvent use the three biggest groups are alkanes 28%, alcohols 24% and aromatics 22%. For transport the shares for the three largest groups of compounds are 41% alkenes, 27% aromatics and 16% alkenes. For stationary combustion the three biggest are aromatics 37%, alkanes 29% and alkenes 19%. For production processes, the shares are alkanes 31%, alcohols 29%, and alkenes 15%.

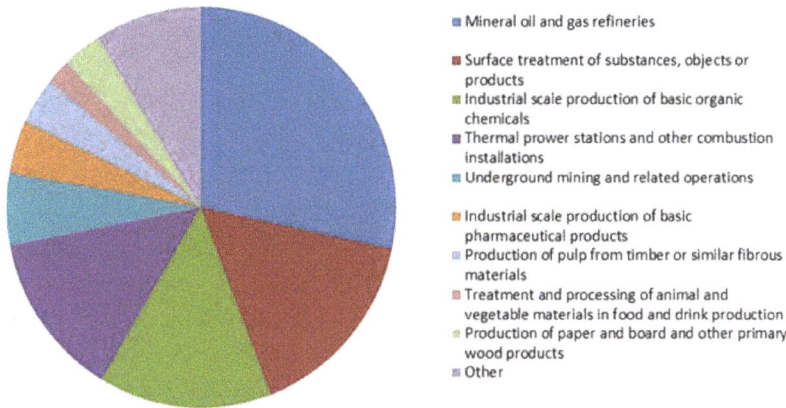

- Mineral oil and gas refineries
- Surface treatment of substances, objects or products
- Industrial scale production of basic organic chemicals
- Thermal prower stations and other combustion installations
- Underground mining and related operations
- Industrial scale production of basic pharmaceutical products
- Production of pulp from timber or similar fibrous materials
- Treatment and processing of animal and vegetable materials in food and drink production
- Production of paper and board and other primary wood products
- Other

Figure 5. NMVOC (Non-Methane VOC) emission shares from European industrial facilities in 2012 (Data collected from: [341]).

Table 8. Division of emissions to substance classes (data from Germany 1998) for different emission sectors [342].

substance class	Solvent use (%)	Transport (%)	Stationary combustion (%)	Production processes (%)
alkanes	28	41	29	31
alkenes	1.3	16	19	15
alkynes	-	3	4	0.03
aromatics	22	27	37	5
alcohols	24	-	-	29
aldehydes	-	9	8	4
esters	9	-	-	-
ethers	1.3	-	-	2.4
ketones	6	1	2	0.2
glycol derivates	7	-	-	-
halogenated hydrocarbons	1.3	-	-	2.7
carbonic acids	0.1	-	-	-
not allocated	-	3	1	11

To get more global view, United States and China are important emission inventory areas as well. U.S. EPA (United States Environmental Protection Agency) categorizes the anthropogenic VOC sources to five different groups [343]:

(1) Stationary fuel combustion
(2) Industrial processes
(3) Highway vehicles
(4) Non-road mobile
(5) Miscellaneous excluding wildfires

231

In the U.S. the VOC emission data is collected and estimated by the National Emission Inventory (NEI). According to NEI data, total estimated VOC emissions from anthropogenic sources was 18,169,000 t in year 2011. Compared to 1990 the decrease is 35% [343]. In 2005 the NMVOC emissions in China were estimated to be 20.1 Tg. The dominated emission sectors were industrial and domestic solvent use (28.6%), road transportation (23.4%), and biofuel combustion (18%). The estimation for year 2020 is 25.9 Tg [171,207,344].

The cost of air pollution is also a key point in this discussion whether the companies should change from current situation and to invest to utilization of VOC emissions. To compare the current situation, it is known that the costs of air pollution from the 14,325 largest polluting facilities in Europe were between 59 and 189 billion euros in 2012. 50% of these costs are result of just 191, or 2% of the facilities. Power generation sector was responsible for around 70% of the total damage costs from industry. Of the top 30 facilities causing the highest damage, 29 were power-generating facilities [345]. It shows that the efforts to find industry that could be interested in utilization of emission gases could be started with few companies. It should be interesting that the high costs caused by air pollution could be converted to incomes when converting the emissions in valuable substances.

However, it seems important that the economics of the utilization needs to be considered case by case. There are emission streams available as well as raw materials to utilization processes, but weather the economic benefits will come from energy usage or chemical production is strongly case dependent. The costs caused by the air pollution could be minimized by utilization of the gases. This should be clearly shown to the polluting industry to increase their interest towards sustainable production.

7. Utilization of VOC Emissions and Sustainability

Environmental catalysis is defined as the development of catalytic materials suitable for abatement of environmental pollutants or to be used in more sustainable and environmentally sound production processes compared with the existing ones [346]. Catalysis has been used already for around 30 years in emissions abatement, e.g., in SO_x, VOCs, automotive exhaust gas purification via converting them to less harmful compounds [347]. The new trend that has started with the CO_2 emissions is to convert the harmful compounds by catalytic processes to valuable products, chemicals, materials and fuels or fuel compounds [348]. VOC emissions that are very often just released into the air as such or after oxidation, offer an excellent starting material for these processes. An important opportunity exists in finding a process route that captures VOCs in, e.g., flue, process and exhaust gases and converts these molecules over preferably a single catalyst composition to valuable compounds [349]. There are still many opportunities for innovations and understanding of these new approaches, e.g., catalytic membranes, hybrid materials, catalytic microreactors. The design and use of catalysts for environmental protection along with their use in processing valuable compounds out of, e.g., gaseous emissions like VOCs is a challenge and a highly motivating field for scientists in the future. The parallel design of the whole process for emissions utilization with separation, purification, concentration and catalysis units or their combinations will be done in the future in a sophisticated manner and the use of catalysts for these purposes will probably show the greatest growth in demand.

Sustainability in, e.g., the chemical industry and sustainable production are the 'creation of goods and services using processes and systems that are non-polluting, conserving of energy and natural resources, are economically efficient, safe and healthful for workers, communities, and consumers, and socially and creatively rewarding for all working people' [350]. In sustainable production the goal is to minimize energy use, waste generation and environmental hazards, and to improve process safety [351–353]. When designing new processes or improving the existing ones it is important to guarantee that sustainability assessment is comparable and relevant. There are many ongoing approaches to develop analysis tools for assessing the sustainability of production processes and for defining common criteria for analyzing sustainability [354,355]. Sustainable production concerns products and services, process design and operation, wellbeing of workers in, e.g., the chemical industry [356]. According to the green chemistry and engineering principles products, processes and services should be safe and ecologically sound throughout their life cycle and appropriate, designed to be durable, repairable, readily recycled, compostable, and easily biodegradable [354–356]. Development of 3D multidimensional sustainability assessment methods including environmental, economic and social sustainability aspects and criteria for processes utilizing, e.g., VOCs existing in various gas streams, *i.e.*, flue, process, exhaust gas streams is very important. Significantly improved capabilities to master and increase sustainable utilization of VOCs exist over novel catalysts and wise selection of process units.

Nanostructured catalytic materials are seen to offer excellent ways to enhance the sustainable use of gaseous emissions since they have excellent features such as better control of activity, selectivity, and deactivation of catalysts. Research in catalysis nanoscience has dealt with the research questions raising from the relationship between catalyst synthesis and active site structure on the atomic- and nanoscale, reaction mechanisms, and catalyst activity, selectivity, lifetime. The outcomes including utilization of VOC emissions will have a great impact on chemicals manufacturing in the future both from environmental, societal and economic perspectives. Process integration at the nanoscale offers as well new tools for process design [357–359].

From the traditional areas of VOCs abatement, research activities are extending to a wide range of applications including, e.g., the use of VOCs in chemicals and fuels production. In all these applications catalysis is the essential tool to enable sustainable production. Catalysis offers not only the way of abating the VOCs' odors, their atmospheric reactivity and harms in the working environment but also provides excellent options for novel and sustainable products.

8. Conclusions

As discussed in the review, the utilization of the VOC emissions is an economically interesting, environmentally sound idea that supports the sustainability of the processes. Currently, the most common way of utilization of VOC is in the energy production. Some of the organic compounds studied find also use in fuels production. If we think the future in a short term-production of synthesis gas and hydrogen seems to be the most straightforward way to utilize VOCs. The VOC emissions have high potential in that application area. In the longer term, more specific low temperature processes in the utilization of organic gases will come into the picture. For example, utilization of non-thermal plasma, photocatalysis and biocatalytic processes become more realistic especially in the case of methane. If we consider chemicals production from the VOC, the currently existing possibilities are not numerous. However, the potential of the VOCs in chemicals production is very wide–even high-value products could be made. When we consider the different emission sectors, namely solvent use, transport, stationary combustion and production processes, the emissions of aromatic compounds, alcohols and alkanes are quantitatively highest. Therefore, currently these are the most potential groups of compounds to be considered in the utilization purposes. When considering the utilization of emissions originating from certain industrial sector, it is important to keep in mind the need for the recovery and pre-treatment technologies (separation, purification, concentration). Also, the flexibility of the utilization process would be important in the situations when potential emission streams change or the markets of the certain products change. Careful selection and development of catalytic materials and reactor technologies take an important role in this approach, especially when the emission streams contain catalyst poisons or if the utilized compound itself contains heteroatoms that may cause catalyst deactivation. The utilization of VOC is a very interesting research area and it offers wide possibilities for further studies.

Acknowledgments: This review is done within the framework of No Waste project that has received funding from European Union Seventh Framework Programme, Marie Curie Actions under grant agreement no (PIRSES-GA-2012-317714). The financial aid of Finnish Cultural Foundation, North Ostrobothnia Regional fund is also appreciated. Seelam would like to address his thanks to Harinya Seelam.

Author Contributions: Satu Ojala is the first and corresponding author of the manuscript. She designed the general contents of the manuscript and invited the other writers to join the work. All the other writers have expertise in the topical area they contributed. Satu Ojala also wrote introduction, conclusions and participated writing of VOC utilization section. She is the second supervisor of the doctoral studies of Niina Koivikko, Tiina Laitinen and Anass Mouammine. Niina Koivikko wrote the Chapter 5 and participated to the background survey for designing the contents of the manuscript. Tiina Laitinen and Anass Mouammine participated writing the VOC utilization in chemicals production. The work described here related to the VOC utilization in the chemicals production, is related to the thesis works of Niina Koivikko, Tiina Laitinen and Anass Mouammine. Prem Kumar Seelam took the complete responsibility on Chapter 3, Said Laassiri wrote the Chapter 4.1. Kaisu Ainassaari was responsible on writing the Chapter 2 and Riitta Keiski wrote the Chapter 7. Riitta Keiski is the principal supervisor of Niina Koivikko, Tiina Laitinen, Anass Mouammine and Kaisu Ainassaari. Rachid Brahmi is the supervisor of the doctoral work of Anass Mouammine.

Conflicts of Interest: The authors declare no conflict of interest.

References

1. Ojala, S.; Pitkäaho, S.; Laitinen, T.; Niskala, N.; Brahmi, R.; Gaálová, J.; Matejova, L.; Kucherov, A.; Päivärinta, S.; Hirschmann, C.; *et al.* Catalysis in VOC abatement–A review. *Top. Catal.* **2011**, *54*, 1224–1256.

2. Council Directive 1999/13/EC of 11 March 1999 on the limitation of emissions of volatile organic compounds due to the use of organic solvents in certain activities and installations. Available online: http://eur-lex.europa.eu/legal-content/EN/TXT/?uri= celex:31999L0013 (accessed on 28 February 2015).

3. Directive 2010/75/EU of the European Parliament and of the Council of 24 November 2010 on industrial emissions (integrated pollution prevention and control). Available online: http://europa.eu/legislation_summaries/environment/soil_ protection/ev0027_en.htm (accessed on 28 February 2015).

4. Moretti, E.C. Reduce VOC and HAP Emissions. *CEP Magazine.* 2002, pp. 30–40. Available online: http://www.cepmagazine.org (accessed on 29 June 2015).

5. Smith, N. Energy, Resources & Public Policy, Paper #2. 2003. Available online: http: //www.brynmawr.edu/geology/206/smith2.htm (accessed on 29 June 2015).

6. Su, S.; Beath, A.; Huo, H.; Mallett, C. An assessment of mine methane mitigation and utilization technologies. *Prog. Energy Combust. Sci.* **2005**, *31*, 123–170.

7. Belis-Bergouignan, M.-C.; Oltra, V.; Saint Jean, M. Trajectories towards clean technology: Example of volatile organic compound emission reductions. *Ecol. Econ.* **2004**, *48*, 201–220.

8. Melhus, Ö. Utilization of VOC in Diesel Engines. Ph.D. Thesis, Department of Marine Engineering, Norwegian University of Science and Technology, Trondheim, Norway, March 2002.

9. US Department of Energy. Mid-Atlantic green energy application center, Penn State University, Clean energy–opportunity fuels–industrial by-products. Available online: http://www.maceac.psu.edu/clean_energy_opportunity_fuels_industrial.htm (accessed on 1 March 2015).

10. Natural Resources Defense Council, NFDC. Renewable Energy for America, Harvesting the benefits of homegrown, renewable energy. Biogas Energy. Available online: http: //www.nrdc.org/energy/renewables/biogas.asp (accessed on 29 June 2015).

11. Nunez, C. VOCs: Sources, Definitions, and Considerations for Recovery, US EPA Seminar, EPA625/R-99/005, presented in 16 September 1998. Available online: http://nepis.epa. gov/ (accessed on 27 February 2015).

12. BAT reference document for Common Waste Water and Waste Gas Treatment/Management Systems in the Chemical Sector. Available online: http://eippcb.jrc.ec.europa.eu/reference/BREF/CWW_Final_Draft_07_2014.pdf (accessed on 20 February 2015).

13. Schenk, E.; Mieog, J.; Evers, D. *Fact Sheets on Air Emission Abatement Techniques*; DHV B.V.: Amersfoort, The Netherlands, 2009.

14. Wu, J. Modeling Adsorption of Organic Compounds on Activated Carbon. Ph.D. Thesis, University of Umeå, Umeå, Sweden, 24 September 2004.

15. Cruz, G.; Pirilä, M.; Huuhtanen, M.; Carrión, L.; Alvarenga, E.; Keiski, R.L. Production of Activated Carbon from Cocoa (Theobroma cacao) Pod Husk. *J. Civ. Environ. Eng.* **2012**, *2*, 109.

16. Girgis, B.S.; El-Hendawy, A.N.A. Porosity development in activated carbons obtained from date pits under chemical activation with phosphoric acid. *Microporous Mesoporous Mater.* **2002**, *52*, 105–117.

17. Olivares-Marín, M.; Fernández-González, C.; Macías-García, A.; Gómez-Serrano, V. Preparation of activated carbon from cherry stones by chemical activation with $ZnCl_2$. *Appl. Surf. Sci.* **2006**, *252*, 5967–5971.

18. Okman, I.; Karagöz, S.; Tay, T.; Erdem, M. Activated Carbons from Grape Seeds by Chemical Activation with Potassium Carbonate and Potassium Hydroxide. *Appl. Surf. Sci.* **2014**, *293*, 138–142.

19. Subrenat, A.S.; le Cloirec, P.A. Volatile organic compound (VOC) removal by adsorption onto activated carbon fiber cloth and electrothermal desorption: An industrial application. *Chem. Eng. Commun.* **2006**, *193*, 478–486.

20. Khan, F.I.; Ghoshal, A.K. Removal of volatile organic compounds from polluted air. *J. Loss Prev. Process Ind.* **2000**, *13*, 527–545.

21. Anonymous. Hydrocarbon Processing's Environmental Processes' 98. *Hydrocarbon Processing* **1998**, *77*, 69–112.

22. Engleman, V.S. Updates on choices of appropriate technology for control of VOC emissions. *Met. Finish.* **2000**, *98*, 433–445.

23. Reference Document on Best Available Techniques in the Large Volume Organic Chemical Industry. Available online: http://eippcb.jrc.ec.europa.eu/reference/BREF/lvo_bref_ 0203.pdf (accessed on 25 February 2015).

24. Kruger, D.; Schultz, K.A. *Guide for Methane Mitigation Projects: Gas-to-Energy at Coal Mines*; US Environmental Protection Agency, Office of Air and Radiation 6202J: Washington, DC, USA, 1996.

25. Xebec Adsorption Inc. Treatment solutions for landfill gas fuel applications. 2007. Available online: http://www.xebecinc.com/pdf/e_white_paper.pdf (accessed on 26 February 2015).

26. Caballero, A.; Perez, P.J. Methane as raw material in synthetic chemistry: The final frontier. *Chem. Soc. Rev.* **2013**, *42*, 8809.

27. Kerr, R.A. Arctic Armageddon needs More Science, Less Hype. *Science* **2010**, *329*, 620–621.

28. US EPA. *Summary Report: Global Anthropogenic Non-CO$_2$ Greenhouse Gas Emissions: 1990–2030*; EPA 430-S-12-002. EPA: Washington, DC, USA, 2012.

29. Bousquet, P.; Tyler, S.C.; Peylin, P.; van der Werf, G.R.; Prigent, C.; Hauglustaine, D.A.; Dlugokencky, E.J.; Miller, J.B.; Ciais, P.; White, J.; *et al.* Contribution of anthropogenic and natural sources to atmospheric methane variability. *Nature* **2006**, *443*, 439–443.

30. McFarland, E. Unconventional Chemistry for Unconventional Natural Gas. *Science* **2012**, *338*, 340–342.

31. GGFR. Available online: http://www.flaringreductionforum.org/agenda.html (accessed on 1 March 2015).

32. Ibitoye, F.I. Ending Natural Gas Flaring in Nigeria's Oil Fields. *J. Sustain. Dev.* **2014**, *7*, 3.

33. Yang, Z.; Liu, J.; Zhang, L.; Zheng, S.; Guo, M.; Yana, Y. Catalytic combustion of low-concentration coal bed methane over CuO/γ-Al$_2$O$_3$ catalyst: Effect of SO$_2$. *RSC Adv.* **2014**, *4*, 39394–39399.

34. Horn, R.; Schlögl, R. Methane activation by heterogeneous catalysis. *Catal. Lett.* **2014**, *145*, 23–39.

35. Alvarez-Galvan, M.C.; Mota, N.; Ojeda, M.; Rojas, S.; Navarro, R.M.; Fierro, J.L.G. Direct methane conversion routes to chemicals and fuels. *Catal. Today* **2011**, *171*, 15–23.

36. Iglesia, E. Challenges and progress in the conversion of natural gas to fuels and chemicals. *Fuel Chem. Div. Prepr.* **2002**, *47*, 128.

37. Lunsford, J.H. Catalytic conversion of methane to more useful chemicals and fuels: A challenge for the 21st century. *Catal. Today* **2000**, *63*, 165–174.

38. Tang, P.; Zhu, Q.; Wu, Z.; Ma, D. Methane activation: The past and future. *Energy Environ. Sci.* **2014**, *7*, 2580–2591.

39. Eliasson, B.; Liu, C.; Kogelschatz, U. Direct conversion of methane and carbon dioxide to higher hydrocarbons using catalytic dielectric-barrier discharges with zeolites. *Ind. Eng. Chem. Res.* **2000**, *39*, 1221–1227.

40. Lavoie, J.-M. Review on dry reforming of methane, a potentially more environmentally-friendly approach to the increasing natural gas exploitation. *Front. Chem.* **2014**, *2*, 81.

41. Pakhare, D.; Spivey, J. A review of dry (CO$_2$) reforming of methane over noble metal catalysts. *Chem. Soc. Rev.* **2014**, *43*, 7813–7837.

42. Wu, H.; La Parola, V.; Pantaleo, G.; Puleo, F.; Venezia, A.M.; Liotta, L.F. Ni-Based Catalysts for Low Temperature Methane Steam Reforming: Recent Results on Ni-Au and Comparison with Other Bi-Metallic Systems. *Catalysts* **2013**, *3*, 563–583.

43. Budiman, A.W.; Song, S.H.; Chang, T.S.; Shin, C.H.; Choi, M.J. Dry reforming of methane over cobalt catalysts: A literature review of catalyst development. *Catal. Surv. Asia* **2012**, *16*, 183–197.

44. Fan, M.S.; Abdullah, A.Z.; Bhatia, S. Catalytic technology for carbon dioxide reforming of methane to synthesis gas. *ChemCatChem* **2009**, *1*, 192–208.

45. Ding, R.G.; Yan, Z.F.; Song, L.H.; Liu, X.M. A review of dry reforming of methane over various catalysts. *J. Nat. Gas Chem.* **2001**, *10*, 237–255.

46. Al-Sayari, S.A. Recent Developments in the Partial Oxidation of Methane to Syngas. *Open Catal. J.* **2013**, *6*, 17–28.

47. Enger, B.C.; Lødeng, R.; Holmen, A. A review of catalytic partial oxidation of methane to synthesis gas with emphasis on reaction mechanisms over transition metal catalysts. *Appl. Catal. A* **2008**, *346*, 1–27.

48. Zhu, Q.; Zhao, X.; Deng, Y. Advances in the partial oxidation of methane to synthesis gas. *J. Nat. Gas Chem.* **2004**, *13*, 191–203.

49. Zheng, X.; Tan, S.; Dong, L.; Li, S.; Chen, H. Silica-coated LaNiO$_3$ nanoparticles for non-thermal plasma assisted dry reforming of methane: Experimental and kinetic studies. *Chem. Eng. J.* **2015**, *265*, 147–156.

50. Ananth, A.; Gandhi, M.S.; Mok, Y.S. A dielectric barrier discharge (DBD) plasma reactor: An efficient tool to prepare novel RuO$_2$ nanorods. *J. Phys. D* **2013**, *46*, 155–202.

51. Kaydouh, M.N.; El Hassan, N.; Davidson, A.; Casale, S.; El Zakhem, H.; Massiani, P. Effect of the order of Ni and Ce addition in SBA-15 on the activity in dry reforming of methane. *Comptes Rendus Chim.* **2015**, *18*, 293–301.

52. Gould, T.D.; Montemore, M.M.; Lubers, A.M.; Ellis, L.D.; Weimer, A.W.; Falconer, J.L.; Medlin, J.W. Enhanced dry reforming of methane on Ni and Ni-Pt catalysts synthesized by atomic layer deposition. *Appl. Catal. A* **2015**, *492*, 107–116.

53. Aw, M.S.; Črnivec, I.G.O.; Djinović, P.; Pintar, A. Strategies to enhance dry reforming of methane: Synthesis of ceria-zirconia/nickel-cobalt catalysts by freeze-drying and NO calcination. *Int. J. Hydrogen Energy* **2014**, *39*, 12636–12647.

54. Cai, W.-J.; Qian, L.-P.; Yue, B.; He, H.-Y. Rh doping effect on coking resistance of Ni/SBA-15 catalysts in dry reforming of methane. *Chin. Chem. Lett.* **2014**, *25*, 1411–1415.

55. Argyle, M.D.; Bartholomew, C.H. Heterogeneous Catalyst Deactivation and Regeneration: A Review. *Catalysts* **2015**, *5*, 145–269.

56. De Miguel, S.R.; Vilella, I.M.J.; Maina, S.P.; San José-Alonso, D.; Roman-Martinez, M.C.; Illan-Gomez, M.J. Influence of Pt addition to Ni catalysts on the catalytic performance for long term dry reforming of methane. *Appl. Catal. A* **2012**, *435*, 10–18.

57. Silva, P.P.; Silva, F.A.; Portela, L.S.; Mattos, L.V.; Noronha, F.B.; Hori, C.E. Effect of Ce/Zr ratio on the performance of Pt/CeZrO$_2$/Al$_2$O$_3$ catalysts for methane partial oxidation. *Catal. Today* **2005**, *108*, 734–740.

58. Bao, Z.; Lu, Y.; Han, J.; Li, Y.; Yu, F. Highly active and stable Ni-based bimodal pore catalyst for dry reforming of methane. *Appl. Catal. A* **2015**, *491*, 116–126.

59. Littlewood, P.; Xie, X.; Bernicke, M.; Thomas, A.; Schomäcker, R. Ni$_{0.05}$Mn$_{0.95}$O catalysts for the dry reforming of methane. *Catal. Today* **2015**, *242*, 111–118.

60. Yonggang, W.; Hua, W.; Kongzhai, L. Ce-Fe-O mixed oxide as oxygen carrier for the direct partial oxidation of methane to syngas. *J. Rare Earths* **2010**, *28*, 560.

61. Talkhoncheh, S.K.; Haghighi, M. Syngas production via dry reforming of methane over Ni-based nanocatalyst over various supports of clinoptilolite, ceria and alumina. *J. Nat. Gas Sci. Eng.* **2015**, *23*, 16–25.

62. Jabbour, K.; El Hassan, N.; Davidson, A.; Massiani, P.; Casale, S. Characterizations and performances of Ni/diatomite catalysts for dry reforming of methane. *Chem. Eng. J.* **2015**, *264*, 351–358.

63. Alenazey, F.S. Utilizing carbon dioxide as a regenerative agent in methane dry reforming to improve hydrogen production and catalyst activity and longevity. *Int. J. Hydrogen Energy* **2014**, *39*, 18632–18641.

64. Sokolov, S.; Kondratenko, E.V.; Pohl, M.M.; Rodemerck, U. Effect of calcination conditions on time on-stream performance of Ni/La$_2$O$_3$-ZrO$_2$ in low-temperature dry reforming of methane. *Int. J. Hydrogen Energy* **2013**, *38*, 16121–16132.

65. Benrabaa, R.; Löfberg, A.; Caballero, J.G.; Bordes-Richard, E.; Rubbens, A.; Vannier, R.N.; Barama, A. Sol-gel synthesis and characterization of silica supported nickel ferrite catalysts for dry reforming of methane. *Catal. Commun.* **2015**, *58*, 127–131.

66. Ma, Q.; Wang, D.; Wu, M.; Zhao, T.; Yoneyama, Y.; Tsubaki, N. Effect of catalytic site position: Nickel nanocatalyst selectively loaded inside or outside carbon nanotubes for methane dry reforming. *Fuel* **2013**, *108*, 430–438.

67. Shinde, V.M.; Madras, G. Catalytic performance of highly dispersed Ni/TiO$_2$ for dry and steam reforming of methane. *RSC Adv.* **2014**, *4*, 4817–4826.

68. Drif, A.; Bion, N.; Brahmi, R.; Ojala, S.; Turpeinen, E.; Seelam, P.K.; Pirault-Roy, L.; Keiski, R.; Epron, F. Study of the dry reforming of methane and ethanol using Rh catalysts supported on doped alumina. *Appl. Catal. A* **2015**. in press.

69. Zhu, Y.; Zhang, S.; Shan, J.J.; Nguyen, L.; Zhan, S.; Gu, X.; Tao, F. *In situ* surface chemistries and catalytic performances of ceria doped with palladium, platinum, and rhodium in methane partial oxidation for the production of syngas. *ACS Catal.* **2013**, *3*, 2627–2639.

70. Kondratenko, V.A.; Berger-Karin, C.; Kondratenko, E.V. Partial Oxidation of Methane to Syngas Over γ-Al2O3-Supported Rh Nanoparticles: Kinetic and Mechanistic Origins of Size Effect on Selectivity and Activity. *ACS Catal.* **2014**, *4*, 3136–3144.

71. Fichtner, M.; Mayer, J.; Wolf, D.; Schubert, K. Microstructured rhodium catalysts for the partial oxidation of methane to syngas under pressure. *Ind. Eng. Chem. Res.* **2001**, *40*, 3475–3483.

72. Branco, J.B.; Ferreira, A.C.; Botelho do Rego, A.M.; Ferraria, A.M.; Almeida-Gasche, T. Conversion of Methane over Bimetallic Copper and Nickel Actinide Oxides (Th, U) Using Nitrous Oxide As Oxidant. *ACS Catal.* **2012**, *2*, 2482–2489.

73. Makarshin, L.L.; Sadykov, V.A.; Andreev, D.V.; Gribovskii, A.G.; Privezentsev, V.V.; Parmon, V.N. Syngas production by partial oxidation of methane in a microchannel reactor over a Ni-Pt/La$_{0.2}$Zr$_{0.4}$Ce$_{0.4}$O$_x$ catalyst. *Fuel Process. Technol.* **2015**, *131*, 21–28.

74. Requies, J.; Cabrero, M.A.; Barrio, V.L.; Güemez, M.B.; Cambra, J.F.; Arias, P.L.; Pérez-Alonso, F.J.; Ojeda, M.; Peña, M.A.; Fierro, J.L.G. Partial oxidation of methane to syngas over Ni/MgO and Ni/La$_2$O$_3$ catalysts. *Appl. Catal. A* **2005**, *289*, 214–223.

75. Vita, A.; Cristiano, G.; Italiano, C.; Specchia, S.; Cipití, F.; Specchia, V. Methane oxy-steam reforming reaction: Performances of Ru/γ-Al$_2$O$_3$ catalysts loaded on structured cordierite monoliths. *Int. J. Hydrogen Energy* **2014**, *32*, 18592–18603.

76. Frind, R.; Borchardt, L.; Kockrick, E.; Mammitzsch, L.; Petasch, U.; Herrmann, M.; Kaskel, S. Complete and partial oxidation of methane on ceria/platinum silicon carbide nanocomposites. *Catal. Sci. Technol.* **2012**, *2*, 139–146.

77. Dong, D.; Shao, X.; Wang, Z.; Lievens, C.; Yao, J.; Wang, H.; Li, C.-Z. Fibrous NiO/CeO$_2$ nanocatalysts for the partial oxidation of methane at microsecond contact times. *RSC Adv.* **2013**, *3*, 1341–1345.

78. Hu, J.; Yu, C.; Bi, Y.; Wei, L.; Chen, J.; Chen, X. Preparation and characterization of Ni/CeO$_2$-SiO$_2$ catalysts and their performance in catalytic partial oxidation of methane to syngas. *Chin. J. Catal.* **2014**, *35*, 8–20.

79. Silva, F.A.; Resende, K.A.; da Silva, A.M.; de Souza, K.R.; Mattos, L.V.; Montes, M.; Souza-Aguiar, E.F.; Noronha, F.B.; Hori, C.E. Syngas production by partial oxidation of methane over Pt/CeZrO$_2$/Al$_2$O$_3$ catalysts. *Catal. Today* **2012**, *180*, 111–116.

80. Holladay, J.D.; Hu, J.; King, D.L.; Wang, Y. An overview of hydrogen production technologies. *Catal. Today* **2009**, *139*, 244–260.

81. Bej, B.; Pradhan, N.C.; Neogi, S. Production of hydrogen by steam reforming of methane over alumina supported nano-NiO/SiO$_2$ catalyst. *Catal. Today* **2013**, *207*, 28–35.

82. Seelam, P.K.; Liguori, S.; Iulianelli, A.; Pinacci, P.; Calabrò, V.; Huuhtanen, M.; Keiski, R.; Piemonte, V.; Tosti, S.; Falco, M.; *et al.* Hydrogen production from bio-ethanol steam reforming reaction in a Pd/PSS membrane reactor. *Catal. Today* **2012**, *193*, 42–48.

83. Iulianelli, A.; Seelam, P.K.; Liguori, S.; Longo, T.; Keiski, R.; Calabrò, V.; Basile, A. Hydrogen production for PEM fuel cell by gas phase reforming of glycerol as byproduct of bio-diesel. The use of a Pd-Ag membrane reactor at middle reaction temperature. *Int. J. Hydrogen Energy* **2011**, *36*, 3827–3834.

84. Saraswat, S.K.; Pant, K.K. Ni$_e$Cu$_e$Zn/MCM-22 catalysts for simultaneous production of hydrogen and multiwall carbon nanotubes via thermo-catalytic decomposition of methane. *Int. J. Hydrogen Energy* **2011**, *36*, 13352–13360.

85. Angeli, S.D.; Pilitsis, F.G.; Lemonidou, A.A. Methane steam reforming at low temperature: Effect of light alkanes' presence on coke formation. *Catal. Today* **2015**, *242*, 119–128.

86. Lee, S.Y.; Lim, H.; Woo, H.C. Catalytic activity and characterizations of Ni/K$_2$Ti$_x$O$_y$-Al$_2$O$_3$ catalyst for steam methane reforming. *Int. J. Hydrogen Energy* **2014**, *39*, 17645–17655.

87. Harshini, D.; Hyung, D.; Lee, J.; Jeong, Y.; Kim, S.; Woo Nam, H.C.; Ham, J.H.H.; Lim, T.-H.; Yoon, C.W. Enhanced oxygen storage capacity of Ce$_{0.65}$Hf$_{0.25}$M$_{0.1}$O$_{2\delta}$ (*M* = rare earth elements): Applications to methane steam reforming with high coking resistance. *Appl. Catal. B* **2014**, *148–149*, 415–423.

88. Homsi, D.; Aouad, S.; Gennequin, C.; El Nakat, J.; Aboukaïs, A.; Abi-Aad, E. The effect of copper content on the reactivity of Cu/Co_6Al_2 solids in the catalytic steam reforming of methane reaction. *Comptes Rendus Chim.* **2014**, *17*, 454–458.

89. Zhai, X.; Ding, S.; Liu, Z.; Jin, Y.; Cheng, Y. Catalytic performance of Ni catalysts for steam reforming of methane at high space velocity. *Int. J. Hydrogen Energy* **2011**, *36*, 482–489.

90. Homsi, D.; Aouad, S.; Gennequin, C.; Aboukaïs, A.; Abi-Aad, E. A highly reactive and stable $Ru/Co_{6−x}Mg_xAl_2$ catalyst for hydrogen production via methane steam reforming. *Int. J. Hydrogen Energy* **2014b**, *39*, 10101–10107.

91. Lim, M.-W.; Yong, S.-T.; Chai, S.-P. Combustion-synthesized Nickel-based Catalysts for the Production of hydrogen from steam reforming of methane. *Energy Procedia* **2014**, *61*, 910–913.

92. Mei, D.; Glezakou, V.-A.; Lebarbier, V.; Kovarik, L.; Wan, H.; Albrecht, K.O.; Gerber, M.; Rousseau, R.; Dagle, R.A. Highly active and stable $MgAl_2O_4$-supported Rh and Ir catalysts for methane steam reforming: A combined experimental and theoretical study. *J. Catal.* **2014**, *316*, 11–23.

93. Palma, S.; Bobadilla, L.F.; Corrales, A.; Ivanova, S.; Romero-Sarria, F.; Centeno, M.A.; Odriozola, J.A. Effect of gold on a $NiLaO_3$ perovskite catalyst for methane steam reforming. *Appl. Catal. B* **2014**, *144*, 846–854.

94. Kondratenko, V.A. Mechanistic aspects of the Andrussow process over Pt-Rh gauzes. Pathways of formation and consumption of HCN. *Appl. Catal. A* **2010**, *381*, 74–82.

95. Guo, Z.; Liu, B.; Zhang, Q.; Deng, W.; Wang, Y.; Yang, Y. Recent advances in heterogeneous selective oxidation catalysis for sustainable chemistry. *Chem. Soc. Rev.* **2014**, *43*, 3480–3524.

96. Schwartz, W.R.; Ciuparu, D.; Pfefferle, L.D. Combustion of Methane over palladium-based catalysts: Catalytic deactivation and Role of the Support. *J. Phys. Chem. C* **2012**, *116*, 8587–8593.

97. Cargnello, M.; Delgado Jaén, J.J.; Hernández Garrido, J.C.; Bakhmutsky, K.; Montini, T.; Calvino Gámez, J.J.; Gorte, R.J.; Fornasiero, P. Exceptional Activity for Methane Combustion over Modular $Pd@CeO_2$ Subunits on Functionalized Al_2O_3. *Science* **2012**, *337*, 713–717.

98. Nozakia, T.; Agıral, A.; Yuzawa, S.; Han Gardeniers, J.G.E.; Okazaki, K. A single step methane conversion into synthetic fuels using microplasma reactor. *Chem. Eng. J.* **2011**, *166*, 288–293.

99. Holmen, A. Direct conversion of methane to fuels and chemicals. *Catal. Today* **2009**, *142*, 2–8.

100. Oshima, K.; Tanaka, K.; Yabe, T.; Kikuchi, E.; Sekine, Y. Oxidative coupling of methane using carbon dioxide in an electric field over $La–ZrO_2$ catalyst at low external temperature. *Fuel* **2013**, *107*, 879–881.

101. Alayon, E.M.C. Copper Cores for the Conversion of Methane to Chemicals. Ph.D. Thesis, ETH Zurich for the degree of Doctor of Sciences, Zürich, Switzerland, 2012.

102. Wei, X.; Ye, L.; Yuan, Y. Low temperature catalytic conversion of methane to formic acid by simple vanadium compound with use of H_2O_2. *J. Nat. Gas Chem.* **2009**, *18*, 295–299.

103. Villa, K.; Murcia-López, S.; Andreu, T.; Ramón Morante, J. Mesoporous WO₃ photocatalyst for the partial oxidation of methane to methanol using electron scavengers. *Appl. Catal. B* **2015**, *163*, 150–155.

104. Li, T.; Wang, S.J.; Yu, C.S.; Ma, Y.C.; Li, K.L.; Lin, L.W. Direct conversion of methane to methanol over nano-[Au/SiO₂] in [Bmim]Cl ionic liquid. *Appl. Catal. A* **2011**, *398*, 150–154.

105. Chen, L.; Zhang, X.-W.; Huang, L.; Lei, L.-C. Partial oxidation of methane with air for methanol production in a post-plasma catalytic system. *Chem. Eng. Process.* **2009**, *48*, 1333–1340.

106. Indarto, A.; Ryook Yang, D.; Palgunadi, J.; Choi, J.-W.; Lee, H.; Song, H.K. Partial oxidation of methane with Cu-Zn-Al catalyst in a dielectric barrier discharge. *Chem. Eng. Process.* **2008**, *47*, 780–786.

107. Adebajo, M.O.; Frost, R.L. Recent Advances in Catalytic/Biocatalytic Conversion of Greenhouse Methane and Carbon Dioxide to Methanol and Other Oxygenates. In *Greenhouse Gases-Capturing, Utilization and Reduction*; Liu, G., Ed.; InTech Europe: Rijeka, Croatia, 2012; pp. 31–56.

108. Hameed, A.; Ismail, I.M.I.; Aslam, M.; Gondal, M.A. Photocatalytic conversion of methane into methanol: Performance of silver impregnated WO₃. *Appl. Catal. A* **2014**, *470*, 327–335.

109. Yuan, Q.; Deng, W.; Zhang, Q.; Wang, Y. Osmium-Catalyzed Selective Oxidations of Methane and Ethane with Hydrogen Peroxide in Aqueous Medium. *Adv. Synth. Catal.* **2007**, *349*, 1199–1209.

110. Polnišer, R.; Štolcová, M.; Hronec, M.; Mikula, M. Structure and reactivity of copper iron pyrophosphate catalysts for selective oxidation of methane to formaldehyde and methanol. *Appl. Catal. A* **2011**, *400*, 122–130.

111. Barbero, J.A.; Alvarez, M.C.; Bañares, M.A.; Peña, M.A.; Fierro, J.L.G. Breakthrough in the direct conversion of methane into C1-oxygenates. *Chem. Commun.* **2002**, 1184–1185.

112. Nagiev, T.M.; Abbasova, M.T. Oxidation of methane to methanol by hydrogen peroxide on a supported hematin catalyst. *Stud. Surf. Sci. Catal.* **2000**, *130*, 3837–3842.

113. Aoki, K.; Ohmae, M.; Nanba, T.; Takeishi, K.; Azuma, N.; Ueno, A.; Ohfune, H.; Hayashi, H.; Udagawa, Y. Direct conversion of methane into methanol over MoO₃/SiO₂ catalyst in an excess amount of water vapor. *Catal. Today* **1998**, *45*, 29–33.

114. Hiyoshi, N.; Ikeda, T. Oxidative coupling of methane over alkali chloride-Mn-Na₂WO₄/SiO₂ catalysts: Promoting effect of molten alkali chloride. *Fuel Process. Technol.* **2015**, *133*, 29–34.

115. Vatani, A.; Jabbari, E.; Askarieh, M.; Torangi, A.M. Kinetic modeling of oxidative coupling of methane over Li/MgO catalyst by genetic algorithm. *J. Nat. Gas Sci. Eng.* **2014**, *20*, 347–356.

116. Ivanov, D.V.; Isupova, L.A.; Gerasimov, E.Y.; Dovlitova, L.S.; Glazneva, T.S.; Prosvirin, I.P. Oxidative methane coupling over Mg, Al, Ca, Ba, Pb-promoted SrTiO₃ and Sr₂TiO₄: Influence of surface composition and microstructure. *Appl. Catal. A* **2014**, *485*, 10–19.

117. Ferreira, V.J.; Tavares, P.; Figueiredo, J.L.; Faria, J.L. Ce-Doped La_2O_3 based catalyst for the oxidative coupling of methane. *Catal. Commun.* **2013**, *42*, 50–53.

118. Wang, Z.; Zou, G.; Luo, X.; Liu, H.; Gao, R.; Chou, L.; Wang, X. Oxidative coupling of methane over $BaCl_2$-TiO_2-SnO_2 catalyst. *J. Nat. Gas Chem.* **2012**, *21*, 49–55.

119. Farsi, A.; Ghader, S.; Moradi, A.; Mansouri, S.S.; Shadravan, V. A simple kinetic model for oxidative coupling of methane over $La_{0.6}Sr_{0.4}Co_{0.8}Fe_{0.2}O_{3-\delta}$ nanocatalyst. *J. Nat. Gas Chem.* **2011**, *20*, 325–333.

120. Godini, H.R.; Gili, A.; Görke, O.; Arndt, S.; Simon, U.; Thomas, A.; Schomäcker, R.; Wozny, G. Sol–gel method for synthesis of Mn-Na_2WO_4/SiO_2 catalyst for methane oxidative coupling. *Catal. Today* **2014**, *236*, 12–22.

121. Majhi, S.; Dalai, A.K.; Pant, K.K. Methanol assisted methane conversion for higher hydrocarbon over bifunctional Zn-modified Mo/HZSM-5 catalyst. *J. Mol. Catal. A* **2015**, *398*, 368–375.

122. Abdelsayed, V.; Shekhawat, D.; Smith, M.W. Effect of Fe and Zn promoters on Mo/HZSM-5 catalyst for methane dehydroaromatization. *Fuel* **2015**, *139*, 401–410.

123. Kenarsari, S.D.; Jiang, D.Y.G.; Zhang, S.; Wang, J.; Russell, A.G.; Wei, Q.; Fan, M. Review of recent advances in carbon dioxide separation and capture. *RSC Adv.* **2013**, *3*, 22739–22773.

124. Cao, Z.; Jiang, H.; Luo, H.; Baumann, S.; Meulenberg, W.A.; Assmann, J.; Mleczko, L.; Liu, Y.; Caro, J. Natural Gas to Fuels and Chemicals: Improved Methane Aromatization in an Oxygen Permeable Membrane Reactor. *Angew. Chem. Int. Ed.* **2013**, *52*, 13794–13797.

125. Schädel, B.T.; Duisberg, M.; Deutschmann, O. Steam reforming of methane, ethane, propane, butane, and natural gas over a rhodium-based catalyst. *Catal. Today* **2009**, *142*, 42–51.

126. Huang, X.; Reimert, R. Kinetics of steam reforming of ethane on Ni/YSZ (yttria-stabilised zirconia) catalyst. *Fuel* **2013**, *106*, 380–387.

127. Olafsen, A.; Slagtern, Å.; Dahl, I.M.; Olsbye, U.; Schuurman, Y.; Mirodatos, C. Mechanistic features for propane reforming by carbon dioxide over a Ni/Mg (Al) O hydrotalcite-derived catalyst. *J. Catal.* **2005**, *229*, 163–175.

128. Resini, C.; Delgado, M.C.H.; Arrighi, L.; Alemany, L.J.; Marazza, R.; Busca, G. Propene *versus* propane steam reforming for hydrogen production over Pd-based and Ni-based catalysts. *Catal. Commun.* **2005**, *6*, 441–445.

129. Faria, W.L.; Dieguez, L.C.; Schmal, M. Autothermal reforming of propane for hydrogen production over Pd/CeO2/Al2O3 catalysts. *Appl. Catal. B* **2008**, *85*, 77–85.

130. Jensen, M.B.; Råberg, L.B.; Sjåstad, A.O.; Olsbye, U. Mechanistic study of the dry reforming of propane to synthesis gas over a Ni/Mg (Al) O catalyst. *Catal. Today* **2009**, *145*, 114–120.

131. Igarashi, A.; Ohtaka, T.; Motoki, S. Low-temperature steam reforming ofn-butane over Rh and Ru catalysts supported on ZrO2. *Catal. Lett.* **1992**, *13*, 189–194.

132. Wang, X.; Gorte, R.J. Steam reforming of n-butane on Pd/ceria. *Catal. Lett.* **2001**, *73*, 15–19.

133. Avcı, A.K.; Trimm, D.L.; Aksoylu, A.E.; Önsan, Z.I. Hydrogen production by steam reforming of n-butane over supported Ni and Pt-Ni catalysts. *Appl. Catal. A* **2004**, *258*, 235–240.

134. Laosiripojana, N.; Sangtongkitcharoen, W.; Assabumrungrat, S. Catalytic steam reforming of ethane and propane over CeO 2-doped Ni/Al2O3 at SOFC temperature: Improvement of resistance toward carbon formation by the redox property of doping CeO_2. *Fuel* **2006**, *85*, 323–332.

135. Nagaoka, K.; Sato, K.; Nishiguchi, H.; Takita, Y. Highly active Ni/MgO in oxidative steam pre-reforming of n-butane for fuel cell application. *Catal. Commun.* **2007**, *8*, 1807–1810.

136. Osipovs, S. Sampling of benzene in tar matrices from biomass gasification using two different solid-phase sorbents. *Anal. Bioanal. Chem.* **2008**, *391*, 1409–1417.

137. Williams, P.T.; Williams, E.A. Recycling plastic waste by pyrolysis. *J. Inst. Energy* **1998**, *71*, 81–93.

138. Liu, Y.; Qian, J.; Wang, J. Pyrolysis of polystyrene waste in a fluidized-bed reactor to obtain styrene monomer and gasoline fraction. *Fuel Process. Technol.* **2000**, *63*, 45–55.

139. Li, B.; Chen, H.P.; Yang, H.P.; Yang, G.L.; Wang, X.H.; Zhang, S.H. Ni/γ-Al_2O_3 Catalyst for CO_2 Reforming of Benzene as a Model Compound of Biomass Gasification Tar: Promotional Effect of Ultrasonic Treatment on Catalytic Performance. In Proceedings of the 20th International Conference on Fluidized Bed Combustion; Springer Berlin Heidelberg: Berlin, Germany, 2010; pp. 576–582.

140. Park, H.J.; Park, S.H.; Sohn, J.M.; Park, J.; Jeon, J.K.; Kim, S.S.; Park, Y.K. Steam reforming of biomass gasification tar using benzene as a model compound over various Ni supported metal oxide catalysts. *Bioresour. Technol.* **2010**, *101*, S101–S103.

141. Sarvaramini, A.; Larachi, F. Mossbauer spectroscopy and catalytic reaction studies of chrysotile-catalyzed steam reforming of benzene. *J. Phys. Chem. C* **2011**, *115*, 6841–6848.

142. Swierczynski, D.; Courson, C.; Kiennemann, A. Study of steam reforming of toluene used as model compound of tar produced by biomass gasification. *Chem. Eng. Process.* **2008**, *47*, 508–513.

143. Zhao, B.; Zhang, X.; Chen, L.; Qu, R.; Meng, G.; Yi, X.; Sun, L. Steam reforming of toluene as model compound of biomass pyrolysis tar for hydrogen. *Biomass Bioenergy* **2010**, *34*, 140–144.

144. Grenoble, D.C. The chemistry and catalysis of the water/toluene reaction: 2. The role of support and kinetic analysis. *J. Catal.* **1978**, *51*, 212–220.

145. Duprez, D.; Miloudi, A.; Little, J.; Bousquet, J. The role of the metal/support interface in toluene steam reforming over rhodium-alumina catalysts. *Appl. Catal.* **1983**, *5*, 219–226.

146. Colby, J.L.; Wang, T.; Schmidt, L.D. Steam reforming of benzene as a model for biomass-derived syngas tars over Rh-based catalysts. *Energy Fuels* **2009**, *24*, 1341–1346.

147. Mei, D.; Lebarbier, V.M.; Rousseau, R.; Glezakou, V.A.; Albrecht, K.O.; Kovarik, L.; Dagle, R.A. Comparative investigation of benzene steam reforming over spinel supported Rh and Ir catalysts. *ACS Catal.* **2013**, *3*, 1133–1143.

148. Quitete, C.P.; Bittencourt, R.C. P.; Souza, M.M. Steam reforming of tar using toluene as a model compound with nickel catalysts supported on hexaaluminates. *Appl. Catal. A* **2014**, *478*, 234–240.

149. Jackson, S.D.; Thomson, S.J.; Webb, G. Carbonaceous deposition associated with the catalytic steam-reforming of hydrocarbons over nickel alumina catalysts. *J. Catal.* **1981**, *70*, 249–263.

150. Asadullah, M.; Ito, S.I.; Kunimori, K.; Yamada, M.; Tomishige, K. Energy efficient production of hydrogen and syngas from biomass: Development of low-temperature catalytic process for cellulose gasification. *Environ. Sci. Technol.* **2002**, *36*, 4476–4481.

151. Świerczyński, D.; Libs, S.; Courson, C.; Kiennemann, A. Steam reforming of tar from a biomass gasification process over Ni/olivine catalyst using toluene as a model compound. *Appl. Catal. B* **2007**, *74*, 211–222.

152. Li, C.; Hirabayashi, D.; Suzuki, K. Development of new nickel based catalyst for biomass tar steam reforming producing H_2-rich syngas. *Fuel Process. Technol.* **2009**, *90*, 790–796.

153. Li, C.; Hirabayashi, D.; Suzuki, K. Steam reforming of biomass tar producing H_2-rich gases over $Ni/MgO_x/CaO_{1-x}$ catalyst. *Bioresour. Technol.* **2010**, *101*, S97–S100.

154. Mukai, D.; Tochiya, S.; Murai, Y.; Imori, M.; Hashimoto, T.; Sugiura, Y.; Sekine, Y. Role of support lattice oxygen on steam reforming of toluene for hydrogen production over $Ni/La_{0.7}Sr_{0.3}AlO_{3-\delta}$ catalyst. *Appl. Catal. A* **2013**, *453*, 60–70.

155. Oemar, U.; Ang, P.S.; Hidajat, K.; Kawi, S. Promotional effect of Fe on perovskite $LaNi_xFe_{1-y}O_3$ catalyst for hydrogen production via steam reforming of toluene. *Int. J. Hydrogen Energy* **2013**, *38*, 5525–5534.

156. Oemar, U.; Ang, M.L.; Hee, W.F.; Hidajat, K.; Kawi, S. Perovskite $La_xM_{1-x}Ni_{0.8}Fe_{0.2}O_3$ catalyst for steam reforming of toluene: Crucial role of alkaline earth metal at low steam condition. *Appl. Catal. B* **2014**, *148*, 231–242.

157. Coll, R.; Salvado, J.; Farriol, X.; Montane, D. Steam reforming model compounds of biomass gasification tars: Conversion at different operating conditions and tendency towards coke formation. *Fuel Process. Technol.* **2001**, *74*, 19–31.

158. Bradford, M.C.; Vannice, M.A. Catalytic reforming of methane with carbon dioxide over nickel catalysts I. Catalyst characterization and activity. *Appl. Catal. A* **1996**, *142*, 73–96.

159. Vagia, E.C.; Lemonidou, A.A. Thermodynamic analysis of hydrogen production via steam reforming of selected components of aqueous bio-oil fraction. *Int. J. Hydrogen Energy* **2007**, *32*, 212–223.

160. Palmeri, N.; Chiodo, V.; Freni, S.; Frusteri, F.; Bart, J.C.J.; Cavallaro, S. Hydrogen from oxygenated solvents by steam reforming on Ni/Al_2O_3 catalyst. *Int. J. Hydrogen Energy* **2008**, *33*, 6627–6634.

161. Hu, X.; Zhang, L.; Lu, G. Pruning of the surface species on Ni/Al_2O_3 catalyst to selective production of hydrogen via acetone and acetic acid steam reforming. *Appl. Catal. A* **2012**, *427*, 49–57.

162. Trane-Restrup, R.; Resasco, D.E.; Jensen, A.D. Steam reforming of light oxygenates. *Catal. Sci. Technol.* **2013**, *3*, 3292–3302.

163. Navarro, R.M.; Guil-Lopez, R.; Gonzalez-Carballo, J.M.; Cubero, A.; Ismail, A.A.; Al-Sayari, S.A.; Fierro, J.L.G. Bimetallic MNi/Al$_2$O$_3$-La catalysts (M = Pt, Cu) for acetone steam reforming: Role of M on catalyst structure and activity. *Appl. Catal. A* **2014**, *474*, 168–177.

164. Navarro, R.M.; Guil-Lopez, R.; Ismail, A.A.; Al-Sayari, S.A.; Fierro, J.L.G. Ni- and PtNi-catalysts supported on Al$_2$O$_3$ for acetone steam reforming: Effect of the modification of support with Ce, La and Mg. *Catal. Today* **2015**, *242*, 60–70.

165. Güell, B.M.; Babich, I.; Nichols, K.P.; Gardeniers, J.G.E.; Lefferts, L.; Seshan, K. Design of a stable steam reforming catalyst—A promising route to sustainable hydrogen from biomass oxygenates. *Appl. Catal. B* **2009**, *90*, 38–44.

166. Devianto, H.; Han, J.; Yoon, S.P.; Nam, S.W.; Lim, T.-H.; Oh, I.-H.; Hong, S.-A.; Lee, H.-I. The effect of impurities on the performance of bioethanol-used internal reforming molten carbonate fuel cell. *Int. J. Hydrogen Energy* **2011**, *36*, 10346–10354.

167. Rass-Hansen, J.; Johansson, R.; Moller, M.; Christensen, C.H. Steam reforming of technical bioethanol for hydrogen production. *Int. J. Hydrogen Energy* **2008**, *33*, 4547–4554.

168. Medrano, J.A.; Oliva, M.; Ruiz, J.; García, L.; Arauzo, J. Catalytic steam reforming of butanol in a fluidized bed and comparison with other oxygenated compounds. *Fuel Process. Technol.* **2014**, *124*, 123–133.

169. Nahar, G.A.; Madhani, S.S. Thermodynamics of hydrogen production by the steam reforming of butanol: Analysis of inorganic gases and light hydrocarbons. *Int. J. Hydrogen Energy* **2010**, *35*, 98–109.

170. Piccot, D.S.; Watson, J.J.; Jones, W.J. A Global inventory of organic compounds emissions from anthropogenic sources. *J. Geophys. Res.* **1992**, *97*, 9897–9912.

171. Wei, W.; Wang, S.; Hao, J.; Cheng, S. Trends of chemical speciation profiles of anthropogenic volatile organic compounds emission in China, 2005–2020. *Front. Environ. Sci. Eng.* **2014**, *8*, 27–41.

172. Katzenstein, A.S.; Doezema, L.A.; Simpson, I.J.; Blake, D.R.; Rowland, F.S. Extensive regional atmospheric hydrocarbon pollution in the southwestern United States. *Proc. Natl. Acad. Sci.* **2003**, *100*, 11975–11979.

173. Birch, L.; Bachofen, R. Microbial production of hydrocarbons. In *Biotechnology*; Rehm, H.-J., Ed.; VCH: Weinheim, Germany, 1988; Volume 6b, pp. 71–99.

174. Tissot, B.; Welte, D. *Petroluem Formation and Occurrence*; Springer Verlag: Berlin, Germany, 1984.

175. Van Zyl, R.L.; Seatlholo, S.T.; van Vuuren, S.F.; Viljoen, A.M. The biological activities of 20 nature identical essential oil constituents. *J. Essent. Oil Res.* **2006**, *18*, 129–133.

176. Ferris, S.W. *Handbook of Hydrocarbons*; Academic Press: Yew York, NY, USA, 1955; p. 324.

177. Wert, B.P.; Trainer, M.; Fried, A.; Ryerson, T.B.; Henry, B.; Potter, W.; Wisthaler, A. (2003). Signatures of terminal alkene oxidation in airborne formaldehyde measurements during TexAQS. *J. Geophys. Res.: Atmos.* **2000**, *108*.

178. Johansson, J.K.; Mellqvist, J.; Samuelsson, J.; Offerle, B.; Lefer, B.; Rappenglück, B.; Yarwood, G. Emission measurements of alkenes, alkanes, SO$_2$, and NO$_2$ from stationary sources in Southeast Texas over a 5 year period using SOF and mobile DOAS. *J. Geophys. Res.* **2014**, *119*, 1973–1991.

179. Bruice, P.Y. *Organic Chemistry*, 5th ed.; Pearson Education: Upper Saddle River, NJ, USA, 2007; pp. 258–286.

180. Schobert, H. Production of Acetylene and Acetylene-based Chemicals from Coal. *Chem. Rev.* **2014**, *114*, 1743–1760.

181. Williams, A.; Smith, D.B. The combustion and oxidation of acetylene. *Chem. Rev.* **1970**, *70*, 267–293.

182. Liu, Y.; Shao, M.; Fu, L.; Lu, S.; Zeng, L.; Tang, D. Source profiles of volatile organic compounds (VOCs) measured in China: Part I. *Atmos. Environ.* **2008**, *42*, 6247–6260.

183. Parrish, D.D.; Kuster, W.C.; Shao, M.; Yokouchi, Y.; Kondo, Y.; Goldan, P.D.; de Gouw, J.A.; Koike, M.; Shirai, T. Comparison of air pollutant emissions among mega-cities. *Atmos. Environ.* **2009**, *43*, 6435–6441.

184. Lin, T.J.; Meng, X.; Shi, L. Catalytic hydrocarboxylation of acetylene to acrylic acid using Ni$_2$O$_3$ and cupric bromide as combined catalysts. *J. Mol. Catal. A: Chem.* **2015**, *396*, 77–83.

185. Matar, S.; Hatch, L.F. *Chemistry of Petrochemical Processes*, 2nd ed.; Gulf Professional Publishing: Boston, MA, USA, 2001; pp. 262–300.

186. Wang, H.; Lie, N.; Li, L.; Wang, Y.; Wang, G.; Wang, J.; Hao, Z. Characterization and assessment of volatile organic compounds (VOCs) emissions from typical industries. *Chin. Sci. Bull.* **2013**, *58*, 724–730.

187. El-Naas, M.; Acio, J.A.; El Telib, A.E. Aerobic biodegradation of BTEX: Progresses and Prospects. *J. Environ. Chem. Eng.* **2014**, *2*, 1104–1122.

188. United States Environmental Protection Agency (EPA). Available online: http://www.epa.gov/reg3hwmd/bf-lr/regional/analytical/semi-volatile.htm (accessed on 30 January 2015).

189. Yassaa, N.; Brancaleoni, E.; Frattoni, M.; Ciccioli, P. Isomeric analysis of BTEXs in the atmosphere using b-cyclodextrin capillary chromatography coupled with thermal desorption and mass spectrometry. *Chemosphere* **2006**, *63*, 502–508.

190. Durmusoglu, E.; Taspinar, F.; Karademir, A. Health risk assessment of BTEX emissions in the landfill environment. *J. Hazard. Mater.* **2010**, *176*, 870–877.

191. Tsai, J.-H.; Lin, K.-H.; Chen, C.-Y.; Lai, N.; Ma, S.-Y.; Chiang, H.-L. Volatile organic compound constituents from an integrated iron and steel facility. *J. Hazard. Mater.* **2008**, *157*, 569–578.

192. Cai, C.; Geng, F.; Tie, X.; Yu, Q.; An, J. Characteristics and source apportionment of VOCs measured in Shanghai, China. *Atmos. Environ.* **2010**, *44*, 5005–5014.

193. Atkinson, R. Gas phase tropospheric chemistry of organic compounds. In *Volatile Organic Compounds in the Atmosphere, Issues in Environmental Science and Technology 4*; Hester, R.E., Harrison, R.M., Eds.; The Royal Society of Chemistry: Letchworth, UK, 1995; pp. 65–89.

194. Huang, D.-Y.; Zhou, S.-G.; Hong, W.; Feng, W.-F.; Tao, L. Pollution characteristics of volatile organic compounds, polycyclic aromatic hydrocarbons and phthalate esters emitted from plastic wastes recycling granulation plants in Xingtan Town, South China. *Atmos. Environ.* **2013**, *71*, 327–334.

195. Huang, H.; Ye, X.; Huang, W.; Chen, J.; Xu, Y.; Wu, M.; Shao, Q.; Peng, Z.; Ou, G.; Shi, J.; *et al.* Ozone-catalytic oxidation of gaseous benzene over MnO_2/ZSM-5 at ambient temperature: Catalytic deactivation and its suppression. *Chem. Eng. J.* **2015**, *264*, 24–31.

196. Scirè, S.; Liotta, L.F. Supported gold catalysts for the total oxidation of volatile organic compounds. *Appl. Catal. B* **2012**, *125*, 222–246.

197. Einaga, H.; Teraoka, Y.; Ogata, A. Catalytic oxidation of benzene by ozone over manganese oxides supported on USY zeolite. *J. Catal.* **2013**, *305*, 227–237.

198. Ren, C.; Liu, X.; Wang, G.; Miao, S.; Chen, Y. Thermo-photocatalytic degradation of benzene on Pt-loaded TiO_2/ZrO_2. *J. Mol. Catal. A* **2012**, *358*, 31–37.

199. Rezaei, E.; Soltan, J. EXAFS and kinetic study of MnO_x/γ-alumina in gas phase catalytic oxidation of toluene by ozone. *Appl. Catal. B* **2014**, *148–149*, 70–79.

200. Legreid, G.; Lööv, J.B.; Staehelin, J.; Hueglinh, C.; Hill, M.; Buchmann, B.; Prevot, A.S.H.; Reimann, S. Oxygenated volatile organic compounds (OVOCs) at an urban background site in Zürich (Europe): Seasonal variation and source allocation. *Atmos. Environ.* **2007**, *41*, 8409–8423.

201. Placet, M.; Mann, C.O.; Gilbert, R.O.; Niefer, M.J. Emissions of ozone precursors from stationary sources: A critical review. *Atmos. Environ.* **2000**, *34*, 2183–2204.

202. Sawyer, R.F.; Harley, R.A.M.; Cadle, S.H.; Norbeck, J.M.; Slott, R.; Bravo, H.A. Mobile sources critical review: 1998 NARSTO assessment. *Atmos. Environ.* **2000**, *34*, 2161–2181.

203. AMEC Earth & Environmental Limited and University of Calgary. Assessment report on reduced sulphur compounds for developing ambient air quality objectives, Alberta environment. 2004. Available online: http://environment.gov.ab.ca/info/library/6664.pdf (accessed on 3 March 2015).

204. Kulkarni, S. Alkyl Halides. In *Encyclopedia of Toxicology*; Elsevier: Berlin, Germany, 2014; pp. 144–145.

205. Cram101 Textbook Reviews. In *Study Guide for Introductory Chemistry: Atoms First by Russo, Steve*, 5th ed.; Cram101: Bellevue, WA, USA, 2014.

206. Huang, B.; Lei, C.; Wei, C.; Zeng, G. Chlorinated volatile organic compounds (Cl-VOCs) in environment—Sources, potential human health impacts, and current remediation technologies. *Environ. Int.* **2014**, *71*, 118–138.

207. Wei, W.; Wang, S.; Chatani, S.; Klimont, Z.; Cofala, J.; Hao, J. Emission and speciation of non-methane volatile organic compounds from anthropogenic sources in China. *Atmos. Environ.* **2008**, *42*, 4976–4988.

208. US EPA. *EPA Coalbed Methane Outreach Program Technical Options Series*; United States Environmental Protection Agency, Office of Air and Radiation 6202J: Washington, DC, USA, 1998.

209. Rojo, F. Enzymes for aerobic degradation of alkanes. In *Handbook of Hydrocarbon and Lipid Microbiology*; Springer: Berlin, Germany, 2010; pp. 781–797.

210. Singh, S.N.; Kumari, B.; Mishra, S. Microbial degradation of alkanes. In *Microbial Degradation of Xenobiotics*; Springer: Berlin, Germany, 2012; pp. 439–469.

211. Wentzel, A.; Ellingsen, T.E.; Kotlar, H.K. Bacterial metabolism of long-chain *n*-alkanes. *Appl. Microbial. Biotechnol.* **2007**, *76*, 1209–1221.

212. Heider, J.; Spormann, A.M.; Beller, H.R. Anaerobic bacterial metabolism of hydrocarbons. *FEMS Microbiol. Rev.* **1998**, *22*, 459–473.

213. Rojo, F. Degradation of alkanes by bacteria. *Environ. Microbial.* **2009**, *11*, 2477–2490.

214. Reetz, M.T.; Daligault, F.; Brunner, B.; Hinrichs, H.; Deege, A. Directed evolution of cyclohexanone monooxygenases: Enantioselective biocatalysts for the oxidation of prochiral thioethers. *Angew. Chem.* **2004**, *116*, 4170–4173.

215. De Gonzalo, G.; Mihovilovic, M.D.; Fraaije, M.W. Recent developments in the application of Baeyer–Villiger monooxygenases as biocatalysts. *ChemBioChem* **2010**, *11*, 2208–2231.

216. Smith, T.J.; Dalton, H. Biocatalysis by methane monooxygenase and its implications for the petroleum industry. *Stud. Surf. Sci. Catal.* **2004**, *151*, 177–192.

217. Akella, S.; Pothini, S.; Veerashekhar, G.E. Recent progress in oxidation of n-alkanes by heterogeneous catalysis. *Res. Rev. Mater. Sci. Chem.* **2012**, *1*, 75–103.

218. Kirillov, A.M.; Kirillova, M.V.; Pombeiro, A.J. Multicopper complexes and coordination polymers for mild oxidative functionalization of alkanes. *Coord. Chem. Rev.* **2012**, *256*, 2741–2759.

219. Chepaikin, E.G. Oxidative functionalization of alkanes under dioxygen in the presence of homogeneous noble metal catalysts. *J. Mol. Catal. A* **2014**, *385*, 160–174.

220. Jia, C.; Kitamura, T.; Fujiwara, Y. Catalytic functionalization of arenes and alkanes via CH bond activation. *Acc. Chem. Res.* **2001**, *34*, 633–639.

221. Maihom, R.; Khongpracha, P.; Sirijaraensre, J.; Limytakul, J. Mechanistic studies on the transformation of ethanol into ethene over Fe-ZSM-5 zeolite. *ChemPhysChem* **2013**, *14*, 101–107.

222. Szymanski, G.S.; Rychlicki, G.; Terzyk, A.P. Catalytic conversion of ethanol on carbon catalysts. *Carbon* **1994**, *32*, 265–271.

223. Eastham, G.R.; Heaton, B.T.; Iggom, J.A.M.; Tooze, R.P.; Whyman, R.; Zacchini, S. Synthesis and spectroscopic characterization of all the intermediates in the Pd-catalysed methoxycarbonylation of ethene. *Chem. Commun.* **2000**, *7*, 609–910.

224. Samel, U.-R.; Kohler, W.; Gamer, A.O.; Keuser, U. Propionic acid and derivatives. *Ullman's Encycl. Ind. Chem.* **2000**.

225. OECD SIDS–Ethylene, UNEP Publications. Available online: http://en.wikipedia.org/wiki/Ethylene (accessed on 2 March 2015).

226. Material Safety Data Sheets, 1-Pentene, Acros Organics. Available online: http://www.fishersci.com (accessed on 30 June 2015).

227. Hashimi, S.K.; Sehwarz, L. Switch from Palladium-Catalyzed Cycloisomerization/Dimerization of Terminal Allenyl Ketones to Preferential Formation of Monomers by a 5-Palladatricyclo[4.1.0.0^{2,4}]heptane Catalyst: Synthesis of Furans from Substrates Incompatible with the Commonly Used Silver Catalysts. *Eur. J. Inorg. Chem.* **2006**.

228. Davarpanah, J.; Kiasat, A.R. Synthesis and characterization of SBA-polyperoxyacid: An efficient heterogeneous solid peroxyacid catalyst for epoxidation of alkenes. *Catal. Commun.* **2014**, *46*, 75–80.

229. Crosthwaite, J.M.; Farmer, V.A.; Hallett, J.P.; Welton, T. Epoxidation of alkenes by Oxone™ using 2-alkyl-3,4-dihydroisoquinolium salts as catalysts in ionic liquids. *J. Mol. Catal. A* **2008**, *279*, 148–152.

230. Pescarmona, P.P.; Jacobs, P.A. A high-throughput experimentation study of the epoxidation of alkenes with transition-metal-free heterogeneous catalysts. *Catal. Today* **2008**, *137*, 52–60.

231. Salmi, T.; Carucci, J.H.; Roche, M.; Eränen, K.; Wärnå, J.; Murzin, D. Microreactors as tools in kinetic investigations: Ethylene oxide formation on silver catalyst. *Chem. Eng. Sci.* **2013**, *87*, 306–314.

232. Hernández Carucci, J.R.; Halonen, V.; Eränen, K.; Wärnå, J.; Ojala, S.; Huuhtanen, M.; Keiski, R.; Salmi, T. Ethylene Oxide Formation in a Microreactor: From Qualitative Kinetics to Detailed Modelling. *Ind. Eng. Chem. Res.* **2010**, *49*, 10897–10907.

233. Smidt, J.; Hafner, W.; Jira, R.; Sedlmeier, J.; Siener, R.; Rüttinger, R.; Kojer, H. Catalytic reactions of olefins on compounds of the platinum group. *Angew. Chem.* **1959**, *71*, 176–182.

234. Cornell, C.N.; Sigman, M.S. Recent progress in Wacker oxidations: Moving toward molecular oxygen as the sole oxidant. *Inorg. Chem.* **2007**, *46*, 1903–1909.

235. Chen, J.; Che, C.-M. A Practical and mild method for the highly selective conversion of terminal alkanes into aldehydes through epoxidation–isomerization with ruthenium (IV)-phorphyrin catalysts. *Angew. Chem.* **2004**, *116*, 5058–5062.

236. Conte, M.; Carley, A.F.; Heirene, C.; Willock, D.J.; Johnston, P.; Herzing, A.; Kiely, C.J.; Hutching, G.J. Hydrochlorination of acetylene using a supported gold catalyst: A study of the reaction mechanism. *J. Catal.* **2007**, *250*, 231–239.

237. Huang, C.; Zhu, M.; Kang, L.; Dai, B. Active carbon supported TiO_2-$AuCl_3$/AC catalyst with excellent stability for acetylene hydrochlorination reaction. *Chem. Eng. J.* **2014**, *242*, 69–75.

238. Lin, T.J.; Meng, X.; Shi, L. Ni-exchanged Y-zeolite: An efficient heterogeneous catalyst for acetylene hydrocarboxylation. *Appl. Catal. A* **2014**, *485*, 163–171.

239. Tang, C.-M.; Zeng, Y.; Yang, X.-G.; Lei, Y.-C.; Wang, G.-Y. The palladium catalyzed hydrocarboxylation of acetylene with carbon monoxide to acrylic acid under mild conditions. *J. Mol. Catal. A* **2009**, *314*, 15–20.

240. Beesu, M.; Periasamy, M. Stereoselective synthesis of α,β-unsaturated carboxylic acids from alkynes using the $Fe(CO)_5$/t-BuOK/AcOH/CH_2Cl_2 reagent system. *J. Organomet. Chem.* **2012**, *705*, 30–33.

241. Zhang, Y.; Riduan, S.N. Catalytic Hydrocarboxylation of Alkenes and Alkynes with CO_2. *Angew. Chem. Int. Ed.* **2011**, *50*, 6210–6212.

242. Periasamy, M.; Radhakrishnan, U.; Rameshkumar, C.; Brumet, J.-J. A New method for the regio and stereoselective hydrocarboxylation of alkynes using $NaHFe(CO)_4$/CH_2Cl_2 System. *Tetrahedron Lett.* **1997**, *38*, 1623–1626.

243. Fujihara, T.; Xu, T.; Semba, K.; Terao, J.; Tsuji, Y. Copper catalyzed hydrocarboxylation of alkynes using carbon dioxide and hydrosilanes. *Angew. Chem. Int. Ed.* **2011**, *50*, 523–527.

244. Dérien, S.; Duñach, E.; Périchon, J. From stoichiometry to catalysis: Electroreductive coupling of alkynes and carbon dioxide with nickel-bipyridine complexes. Magnesium ions as the key for catalysis. *J. Am. Chem. Soc.* **1991**, *113*, 8447–8454.

245. Zhang, H.; Dai, B.; Wang, X.; Xu, L.; Zhu, M. Hydrochlorination of acetylene to vinyl chloride monomer over bimetallic Au-La/SAC catalysts. *J. Ind. Eng. Chem.* **2012**, *18*, 49–54.

246. Li, X.; Pan, X.; Bao, X. Nitrogen doped carbon catalyzing acetylene conversion to vinyl chloride. *J. Energy Chem.* **2014**, *23*, 131–135.

247. Gulyaeva, Y.K.; Kaicher, V.V.; Zaikovskii, V.I.; Kovalyov, E.V.; Suknev, A.P.; Bal'zhinimaev, B.S. Selective hydrogenation of acetylene over novel Pd/fiberglass catalyst. *Catal. Today* **2015**, *245*, 139–146.

248. Yan, X.; Wheeler, J.; Jang, B.; Lin, W.-Y.; Zhao, B. Stable Au catalyst for selective hydrogenation of acetylene in ethylene. *Appl. Catal. A* **2014**, *487*, 36–44.

249. Azizi, Y.; Petit, C.; Pitchon, V. Formation of polymer-grade ethylene by selective hydrogenation of acetylene over Au/CeO$_2$ catalyst. *J. Catal.* **2008**, *256*, 338–344.

250. Teschner, D.; Vass, E.; Hävecker, M.; Zafeiratos, S.; Schnörch, P.; Sauer, H.; Knop-Gericke, A.; Schlögl, R.; Chamam, M.; Wootsch, A.; *et al.* Alkyne hydrogenation over Pd catalysts: A new paradigm. *J. Catal.* **2006**, *242*, 26–37.

251. McCue, A.J.; McRitchie, C.J.; Shepherd, A.M.; Anderson, J.A. Cu/Al$_2$O$_3$ catalysts modified with Pd for selective acetylene hydrogenation. *J. Catal.* **2014**, *319*, 127–135.

252. Liptáková, B.; Báhidský, M.; Hronec, M. Preparation of phenol from benzene by one-step reaction. *Appl. Catal. A* **2004**, *263*, 33–38.

253. Molinary, R.; Argurio, P.; Poerio, T. Vanadyl acetylacetonate filled PVDF membranes as the core of a liquid phase continuous process for pure phenol production from benzene. *J. Membr. Sci.* **2015**, *476*, 490–499.

254. Busca, G.; Berardinelli, S.; Resini, C.; Arrighi, L. Technologies for the removal of phenol from fluid streams: A short review of recent developments. *J. Hazard. Mater.* **2008**, *160*, 265–288.

255. Schmidt, R. Industrial catalytic processes—Phenol production. *Appl. Catal. A* **2005**, *280*, 89–103.

256. Maity, D.; Jagtap, R.; Kaistha, N. Systematic top-down economic plantwide control of thecumene process. *J. Process Control* **2013**, *23*, 1426–1440.

257. Sayyar, M.H.; Wakeman, R.J. Comparing two new routes for benzene hydroxylation. *Chem. Eng. Res. Des.* **2008**, *86*, 517–526.

258. INEOS group AG. Available online: http://www.ineos.com/ (accessed on 1 March 2015).

259. Molinary, R.; Argurio, P.; Lavorato, C. Review on reduction and partial oxidation of organic in photocatalytic (membrane) reactors. *Curr. Org. Chem.* **2013**, *17*, 2516–2537.

260. Baykan, D.; Oztas, N.A. Synthesis of iron orthophosphate catalysts by solution and solution combustion methods for the hydroxylation of benzene to phenol. *Mater. Res. Bull.* **2015**, *64*, 294–300.

261. Vulpescu, G.D.; Ruitenbeek, M.; van Lieshout, L.L; Correia, L.A.; Meyer, D.; Pex, P.P.A.C. One-step selective oxidation over a Pd-based catalytic membrane; evaluation of the oxidation of benzene to phenol as a model reaction. *Catal. Commun.* **2004**, *5*, 347–351.

262. Bahidsky, M.; Hronec, M. Direct gas phase hydroxylation of benzene over phosphate catalysts. *Catal. Today* **2004**, *91–92*, 13–16.

263. Lemke, K.; Ehrich, H.; Lohse, U.; Berndt, H.; Jähnisch, K. Selective hydroxylation of benzene to phenol over supported vanadium oxide catalysts. *Appl. Catal. A* **2003**, *243*, 41–51.

264. Joseph, J.; Singhal, S.; Jain, S.L.; Sivakumaran, R.; Kumar, B.; Sain, B. Studies on vanadium catalyzed direct hydroxylation of aromatic hydrocarbons using hydrogen peroxide as oxidant. *Catal. Today* **2009**, *141*, 211–214.

265. Callanan, L.H.; Burton, R.M.; Mullineux, J.; Engelbrecht, J.M.M.; Rau, U. Effect of semi-batch reactor configuration on aromatic hydroxylation reactions. *Chem. Eng. J.* **2012**, *180*, 255–262.

266. Yashima, T.; Kobayashi, Y.; Komatsu, T.; Namba, S. Hydroxylation of toluene with hydrogen peroxide on HY zeolites with various Si/Al ratios. *Stud. Surf. Sci. Catal.* **1993**, *75*, 1689–1692.

267. Motz, J.L.; Heinichen, H.; Hölderich, W.L. Direct hydroxylation of aromatics to their corresponding phenols catalysed by H-[Al]ZSM-5 zeolite. *J. Mol. Catal. A* **1998**, *136*, 175–184.

268. Bartoli, J.-F.; Lambert, F.; Morgenstern-Badarau, I.; Battioni, P.; Mansuy, D. Unusual efficiency of a non-heme iron complex as catalyst for the hydroxylation of aromatic compounds by hydrogen peroxide: Comparison with iron porphyrins. *C.R. Chim.* **2002**, *5*, 263–266.

269. Balland, V.; Mathieu, D.; Pons-Y-Moll, N.; Bartoli, J.F.; Banse, F.; Battioni, P.; Girerd, J.J.; Daniel Mansuyb, D. Non-heme iron polyazadentate complexes as catalysts for oxidations by H_2O_2: Particular efficiency in aromatic hydroxylations and beneficial effects of a reducing agent. *J. Mol. Catal. A* **2004**, *215*, 81–87.

270. Bianchi, D.; Bertoli, M.; Tassinari, R.; Ricci, M.; Vignola, R. Direct synthesis of phenols by iron-catalyzed biphasic oxidation of aromatic hydrocarbons with hydrogen peroxide. *J. Mol. Catal. A* **2003**, *200*, 111–116.

271. Yashima, T.; Nagase, S.; Komatsu, T.; Namba, S. Liquid-phase hydroxylation of xylene with hydrogen peroxide over zeolite catalysts. *Stud. Surf. Sci. Catal.* **1993**, *77*, 417–420.

272. Kang, Z.; Wang, E.; Mao, B.; Su, Z.; Gao, L.; Niu, L.; Shan, H.; Xu, L. Heterogeneous hydroxylation catalyzed by multi-walled carbon nanotubes at low temperature. *Appl. Catal. A* **2006**, *299*, 212–217.

273. Singh, A.P. Selective chlorination of various aromatics over zeolite catalysts. *Catal. Stud. Surf. Sci. Catal.* **1998**, *113*, 419–423.

274. Singh, A.P.; Mirajkar, S.P.; Sharma, S. Liquid phase bromination of aromatics over zeolite H-beta catalyst. *J. Mol. Catal. A* **1999**, *150*, 241–250.

275. Daou, T.J.; Boltz, M.; Tzanis, L.; Michelin, L.; Louis, B. Gas-phase chlorination of aromatics over FAU- and EMT-type zeolites. *Catal. Commun.* **2013**, *39*, 10–13.

276. Ju, J.; Li, Y.J.; Gao, J.R.; Jia, R.H.; Han, L.; Sheng, W.J.; Jia, Y.X. High selectively oxidative bromination of toluene derivatives by the H_2O_2–HBr system. *Chin. Chem. Lett.* **2011**, *22*, 382–384.

277. Olah, G.A.; Kuhn, S.J.; Flood, S.H.; Hardie, B.A. Aromatic Substitution. XIV. Ferric chloride catalyzed bromination of benzene and alkylbenzenes with bromine in nitromethane solution. *J. Am. Chem. Soc.* **1964**, *86*, 1039–1044.

278. Ratnasamy, P.; Singh, A.P.; Sharma, S. Halogenation over zeolite catalysts. *Appl. Catal. A* **1996**, *135*, 25–55.

279. Wang, P.-C.; Lu, M.; Zhu, J.; Song, Y.-M.; Xiong, X.-F. Regioselective nitration of aromatics under phase-transfer catalysis conditions. *Catal. Commun.* **2011**, *14*, 42–47.

280. Ma, X.M.; Li, B.D.; Chen, L.; Lu, M.; Lv, C.X. Selective nitration of aromatic compounds catalysed by Hβ zeolite using N_2O_5. *Chin. Chem. Lett.* **2012**, *23*, 809–812.

281. Nowrouzi, N.; Mehranpour, A.M.; Bashiri, E.; Shayan, Z. Aromatic nitration under neutral conditions using N-bromosuccinimide/silver(I) nitrate. *Tetrahedron Lett.* **2012**, *53*, 4841–4842.

282. Ott, J.; Gronemann, V.; Pontzen, F.; Fiedler, E.; Grossmann, G.; Kersebohm, D.B.; Weiss, G.; Witte, C. Methanol. In *Ullmann's Encyclopedia of Industrial Chemistry*; Bellussi, G., Bohnet, M., Bus, J., Drauz, K., Faulhammer, H., Greim, H., Jäckel, K.-P., Karst, U., Kleemann, A., Kutscher, B., *et al.*, Eds.; John Wiley and Sons: Weinheim, Germany, 2012; pp. 1–27.

283. Kung, H.H.; Smith, K.J. Methanol to chemical. In *Methanol Production and Use*; Cheng, W.H., Kung, H.H., Eds.; Marcel Dekker: New York, NY, USA, 1994; pp. 175–214.

284. Hordeski, M.F. *Alternative Fuels: The Future of the Hydrogen*, 2nd ed.; The Fairmont Press: Lilburn, GA, USA, 2008; pp. 1–32.

285. Kosaric, N.; Duvnjak, Z.; Farkas, A.; Sahm, H.; Bringer-Meyer, S.; Goebel, O.; Mayer, D. Ethanol. In *Ullmann's Encyclopedia of Industrial Chemistry*; Bellussi, G., Bohnet, M., Bus, J., Drauz, K., Faulhammer, H., Greim, H., Jäckel, K.-P., Karst, U., Kleemann, A., Kutscher, B., *et al.*, Eds.; John Wiley and Sons: Weinheim, Germany, 2011; pp. 333–403.

286. Hahn, H.D.; Dämbkes, G.; Rupprich, N.; Bahl, H.; Frey, G.D. Buthanols. In *Ullmann's Encyclopedia of Industrial Chemistry*; Bellussi, G., Bohnet, M., Bus, J., Drauz, K., Faulhammer, H., Greim, H., Jäckel, K.-P., Karst, U., Kleemann, A., Kutscher, B., *et al.*, Eds.; John Wiley and Sons: Weinheim, Germany, 2013; pp. 1–13.

287. McGrath, K.M.; Prakash, G.K.S.; Olah, G.A. Direct methanol fuel cells. *J. Ind. Eng. Chem.* **2004**, *10*, 1063–1080.

288. Kamarudin, S.K.; Achmad, F.; Daud, W.R.W. Overview on the application of direct methanol fuel cell (DMFC) for portable electronic devices. *Int. J. Hydrogen Energy* **2009**, *34*, 6902–6916.

289. Kundu, A.; Jang, J.H.; Gil, J.H.; Jung, C.R.; Lee, H.R.; Kim, S.-H.; Ku, B.; Oh, Y.S. Micro-fuel cells—Current development and applications. *J. Power Sources* **2007**, *170*, 67–78.

290. Ojala, S.; Laakso, I.; Maunula, T.; Silvonen, R.; Ylönen, R.; Lassi, U.; Keiski, R.-L. Abatement of malodorous pulp mill emissions by catalytic oxidation–Pilot experiments in Stora Enso Pulp Mill Oulu, Finland. *TAPPI J.* **2005**, *4*, 9–13.

291. Koivikko, N.; Laitinen, T.; Ojala, S.; Pitkäaho, S.; Kucherov, A.; Keiski, R.L. Formaldehyde production from methanol and methyl mercaptan over titania and vanadia based catalysts. *Appl. Catal. B* **2011**, *103*, 72–78.

292. Henze, M.; van Loosdrecht, M.; Ekama, G.; Brdjanovic, D. *Biological Wastewater Treatment: Principles, Modeling and Design*; IWA Publishing: London, UK, 2008; pp. 87–154.

293. Sakuth, M.; Mensing, T.; Schuler, J.; Heitmann, W.; Strehlke, G.; Mayer, D. Ethers, Aliphatic. In *Ullmann's Encyclopedia of Industrial Chemistry*; Bellussi, G., Bohnet, M., Bus, J., Drauz, K., Faulhammer, H., Greim, H., Jäckel, K.-P., Karst, U., Kleemann, A., Kutscher, B., *et al.*, Eds.; John Wiley and Sons: Weinheim, Germany, 2010; pp. 433–449.

294. Lewis, R.A. *Lewis' Dictionary of Toxicology*; CRC Press LLC: Boca Raton, MA, USA, 1998.

295. Leisch, H.; Omori, A.T.; Finn, K.J.; Gilmet, J.; Bissett, T.; Ilceski, D.; Hudlicky, T. Chemoenzymatic enantiodivergent total synthesis of (+)- and (+)-cdeine. *Tetrahedron* **2009**, *65*, 9862–9875.

296. Nadim, F.; Zack, P.; hoag, G.E.; Liu, S. United States experience with gasoline additives. *Energy Policy* **2001**, *29*, 1–5.

297. Wang, S.Z.; Ishihara, T.; Takita, Y. Partial oxidation of dimethyl ether over various supported metal catalysts. *Appl. Catal. A* **2002**, *228*, 167–176.

298. Yu, L.; Xu, J.; Sun, M.; Wang, X. Catalytic oxidation of dimethyl ether to hydrocarbons over SnO_2/MgO and SnO_2/CaO catalysts. *J. Nat. Gas Chem.* **2007**, *16*, 200–203.

299. Sun, M.; Yu, L.; Ye, F.; Diao, G.; Yu, Q.; Hao, Z.; Zheng, Y.; Yuan, L. Transition metal doped cryptomelane-type manganese oxide for low temperature catalytic combustion of dimethyl ether. *Chem. Eng. J.* **2013**, *220*, 320–327.

300. Park, S.; Watanabe, Y.; Nishita, Y.; Fukuoka, T.; Inagaki, S.; Kubota, Y. Catalytic conversion of dimethyl ether into propylene over MCM-68 zeolite. *J. Catal.* **2014**, *319*, 265–273.

301. Siegel, H.; Eggersdorfer, M. Ketones. In *Ullmann's Encyclopedia of Industrial Chemistry*; Bellussi, G., Bohnet, M., Bus, J., Drauz, K., Faulhammer, H., Greim, H., Jäckel, K.-P., Karst, U., Kleemann, A., Kutscher, B., *et al.*, Eds.; John Wiley and Sons: Weinheim, Germany, 2000; pp. 187–207.

302. Weber, M.; Pompetzki, W.; Bonmann, R.; Weber, M. Acetone. In *Ullmann's Encyclopedia of Industrial Chemistry*; Bellussi, G., Bohnet, M., Bus, J., Drauz, K., Faulhammer, H., Greim, H., Jäckel, K.-P., Karst, U., Kleemann, A., Kutscher, B., *et al.*, Eds.; John Wiley and Sons: Weinheim, Germany, 2014; pp. 1–19.

303. Suzuki, T.; Yamagiwa, N.; Matsuo, Y.; Sakamoto, S.; Yamaguchi, K.; Shibasaki, M.; Noyori, R. Catalytic asymmetric aldol reaction of ketones and aldehydes using chiral alkoxides. *Tetrahedron Lett.* **2001**, *42*, 4669–4671.

304. Vetere, V.; Merlo, A.B.; Casella, M.L. Chemoselective hydrogenation of aromatic ketones with Pt-based heterogeneous catalysts. Substituent effects. *Appl. Catal. A* **2015**, *491*, 70–77.

305. Malyala, R.V.; Rode, C.V.; Arai, M.; Hegde, S.G.; Chaudhari, R.V. Activity, selectivity and stability of Ni and bimetallic Ni-Pt supported on zeolite Y catalysts for hydrogenation of acetophenone and its substituted derivatives. *Appl. Catal. A* **2000**, *193*, 76–86.

306. Trasarti, A.F.; Bertero, N.M.; Apesteguia, C.R.; Marchi, A.J. Liquid-phase hydrogenation of acetophenone over silica-supported Ni, Co and Cu catalysts: Influence of metal and solvent. *Appl. Catal. A* **2014**, *475*, 282–291.

307. Mäki-Arvela, P.; Hajek, J.; Salmi, T.; Murzin, D.Y. Chemoselective hydrogenation of carbonyl compounds over heterogeneous catalysts. *Appl. Catal. A* **2005**, *292*, 1–49.

308. Bergault, I.; Fouilloux, P.; Joly-Vuillemin, C.; Delmas, H. Kinetics and intraparticle diffusion modelling of a complex multistep reaction: Hydrogenation of acetophenone over Rhodium catalyst. *J. Catal.* **1998**, *175*, 328–337.

309. Deshmukh, A.; Kinage, A.; Kumar, R.; Meijboom, R. An efficient heterogeneous catalytic system for chemoselective hydrogenation of unsaturated ketones in aqueous medium. *Polyhedron* **2010**, *29*, 3262–3268.

310. Madduri, A.V.R.; Harutyunyan, S.R.; Minnaard, A.J. Catalytic asymmetric alkylation of ketones using organometallic reagents. *Drug Discov. Today: Technol.* **2013**, *10*, e21–e27.

311. Anneken, D.J.; Both, S.; Christoph, R.; Fieg, G.; Steinberger, U.; Westfechtel, A. Fatty acids. In *Ullmann's Encyclopedia of Industrial Chemistry*; Bellussi, G., Bohnet, M., Bus, J., Drauz, K., Faulhammer, H., Greim, H., Jäckel, K.-P., Karst, U., Kleemann, A., Kutscher, B., *et al.*, Eds.; John Wiley and Sons: Weinheim, Germany, 2006; pp. 73–116.

312. Kubitschke, J.; Lange, H.; Strutz. Carboxylic acids, aliphatic. In *Ullmann's Encyclopedia of Industrial Chemistry*; Bellussi, G., Bohnet, M., Bus, J., Drauz, K., Faulhammer, H., Greim, H., Jäckel, K.-P., Karst, U., Kleemann, A., Kutscher, B., *et al.*, Eds.; John Wiley and Sons: Weinheim, Germany, 2014; pp. 1–18.

313. Mérel, D.S.; Do, M.L.T.; Gaillard, S.; Dupau, P.; Renaud, J.-L. Iron-catalyzed reduction of carboxylic and carbonic acid derivatives. *Coord. Chem. Rev.* **2015**, *288*, 50–68.

314. Misal Castro, L.C.; Li, H.; Sortais, J.-B.; Darcel, C. Selective switchable iron-catalyzed hydrosilylation of carboxylic acids. *Chem. Commun.* **2012**, *48*, 10514–10516.

315. Liu, H.-Q.; Liu, J.; Zhang, Y.-H.; Shao, C.D.; Yu, J.-X. Copper-catalyzed amide bond formation from formamides and carboxylic acids. *Chin. Chem. Lett.* **2015**, *26*, 11–14.

316. Riemenschneider, W.; Bolt, H. Esters, Organic. In *Ullmann's Encyclopedia of Industrial Chemistry*; Bellussi, G., Bohnet, M., Bus, J., Drauz, K., Faulhammer, H., Greim, H., Jäckel, K.-P., Karst, U., Kleemann, A., Kutscher, B., *et al.*, Eds.; John Wiley and Sons: Weinheim, Germany, 2005; pp. 245–266.

317. Koritala, S. Hydrogenation of unsaturated fatty esters with copper-chromite catalyst: Kinetics, mechanism and isomerization. *J. Am. Oil Chem. Soc.* **1970**, *47*, 463–466.

318. Niczke, L.; Czechowski, F.; Gawel, I. Oxidized rapeseed oil methyl ester as a bitumen flux: Structural changes in the ester during catalytic oxidation. *Prog. Org. Coat.* **2007**, *59*, 304–311.

319. Nguyen, B.N.T.; Leclerc, C.A. Catalytic partial oxidation of methyl acetate as a model to investigate the conversion of methyl esters to hydrogen. *Int. J. Hydrog. Energy* **2008**, *33*, 1295–1303.

320. Wachs, I.E. Production of Formaldehyde from Methyl Mercaptans. US Patent N: 5969191, 6 September 1999.

321. Mouammine, A.; Ojala, S.; Pirault-Roy, L.; Bensitel, M.; Keiski, R.; Brahmi, R. Catalytic Partial Oxidation of Methanol and Methyl Mercaptan: Studies on the Selectivity of TiO_2 and CeO_2 Supported V_2O_5 Catalysts. *Top. Catal.* **2013**, *56*, 650–657.

322. Krause, E.; Rroka, K. Making Formaldahyde from Methylene Chloride. US Patent N: 1616533, 7 August 1927.

323. Stauffer, J.E. Formaldehyde process. Separation. US patent N: 6822123 B2, 02 April 2003.

324. Gong, H.; Chen, Z.; Zhang, M.; Wu, W.; Wang, W.; Wang, W. Study on the deactivation of the deoxygen catalyst during the landfill gas upgrading process. *Fuel* **2015**, *144*, 43–49.

325. Karacan, Ö.; Ruiz, F.A.; Cotè, M.; Phipps, S. Coal mine methane: A review of capture and utilization practices with benefits to mining safety and to greenhouse gas reduction. *Int. J. Coal Geol.* **2011**, *86*, 121–156.

326. Stasińska, B.; Machocki, A. Catalysts for the utilization of methane from the coal mine ventilation air. *Pol. J. Chem. Technol.* **2007**, *9*, 29–32.

327. McMinn, E.; Moates, F.C.; Richardson, J.T. Catalytic Steam Reforming of Chlorocarbons: Catalyst Deactivation. *Appl. Catal. B* **2001**, *31*, 93–105.

328. Zhang, Q.-H.; Li, Y.; Xu, B.-Q. Reforming of Methane and Coalbed Methane over Nanocomposite Ni/ZrO_2. *Prepr. Pap.-Am. Chem. Soc. Div. Fuel Chem.* **2004**, *49*, 137.

329. Ojeifo, E.; Abaa, K.; Orsulak, M.; Pamidimukkala, P.K.; Sircar, S.; Sharma, S.; Vatsa, T. *Coalbed Methane: Recovery & Utilization in North Western San Juan*; Colorado EME 580: Integrative Design; Department of Energy and Mineral Engineering, Penn State University: State College, PA, USA, 2010.

330. Wachs, I.E. Treating Methanol-Containing Waste Gas Streams. U.S. Patent 5907066, 25 May 1999.

331. Burgess, T.L.; Gibson, A.G.; Furstein, S.J.; Wachs, I.E. Converting waste gases from pulp mills into value-added chemicals. *Environ. Prog.* **2002**, *21*, 137–141.

332. Kalantar Neyestanaki, A.; Klingstedt, F.; Salmi, T.; Murzin, D.Y. Deactivation of post combustion catalysts, a review. *Fuel* **2004**, *83*, 395–408.

333. Spivey, J.J.; Butt, J.B. Deactivation of catalysts in the oxidation of volatile organic compounds. *Catal. Today* **1992**, *11*, 465–500.

334. Spivey, J.J. Deactivation of reforming Catalysts. In *Fuel Cells: Technologies for fuel processing*; Shekhawat, D, Spivey, J.J., Berry, D.A., Eds.; Elsevier: Amsterdam, The Netherlands, 2011; pp. 285–315.

335. Hulteberg, C. Sulphur-tolerant catalysts in small-scale hydrogen production, a review. *Int. J. Hydrogen Energy* **2012**, *37*, 3978–3992.

336. Pitkäaho, S.; Ojala, S.; Maunula, T.; Savimäki, A.; Kinnunen, T.; Keiski, R.L. Oxidation of dichloromethane and perchloroethylene as single compounds and in mixtures. *Appl. Catal. B* **2011**, *102*, 395–403.

337. Pitkäaho, S.; Matejova, L.; Jiratova, K.; Ojala, S.; Keiski, R.L. Oxidation of perchloroethylene–Activity and selectivity of Pt, Pd, Rh and V2O5 catalysts supported on Al$_2$O$_3$, Al$_2$O$_3$-TiO$_2$, Al$_2$O$_3$-CeO$_2$, Part 2. *Appl. Catal. B* **2012**, *126*, 215–224.

338. Bartholomew, C.H. Mechanisms of catalyst deactivation. *Appl. Catal. A* **2001**, *212*, 17–60.

339. De Rivas, B.; López-Fonseca, R.; Gutiérrez-Ortiz, M.A.; Gutiérrez-Ortiz, J.I. Impact of induced chlorine-poisoning on the catalytic behavior of Ce$_{0.5}$Zr$_{0.5}$O$_2$ and Ce$_{0.15}$Zr$_{0.85}$O$_2$ in the gas-phase oxidation of chlorinated VOCs. *Appl. Catal. B* **2011**, *104*, 373–381.

340. EEA. Non-methane volatile organic compounds (NMVOC) emissions (Ape 004). European Environment Agency, 2014. Available online: http://www.eea.europa.eu/data-and-maps/indicators/eea-32-non-methane-volatile-1/assessment-4 (accessed on 31 January 2015).

341. E-PRTR (The European Pollutant Release and Transfer Register) 2012 data. Available online: http://prtr.ec.europa.eu (accessed on 28 January 2015).

342. Theloke, J.; Friedrich, R. Compilation of a database on the composition of anthropogenic VOC emissions for atmospheric modeling in Europe. *Atmos. Environ.* **2007**, *41*, 4148–4160.

343. U.S. EPA. Profile of the 2011 National Air emissions inventory U.S. EPA 2011 NEI version1.0. Office of Air Quality Planning&Standards, Emissions inventory&Analysis group, United States Environmental Protection Agency, April 2014. Available online: http://www.epa.gov/ttn/chief/net/lite_finalversion_ver10.pdf (accessed on 31 January 2015).

344. Wei, W.; Wang, S.; Hao, J.; Cheng, S. Projection of anthropogenic volatile organic compounds (VOCs) emissions in China for the period 2010–2020. *Atmos. Environ.* **2011**, *45*, 6863–6871.

345. EEA. Costs of air pollution from European industrial facilities 2008–2012–an updated assessment. European Environment Agency, 2014. Available online: http://www.eea.europa.eu/publications/costs-of-air-pllution-2008--2012 (accessed on 31 January 2015).

346. Centi, G.; Ciambelli, P.; Perathoner, S.; Russo, P. Environmental catalysis: Trends and outlook. *Catal. Today* **2002**, *75*, 3–15.

347. Centi, G.; van Santen, R. *Catalysis for Renewables*; Wiley-VCH Verlag GmbH & Co. KGaA: Weinheim, Germany, 2007; p. 448.

348. Ballivet-Tkatchenko, D.; Chambrey, S.; Keiski, R.; Ligabue, R.; Plasseraud, L.; Richard, P.; Turunen, H. Direct synthesis of dimethyl carbonate with supercritical carbon dioxide: Characterization of a key organotin oxide intermediate. *Catal. Today* **2006**, *115*, 80–87.

349. Jenssen, F.J.J.G.; van Santen, R.A. Environmental Catalysis. *NIOK Catal. Sci. Ser.* **1999**, *1*, 369.

350. Lowell Center for Sustainable Production. Available online: http://sustainableproduction.org/ (accessed on 2 May 2009).

351. Anastas, P.T.; Warner, J.C. *Green Chemistry: Theory and Practice*; Oxford University Press: New York, NY, USA, 1998.

352. Anastas, P.T.; Heine, G.L.; Williamson, T.C. *Green Engineering*; Oxford University Press: Washington, DC, USA, 2001.

353. Anastas, P.T.; Williamson, T.C. *Green Chemistry, Frontiers in Benign Chemical Syntheses and Processes*; Bookcraft Ltd.: Oxford, UK, 1998.

354. Saavalainen, P.; Majala, A.; Pongrácz, E.; Keiski, R.L. Designing for sustainability: Developing a tool for sustainable chemical process design. Manuscript submitted in 2011.

355. Manley, J.B.; Anastas, P.T.; Cue, B.W. Frontiers in Green Chemistry: Meeting the grand challenges for sustainability in R&D and manufacturing. *J. Clean. Prod.* **2008**, *16*, 743–750.

356. Quinn, M.; Kriebel, D.; Geiser, K.; Moure-Eraso, R. Sustainable production. A proposed strategy for the work environment. In *Approaches to Sustainable Development*; Forrant, R., Pyle, J.L., Lazonick, W., Levenstein, C., Eds.; University of Massachusetts Press: Amherst, MA, USA, 2001.

357. Grunes, J.; Zhu, J.; Somorjai, G.A. Catalysis and nanoscience. *Chem. Commun.* **2003**, 2257–2260.

358. Somorjai, G.A. On the move. *Nature* **2004**, *430*, 730.

359. Zhou, B.; Hermans, S.; Somorjai, G.A. *Nanotechnology in Catalysis*; Springer: New York, NY, USA, 2004; Volumes 1–3, p. 555,333.

Removal of Toluene over NaX Zeolite Exchanged with Cu²⁺

Douglas Romero, Dayan Chlala, Madona Labaki, Sébastien Royer,
Jean-Pierre Bellat, Igor Bezverkhyy, Jean-Marc Giraudon and
Jean-François Lamonier

Abstract: Toluene is a major air pollutant emitted from painting and metal coating processes and might have some health effects. Adsorption and catalytic complete oxidation are promising ways to retain or convert toluene into harmless products. The present work aims to develop a bifunctional material which can be used as an adsorbent and catalyst for low-temperature toluene removal. Copper zeolites were obtained by exchanging the sodium in the parent NaX zeolite with copper from aqueous solutions of $Cu(NO_3)_2 \cdot 2.5H_2O$. Several characterization techniques, H_2-TPR, XPS, XRD and N_2 physisorption, were used in order to evaluate the redox, surface, structural and textural properties of the materials, respectively. The various materials were tested in adsorption and catalytic processes. The sample with low copper content (1 wt. %) exhibited promising features in terms of toluene adsorption capacity and total oxidation. The results can be correlated to the presence of micropores and well dispersed CuO species.

Reprinted from *Catalysts*. Cite as: Romero, D.; Chlala, D.; Labaki, M.; Royer, S.; Bellat, J.-P.; Bezverkhyy, I.; Giraudon, J.-M.; Lamonier, J.-F. Removal of Toluene over NaX Zeolite Exchanged with Cu²⁺. *Catalysts* **2015**, *5*, 1479–1497.

1. Introduction

Volatile organic compounds (VOCs) are organic chemicals that have a high vapor pressure at room temperature. VOCs are numerous, varied and ubiquitous. Among them aromatic compounds are widely distributed in the environment and have been found to be a significant cancer risk factor, and to be implicated in neurological problems in urban children [1]. If inhaled or contacted, toluene can cause dermatitis (dry, red, cracked skin) and damage the nervous system and kidneys. Therefore, toluene emission control has become more stringent.

Natural sources of toluene are forest fires (incomplete combustion plants), volcanic eruptions and emissions from vegetation. These sources are minor compared with anthropogenic emissions from various oil conversion processes. Toluene emission results from the transformation of fossil fuels (oil, gas and coal). It is produced in combination with other substances (benzene, xylenes, *etc.*) as a result of various petrochemical processes such as catalytic reforming, steam cracking and dealkylation. Following petrochemical operations, the richest in toluene fractions will be distilled and purified to obtain pure trading toluene at 99%.

Toluene is a constituent of unleaded gasoline and replaced tetramethyl lead in order to improve the octane rating. As a result toluene is emitted during the vaporization of gasoline (petrol station, fuel transport and storage, *etc.*) and is present in vehicle exhaust gases (unburned products). Toluene is also found in industrial exhaust when used as a solvent or produced in the incineration processes.

Generally, the best way to reduce VOC emissions is to replace VOCs or to limit their use in industrial processes. If VOC substitution is not possible, it is necessary to control their emission in air using non-destructive or destructive methods [2]. The first group includes adsorption, absorption and condensation. Absorption and condensation are very useful in recovering expensive solvents and operating at a large capacity, but they require high capital and operating costs. Adsorption is a reliable alternative to eliminate organic compounds from industrial waste gases because of the flexibility of the system, low energy requirements and low operation costs [3]. The second group includes thermal and catalytic oxidation. Thermal oxidation, which requires high capital and operating costs, is the simplest method but it has to be carried out at high temperature, typically over 1000 °C. Also, this method can generate harmful by-products such as dioxin and nitrogen oxides [3].

In the case of toluene, control of its emission is often accomplished by catalytic oxidation or adsorption. Toluene removal by adsorption is the traditional method for cleaning air contaminants. However, the use of adsorbents just transfers pollutants from the gaseous phase to the solid phase and causes a disposal and regeneration problem. Catalytic oxidation is an attractive technique to selectively destroy this compound in CO_2 and H_2O. Catalytic oxidation of toluene can be carried out at temperatures 200–600 °C lower than that for thermal oxidation and has high selectivity for the formation of harmless reaction products such as H_2O and CO_2 [4]. Supported noble metal (Pt, Pd, Rh) materials have been intensively examined in VOC catalytic combustion, but their limited availability and high price have encouraged their replacement by other active phases [5–7]. Transition metal oxides have been found to be very active and present advantages over noble metals associated with their lower cost, higher thermal stability and greater resistance to poisoning [8–10]. The most active and selective transition metal-containing catalysts for VOC total oxidation are based on copper oxides [11]. Usually, the active phase is supported on a porous material in order to expose their active phases in highly dispersed, easily reducible metal ions. Large pore zeolites are often chosen as supports because of their high specific surface area and acid-base properties, allowing high metal loading and generating specific interaction between the active phase and the support. In the literature, zeolites NaX and NaY have become of special interest in the oxidation of toluene, due to their large pore channel system and high specific surface area [12,13].

It has been shown that toluene adsorption performed on faujasites is appealing due to their incombustibility, great stability, hydrophobic character, and recovery at low temperature [14,15]. Benmaamar and Bengueddach [16] studied the capacity of NaY, KY, BaY, BaX and NaX zeolites to adsorb m-xylene and toluene-measuring adsorption isotherms at 298, 308, 318, and 333 K. The NaX and NaY zeolites presented the highest capacity for toluene and m-xylene adsorptions in comparison with the other samples.

For low pollutant concentrations in the atmosphere, VOC catalytic oxidation is unappealing due to high energy consumption. Conversely, the cost-saving adsorption-catalytic oxidation hybrid process is attractive. The adsorption process will first concentrate the VOC, followed by the catalytic process, by increasing the temperature. In this work, the performances of copper-exchanged X zeolites were studied, taking into consideration the individual adsorbent and catalytic properties of the solid in toluene adsorption and catalytic oxidation processes.

2. Results and Discussion

2.1. Characterization of the Materials

2.1.1. Redox Properties

H_2-TPR experiments were carried out to identify and to quantify the copper-species-exchanged zeolites and to characterize their reducibility. The reduction profiles of the copper-exchanged zeolite X samples as well as the one of NaX zeolites after calcination at 400 °C are presented in Figure 1. The reduction profiles for the copper-exchanged samples are characterized by two regions of hydrogen consumption: the low temperature region (LTR) from 200 to 520 °C, and the high temperature region (HTR) from 520 to 900 °C. The NaX sample does not show any reduction peaks in the studied temperature range. The CuX-B1 (see experimental Section 3.1) sample shows unresolved signals in the LTR and HTR regions, with less H_2 consumption in comparison to the other exchanged samples. The reduction profile shows two well-defined peaks in the LTR, with maxima at 300 and 380 °C and at 320 and 365 °C for CuX-B2 and CuX-B3 samples, respectively. In the HTR, one well-resolved peak can be observed, with the maximum at 620 °C for both samples. According to the literature [4,17,18], the reduction of Cu^{2+} ions occurs by one-step and two-step mechanisms. The reactions involved in the reduction process are the following [19]:

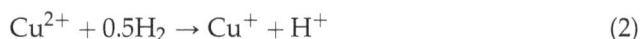

$$CuO + H_2 \rightarrow Cu^0 + H_2O \tag{1}$$

$$Cu^{2+} + 0.5H_2 \rightarrow Cu^+ + H^+ \tag{2}$$

$$Cu^+ + 0.5H_2 \rightarrow Cu^0 + H^+ \tag{3}$$

The reduction of CuO aggregates takes place via a one-step mechanism (Equation (1)) and occurs in the low-temperature region, along with the reduction of Cu^{2+} to Cu^+ (Equation (2)). Equation (3) corresponds to the reduction of the Cu^+ formed during the previous reduction process. Torre-Abreu *et al.* [19] pointed out the presence of CuO species in copper-exchanged MFI and Y zeolites. The authors also reported two reduction peaks, the first one of greater area at 300 °C and the second in the range of 500–900 °C. This means that a significant fraction of the copper species in the solids goes through a one-step mechanism according to Equation (1), instead of a two-step mechanism. It is expected with copper exchange treatment that $[Cu(OH)]^+$ as well as $Cu(OH)_2$ species in solution are transformed into CuO, after the zeolite activation treatment, which is present within zeolite channels along with the exchanged Cu^{2+} ions [20,21].

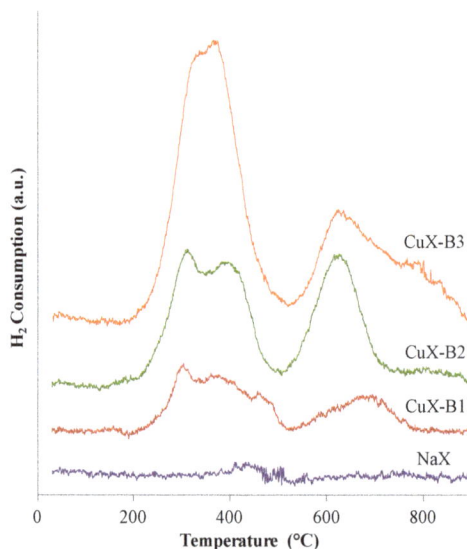

Figure 1. H_2-TPR profiles of the NaX and copper-exchanged zeolites.

Table 1 gathers the H_2 consumption values together with the theoretical copper content, calculated from the H_2 uptake and based in the stoichiometry of Equations (1) to (3). For the calculation, three assumptions are made:

(i)   NaX sample is irreducible;

(ii)   All the copper species are present as CuO and Cu^{2+} species in the copper-exchanged samples;

262

(iii) All the copper species are reduced by H_2 in the copper-exchanged samples into Cu^0 species.

The copper content in the sample increases by repeated exchange treatments from CuX-B1 (1.1 wt. %) to CuX-B3 (4 wt. %). The difference in the H_2 consumption between the low-temperature and high-temperature regions is an indication of CuO formation. The presence of the HT region peak for all samples indicates that a fraction of the copper + II is also present as Cu^{2+} ionic species. Table 1 gives the molar ratio between Cu^{2+} ions and total copper in the sample, calculated from the H_2 uptake of the LT and HT peaks. Whatever the copper content, this ratio is close to 0.6–0.7. It follows that the fraction of CuO species is rather similar in the three materials. The degree of exchange (in %) has been estimated from this ratio and the degree increases with the copper content in the zeolite (Table 1).

Table 1. Quantitative results from H_2-TPR experiments for NaX and copper-exchanged zeolites.

Sample	Consumed H_2 (mmol/g)	LTR Consumed H_2 (mmol/g)	HTR Consumed H_2 (mmol/g)	Copper Content (wt. %) *	Molar Ratio Cu^{2+}/Cu_{total}	Degree of Exchange
NaX	0.02	-	-	-	-	-
CuX-B1	0.17	0.12	0.05	1.1	0.6	2.0
CuX-B2	0.36	0.23	0.13	2.3	0.7	5.3
CuX-B3	0.62	0.41	0.21	4.0	0.7	8.5

* Calculated from the H_2 uptake.

2.1.2. Bulk Structure

The wide angle power XRD patterns of the copper-exchanged zeolite X samples as well as the parent one (NaX) after calcination at 400 °C are displayed in Figure 2. The diffractrogram of the NaX sample shows the characteristic peaks of the crystalline faujasite framework ($Na_{88}Al_{88}Si_{104}O_{384}(H_2O)_{220}$ JCPDS 01-070-2168) [22,23]. The XRD patterns of the CuX-B materials are similar to that of pure NaX. No shift in the peak positions and no new phase have been observed for CuX-B samples. This result suggests that the copper species are probably well dispersed in the zeolite structure. A decrease in the intensities of the diffraction peaks can be also observed, indicating a certain loss of crystallinity.

Abu-Zied [22] obtained similar results in samples prepared by exchanging a NaX zeolite with different concentration of aqueous copper(II) acetate solutions. Benaliouche *et al.* [23] also observed no shift in the peak positions and no significant diffraction lines assigned to any new phase in Ag^+- and Cu^{2+}-exchanged X zeolites prepared from $AgNO_3$ or $CuCl_2$ aqueous solutions. The authors concluded from these results, that both Ag^+ and Cu^{2+} seemed to be well dispersed in the zeolite framework.

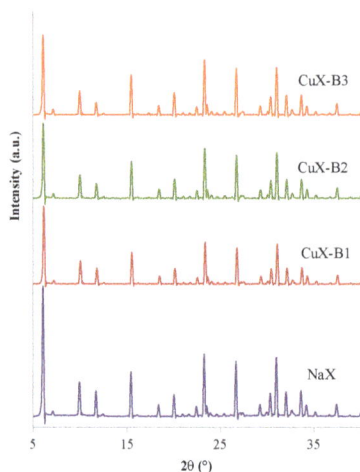

Figure 2. XRD patterns of the NaX and copper-exchanged zeolites.

2.1.3. Specific Surface Area

N$_2$ physisorption isotherms obtained for the samples after calcination at 400 °C are presented in Figure 3. The isotherms of the samples show a type I in Brunauer, Deming, Deming and Teller (BDDT) classification [24], which is typical of microporous materials. Upon copper treatment, the adsorption capacity decreases with the increase in Cu-exchange and a hysteresis loop appeared on the CuX-B2 and CuX-B3 samples, suggesting that ink-bottle type mesopores are formed on these zeolites [22]. According to the shapes of the isotherms the filling of micropores is done for approximately P/P$_0$ = 0.2; beyond that point, adsorption takes place outside micropores. The adsorbed quantity between P/P$_0$ = 0.2 and 0.95 is usually quite small, around 4% to 6% compared to that adsorbed inside the micropores.

The textural properties of the samples are listed in Table 2. The NaX sample presents a specific surface area of 451 m^2/g, which is in the order of the values reported in the literature for this type of zeolite [22,25,26]. However, Cu addition leads to a decrease in the specific surface area of the copper-exchanged samples, the effect being more important for the CuX-B3 sample, which is the sample with the highest degree of exchange (see Table 1). Benaliouche *et al.* [23] also observed a decrease in the specific surface area in Ag$^+$ and Cu^{2+} exchanged X zeolites prepared from AgNO$_3$ or CuCl$_2$ aqueous solutions, and attributed this behavior to the partial blocking of the zeolite pore access by Ag and Cu ions, leading to a reduction in total specific surface area. The exchange treatment do not cause a significant effect on the pore volume of the samples, since a slight decrease is observed in comparison to the NaX zeolite. However, considering the micropore volume (Table 2), the decrease is more pronounced and seems to be correlated to the increase of copper content in the sample. For NaX material, the micropore volume represents 97% of pore volume, whereas the value drops to 85% for the CuX-B3 sample.

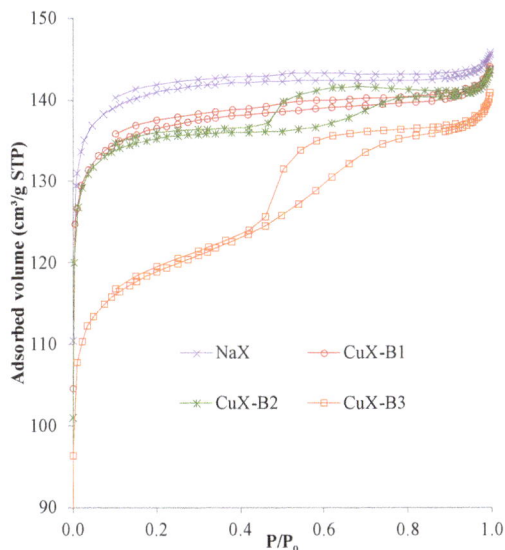

Figure 3. Adsorption-desorption isotherms of the NaX and copper-exchanged zeolites.

Table 2. Textural properties of the NaX and copper-exchanged zeolites.

Sample	Specific Surface Area (m²/g)	Pore Volume * (cm³/g)	Micropore Volume ** (cm³/g)
NaX	451	0.226	0.218
CuX-B1	434	0.223	0.211
CuX-B2	432	0.222	0.201
CuX-B3	384	0.218	0.185

* Calculated at $P/P_0 = 0.98$; ** Calculated from the equation proposed by Dubinin [24].

Figure 4 shows the pore size distribution of the samples. Exchanging the NaX zeolite with solutions of copper results in emergence of a peak in the mesoporous region, centered at 6.5 and 4.5 nm for the CuX-B2 and CuX-B3 samples, respectively. Abu-Zied [22] obtained this behavior for Cu^{2+}-acetate exchanged X zeolites. The author observed two peaks at 3.9 and 6.5 nm and 3.9 and 4.9 nm for the lowest copper content catalysts, 2.6 and 7.3 wt. %, respectively. Hammoudi *et al.* [26] reported a decrease in the N_2 uptake with an increasing number of Cu^{2+} ions in copper-exchanged X zeolites; the authors suggested that the zeolite could endure a structural deformation at higher levels of exchange, that would consist in a modification of the supercage geometry which would impact the adsorption characteristics of the parent zeolite, in terms of the interaction of the adsorbate molecules with the zeolite surface through lattice oxygen atoms and accessible extraframework cations. However, in our case, the degree of Na^+ exchange is not so high (less than 9%) (Table 1). Therefore, in the case of CuX-B2 and CuX-B3 samples, the increase in copper content could explain this additional mesoporosity arising from the CuO species which are less dispersed and could contribute to the porosity at 6.5 and 4.5 nm.

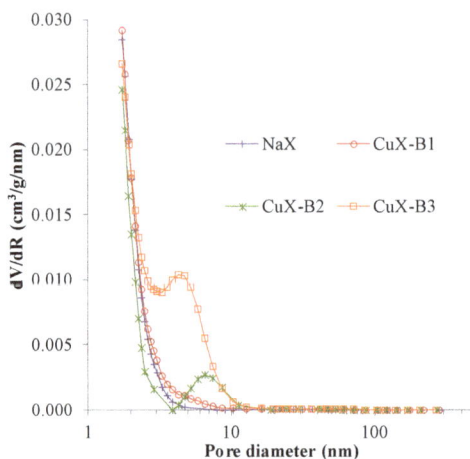

Figure 4. Pore size distribution of the CuX-B materials and the parent NaX zeolite.

2.1.4. XPS Analyses

The XPS spectra of the core levels of the Cu 2p were analyzed to estimate the copper oxidation state. The samples show two peaks in Figure 5, which intensity increase with copper content; the binding energy (BE) of the first one is centered at 934.6–934.9 eV (Cu $2p_{3/2}$) (Table 3) whereas the BE of the second one is at 954.3 eV (Cu $2p_{1/2}$). The presence of a shake-up satellite is also observed at around 944.5 eV, especially for the CuX-B3 sample (Table 3 and Figure 5). The presence of this shake-up satellite peak is characteristic of the presence Cu(II) species [27]. Ning *et al.* [28] reported Cu $2p_{3/2}$ peaks centered at 934.7 eV in samples of Na-ZSM-5 exchanged with copper, which were assigned to the presence of CuO. They also reported other Cu $2p_{3/2}$ peaks centered at 932.52 eV related to Cu_2O. It is worth noting that the Cu $2p_{3/2}$ photopeak BE observed for the materials are located at higher BE than those reported in the literature for bulk copper materials. For example, Morales *et al.* [29] reported that Cu $2p_{3/2}$ BE in CuO is 933.7 eV. The highest BE value of electrons coming from the 2p level of Cu in zeolites in comparison to those in bulk metal oxides has been already reported [30,31]. The shifts reflected the presence of small clusters of ions together with isolated metal ions dispersed in the zeolite [27]. Then, the values of 934.6–934.9 eV for BE Cu $2p_{3/2}$ are in line with the presence of Cu^{2+} ions and CuO species suggested by TPR analyses.

Table 3. Surface composition (at. %) of the NaX and copper-exchanged zeolites, Cu $2p_{3/2}$ BE (eV) and satellite (Sat) Cu $2p_{3/2}$ BE (eV) in the copper-exchanged zeolites.

Sample	Si (at. %)	Al (at. %)	O (at. %)	Na (at. %)	Cu (at. %)	Cu $2p_{3/2}$ (eV)	Sat Cu $2p_{3/2}$ (eV)
NaX	13.4	9.6	61.4	15.6	-	-	-
CuX-B1	14.3	10.8	61.6	12.5	0.8	934.9	944.5
CuX-B2	15.1	10.7	63.9	9.2	1.1	934.6	944.3
CuX-B3	15.0	12.0	63.3	8.0	1.7	934.6	944.7

Figure 6 shows the Na 1s and O 1s core level spectra of the different samples. Figure 6a shows that the Na 1s photopeak decreases in intensity from the parent zeolite (NaX) to the Cu^{2+} exchanged samples, a result that supports the behavior observed in the Cu 2p photopeak. Quantitative results corroborate this trend (Table 3). Interestingly, Na 1s BE is shifted to lower value with copper addition, showing an evidence of change in Na species surrounding. A decrease in O 1s intensity and a change in the O 1s shape are clearly observed from NaX to CuX-B3 materials (Figure 6b). Almeida *et al.* [32] reported that the O 1s peak in the NaX zeolite originates from oxygen atoms belonging to different environments: aluminate anions compensated by sodium cations, NaAlO, SiOSi bonds and silanol groups, SiOH. They determined that the contribution from the NaAlO signals decreased in zeolites exchanged with Cs and ammonium ions in comparison to the original NaX zeolite. Therefore, the change in the shape of the O 1s peak could be interpreted as a decrease in the contribution of NaAlO component resulting from Na^+–Cu^{2+} exchange.

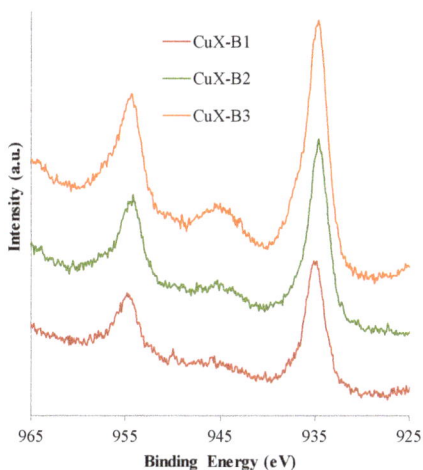

Figure 5. Cu 2p XPS spectra of the copper-exchanged zeolites.

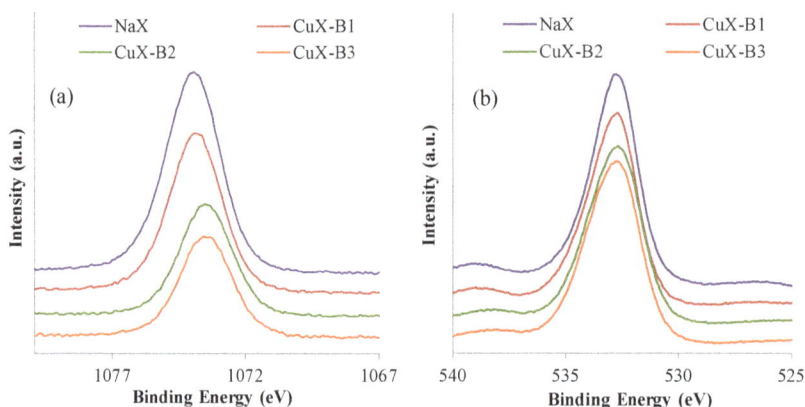

Figure 6. (**a**) Na 1s and (**b**) O 1s XPS spectra of the NaX and copper-exchanged zeolites.

2.2. Adsorption Tests

Figure 7 shows the toluene adsorption breakthrough curves of the NaX and copper-exchanged zeolites depicted by the variation in the C/Co ratio *vs* the test duration, where Co and C are the inlet and outlet concentrations, respectively. All the breakthrough curves reach adsorption equilibrium, but the samples exhibit different behaviors. Table 4 presents the parameters associated with the behavior of the samples in relation with their properties for toluene adsorption.

Among the samples, CuX-B1 shows the higher adsorption capacity and the longest breakthrough time compared with the other materials, reaching 22.2 g toluene/100 g solid and 473 min, respectively. This result for adsorption capacity is in the order of the values that can be found in the literature for other materials. For example it has been reported for activated carbon values between 11 and 34 g toluene/100 g solid [33]. All other samples exhibit a similar breakthrough time at about 260 min but differ in the time needed to reach the saturation point. The samples can be ranked by their ease in reaching adsorption equilibrium: NaX < CuX-B1 < CuX-B2 < CuX-B3, and by their decreasing adsorption capacity: CuX-B1 > NaX > CuX-B2 > CuX-B3 (Figure 7). On the other side, the NaX presents the highest PUB (percentage of unused bed) value, which can be related to the fact that the residence time of toluene molecules in the bed is not enough to benefit from all the extension of the NaX bed.

The difference between samples could be attributed not only to a difference in the copper content, but also to the difference in their textural properties. Archipov *et al.* [34] studied the adsorption of benzene in copper-exchanged Y zeolites, using Fourier transform infrared spectroscopy; the authors found that benzene adsorbed on both copper and brønsted acid sites of the zeolite, but preferentially on copper site. With the assumptions of a similar behavior for the toluene-zeolite system, it can be expected that the copper-exchanged samples exhibit higher capacities of adsorption in comparison to the NaX zeolite. However CuX-B2 and CuX-B3 samples show less capacity for adsorption. These results may be explained by a decrease in the specific surface area affecting mostly the micropore volume (Table 2). It has been pointed out in the literature that the micropore volume of an adsorbent plays an important role in the adsorption of toluene. Zhang *et al.* [35] reported for several types of adsorbents, such as SBA-15, MCM-41, NaY and SiO_2, that the volume of micropores was an important parameter for the adsorption of VOCs. It was found that the amount of adsorbed toluene was linearly correlated to the micropore volume and that the presence of micropores directly led to an increase in the dynamic adsorption capacity of VOCs. Lillo-Ródenas *et al.* [33] also obtained such a correlation between the capacity of toluene and benzene adsorptions for different types of activated carbon and the total volume of micropores.

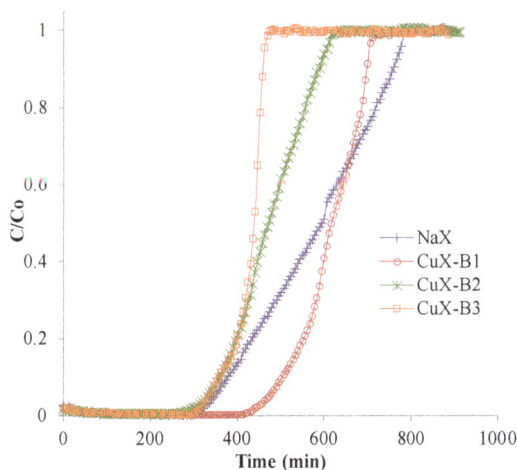

Figure 7. Adsorption breakthrough curves of toluene over NaX and copper-exchanged zeolites.

Table 4. Adsorption parameters of the NaX and copper-exchanged zeolites.

Sample	Adsorption Capacity (g/100 g Solid)	t 5% (min)	t 50% (min)	t 95% (min)	PUB (%)
NaX	20.1	347	594	770	40
CuX-B1	22.2	473	616	704	23
CuX-B2	16.7	330	473	605	31
CuX-B3	15.1	336	440	462	21

2.3. Catalytic Tests

Figure 8 shows the catalytic performances of the three copper-exchanged zeolites and the parent zeolite NaX, in terms of toluene conversion as a function of temperature. The toluene concentration reaches the maximum value (conversion = 0) at 50 °C. After this point, the concentration decreased progressively as the conversion increased, reaching the minimum value (conversion = 98%) at around 263, 320 and 346 °C ($T_{98\%}$) for CuX-B1, CuX-B2 and CuX-B3, respectively (Table 5).

Considering the $T_{50\%}$ as a measure of the catalytic performances of the materials, the activity order can be established as CuX-B1 > CuX-B2 > CuX-B3 > NaX. A remarkable improvement can be observed in the catalytic performance after the copper exchange treatment of the parent zeolite. Among the copper-exchanged samples, CuX-B1 presents the best catalytic performance, even though the copper loading in the sample is the lowest. By contrast, in the literature an increase in the toluene conversion has been reported with the copper content increase in the zeolites. For example, Antunes et al. [4] reported that the oxidation of toluene is dependent on the copper content and the degree of copper exchange in the NaHY zeolite. A very low CuO quantity has been observed in the samples and the sample with the lowest copper content (1.1%) was classified as quasi-inactive for toluene oxidation. A similar result was obtained by Ribeiro et al. [13], who studied the evolution of $T_{50\%}$ as a function of the Cu loading from 0 to 5 wt. % for series of CuHY and CuCsHY zeolites, and showed a decrease of the light-off temperatures when copper content of the catalysts is increasing. It has also been reported [20] for Cu ions exchanged in ZSM-5 zeolites that well dispersed CuO species exhibited higher oxidation activity and reducibility in comparison with the exchanged Cu ions. In comparison with literature data [4] and in similar experimental conditions, our copper-exchanged zeolites are much more active in the toluene oxidation. Therefore we suggest that a high copper content and high Na$^+$ exchange is not required for toluene oxidation. Well dispersed CuO are most probably the active species which are easily reducible by the toluene molecule (redox mechanism). Increasing the copper content in the sample led to the decrease in CuO species dispersion, species less accessible and less active for toluene oxidation (Figure 8).

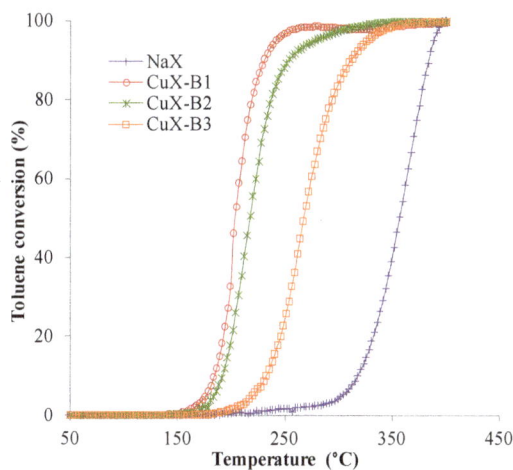

Figure 8. Toluene conversion as a function of the temperature reaction in the presence of different catalysts.

Table 5. Catalytic performances of the NaX and copper-exchanged zeolites.

Sample	$T_{50\%}$ (°C)	$T_{98\%}$ (°C)
NaX	357	394
CuX-B1	203	263
CuX-B2	217	320
CuX-B3	270	346

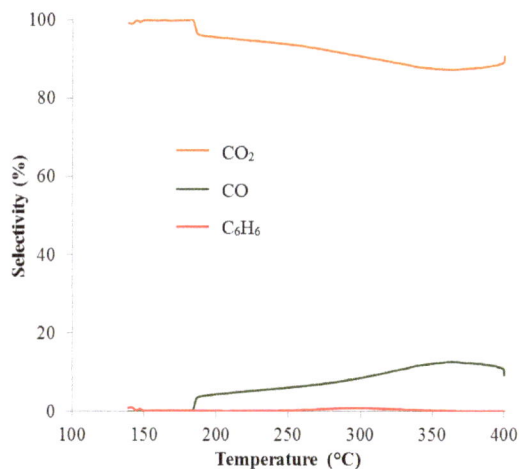

Figure 9. Variation of CO_2, CO and benzene selectivities (%) with the reaction temperature over CuX-B1 catalyst.

271

Boikov *et al.* [36] proposed a mechanism for the selective catalytic oxidation of toluene using V_2O_5 and MoO_3 as catalysts. According to the mechanism, toluene is first oxidized to benzene, through the formation of benzaldehyde as an intermediate. Then, benzene is oxidized to benzoquinone to yield carbon oxides. Figure 9 shows the variation of the selectivities in CO_2, CO and benzene during the catalytic test for the CuX-B1 catalyst. CuX-B2 and CuX-B3 samples presented similar results. At 400 °C, the selectivity towards CO_2 is 90.6%, whereas the selectivity in CO is 9.7%. No benzene is detected, evidencing the high performances of copper-exchanged zeolites in the oxidation of gaseous toluene reactant in carbon monoxide, carbon dioxide and water products.

3. Experimental Section

3.1. Materials Synthesis

CuX materials were synthesized by an ion exchange using a commercial NaX zeolite (Si/Al = 1.3). The zeolite NaX was exchanged with aqueous solutions of $Cu(NO_3)_2 \cdot 2.5H_2O$ (99.99% metal basis, Sigma-Aldrich, St. Louis, MO, USA, as received). 7 g of parent zeolite were stirred at room temperature in 350 mL of aqueous solution containing 5×10^{-3} mol/L of the metal nitrate for 24 h. The sample was filtered, washed thoroughly and dried overnight at 373 K. This material was called CuX-B1. Higher degrees of copper exchange in the zeolite were obtained by repeated treatments. CuX-B2 was obtained by stirring 4 g of the CuX-B1 material with 200 mL of the solution for the same time and applying the same procedure. Finally, CuX-B3 was synthesized from CuX-B2 material, stirring 2 g of the solid with 100 mL of the solution [37,38]. The copper-exchanged level extent was evaluated assuming that one Cu^{2+} ion can replaced two Na^+ ions in the zeolite.

3.2. Characterization Techniques

Powder X-ray diffraction (XRD) patterns were recorded on a Bruker D8 X-ray diffractometer (Champs-sur-Marne, France) at room temperature with CuKα radiation (λ = 1.5418 Å). The diffractograms were recorded for 2θ values comprised between 5° and 80° using a 0.04° step with an integration time of 2 s. The diffraction patterns were indexed by comparison with the Joint Committee on Powder Diffraction Standards files (JCPDS).

XPS experiments were performed using an AXIS Ultra DLD Kratos spectrometer (Manchester, UK) equipped with a monochromatised aluminum source (Al Kα = 1486.69 eV) and charge compensation gun. All binding energies were referenced to the C 1s core level at 285 eV. Simulation of the experimental photopeaks was carried out using a mixed Gaussian/Lorentzian peak fit procedure by means of the CasaXPS software.

Temperature programmed reduction (H$_2$-TPR) measurements were performed on a Micromeritics AutoChem apparatus (Verneuil en Halatte, France). H$_2$-TPR was performed in a quartz microreactor (9 mm). The H$_2$-TPR was conducted in two stages:

1. Degassing treatment to remove the adsorbed species: The sample was heated from room temperature to 200 °C (10 °C/min) under an argon flow (50 mL/min). The sample was then maintained at 200 °C for 1 h and cooled to room temperature under argon atmosphere.
2. Reduction process: The reduction mixture containing 5% v/v H$_2$ in argon was send to the sample at room temperature, the flow was 50 mL/min. After a short stabilization process the temperature was increased from room temperature to 900 °C (with 10 °C/min heating rate). Finally, the sample was cooled to room temperature under argon.

N$_2$ adsorption-desorption isotherms were recorded at −196 °C using an automated ASAP2010 and TRISTAR II apparatus from Micromeritics. Before the adsorption-desorption experiment, the sample was heated at 120 °C under vacuum for 3 h for degassing and the adsorbed species was removed. Specific surface areas were calculated from the linear part of the Brunauer-Emmett-Teller line whereas the micropore volumes were calculated from the Dubinin method commonly used for adsorption description in microporous materials [24]. The desorption branch of the isotherm was used for the evaluation of the pore size distribution.

3.3. Toluene Catalytic Oxidation Test

The catalytic activity of materials for toluene oxidation is typically compared using light off curves, curve of conversion *versus* reaction temperature, where the reactant mixture passes through a catalyst bed while the temperature is increased [26]. The subsequent S-shaped curves give an idea of the performance of different catalysts as well as total conversion minimal temperature. The toluene catalytic setup is presented in Figure 10.

The gaseous toluene was generated using a saturator containing reagent-grade toluene in liquid phase. The saturator is a small and inert capsule containing a pure chemical compound in two phases between its gas and liquid phases. The release of the chemical occurs when a carrier gas (in this case air) passes through the device and reaches a saturated state. The temperature controller of the thermostatic bath maintains the saturator temperature at a set point of 65 °C with an accuracy of 0.01 °C. This gives a toluene concentration in the air stream of 800 ppm.

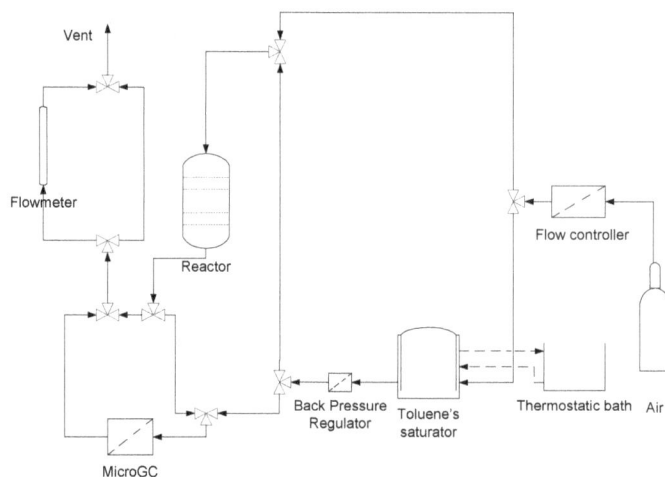

Figure 10. Toluene adsorption and catalytic setup.

Toluene complete catalytic oxidation produces only CO_2 and H_2O. However, incomplete oxidation products can be obtained, such as CO and benzene, among others. Analysis of these compounds was performed using a microGC from SRA Instruments (Marcy l'Etoile, France). Quantification was realized by calibration of peak areas, and for each compound, linear regression was used to determine the response factors.

The toluene conversion was estimated using the following equation:

$$X_{\text{toluene}}(\%) = \left(1 - \frac{[\text{toluene}]}{[\text{toluene}] + [\text{benzene}] + [\text{CO}] + [\text{CO}_2]}\right) \times 100 \qquad (4)$$

The selectivity towards products was estimated with the following equation, which was written for the specific case of CO:

$$S_{\text{CO}}(\%) = \left(\frac{[\text{CO}]}{[\text{toluene}] + [\text{benzene}] + [\text{CO}] + [\text{CO}_2]}\right) \times 100 \qquad (5)$$

The toluene catalytic oxidation was performed in a fixed bed reactor (i.d. 8–10 mm). In a test, 200 mg of the catalyst was loaded in the reactor. The total flow rate was 100 mL/min and the gas hourly space velocity was 30,000 mL/gcat.

- Step 1: Catalyst activation is the first step of the catalyst test. The conditions employed are an air stream with a flow of 70 mL/min. The aim of this activation process is to clean the catalyst surface and remove physisorbed water and surface impurities that can eventually be present on the catalyst.

- Step 2: The toluene catalytic oxidation start, the temperature is increased and the effluent gas from the reactor is analyzed using an on-line microGC analyzer; injections were made every five minutes.

3.4. Toluene Adsorption Test

The adsorption capacity of the samples was measured employing the same system depicted in Figure 10 and measuring the variation of the ratio of outlet (C) and inlet (Co) concentrations with time. The results are typical S-shape curves. The first toluene adsorption is nearly complete and the breakthrough curves are adjacent to a straight line. In the second, the concentration of the toluene slowly reaches the breakthrough point and the outlet toluene concentration increased with the extension of adsorption time. At this end the outlet concentration rise to the inlet concentration [35]. For a given concentration, the longer breakthrough and saturation time means a greater adsorption capacity. The adsorption capacity (AC) was calculated by the numerical integration of the breakthrough curve, using the following equation:

$$AC = \frac{Area \times C_0}{W_s} \tag{6}$$

where:

Area $- \int_{t0}^{tf} C/C_0 dt$, C_0 = Initial concentration, W_s – Sample weight.

The times at which the outlet toluene concentration was 5% (t 5% = breakpoint time), 50% (t 50% = stoichiometric time) and 95% (t 95% = saturation time) of the inlet toluene concentration were also estimated. The percentage of unused bed (PUB) was calculated from the length of unused bed (LUB) by using the following equations:

$$LUB = L\left(1 - \frac{AC_{t5\%}}{AC_{t95\%}}\right) \text{ and } PUB(\%) = \frac{LUB}{L} \times 100 \tag{7}$$

where: L = total length of the bed, $AC_{t\,5\%}$ and $AC_{t\,95\%}$ = adsorption capacity at t 5% and t 95% respectively.

The toluene's concentration, flow of stream and mass of the adsorbent were empirically adjusted in order to have acceptable adsorption times before the breakpoint for comparison purposes between samples. The temperature of the thermostatic bath was maintained at 25 °C, the mass of the sample loaded in the reactor was 200 g and the total flow rate was 13 mL/min. This gives a toluene concentration in the air stream of approximately 1250 ppm. For the adsorption test, activated samples (400 °C with an air flow of 70 mL/min) were exposed to the toluene stream at a temperature of 30 °C. The effluent gas from the reactor was analyzed using on-line microGC analyzer; injections were made every five minutes.

4. Conclusions

In this study the potential of copper-exchanged zeolites as a toluene adsorbent and as a catalyst for toluene total oxidation was evaluated. A low quantity of copper (from 1 wt. % to 4 wt. %) was deposited into the zeolite as copper + II species. Through the consequent exchange, it has been shown that the fraction of copper-exchanged (Cu^{2+} species) *versus* CuO species remained rather constant whatever the copper content. The toluene adsorption studies have shown that when the microporosity is maintained for low-copper-content zeolite, the adsorption capacity of the corresponding material (CuX-B1) is not affected in comparison with the parent zeolite. Interestingly, this material exhibited remarkable toluene oxidation due to the presence of well-dispersed CuO species.

The individual adsorbent and catalytic properties of CuX-B1 sample are promising, and this material can be considered a potential hybrid system for the treatment of toluene in low concentrations in air.

Acknowledgments: The Chevreul institute (FR 2638), Ministère de l'Enseignement Supérieur et de la Recherche, Région Nord–Pas de Calais and FEDER are acknowledged for supporting and funding this work. The French Agency for Sustainable Development, ADEME, is also acknowledged for its financial support through CORTEA program n°11-81-C0108. This work was also supported by a grant from the project PHC CEDRE 2015 N° 32933QE. D. C. acknowledges the award of a doctoral fellowship from the Agence Universitaire de la Francophonie (AUF)—Région du Moyen-Orient.

Author Contributions: D.R. and D.C. performed the experiments and analyzed the data; M.L. and S.R. contributed to the interpretation of the materials characterization results. J.P.B. and I.B. contributed towards the interpretation of the adsorption tests results. J.M.G. and. J.F.L. contributed towards the interpretation of the catalytic results. D.R., J.M.G. and J.F.L. contributed to the writing of the manuscript.

Conflicts of Interest: The authors declare no conflict of interest.

References

1. Kostiainen, R. Volatile organic compounds in the indoor air of normal and sick houses. *Atmos. Environ.* **1995**, *29*, 693–702.
2. Soylu, G.; Özcelik, Z.; Boz, I. Total oxidation of toluene over metal oxides supported on a natural clinoptilolite-type zeolite. *Chem. Eng. J.* **2010**, *162*, 380–387.
3. Kima, H.; Kimb, T.; Koh, H.; Lee, S.; Mina, B. Complete benzene oxidation over Pt-Pd bimetal catalyst supported on γ-alumina: Influence of Pt-Pd ratio on the catalytic activity. *Appl. Catal. A* **2005**, *280*, 125–131.
4. Antunes, A.; Ribeiro, M.; Silva, J.; Ribeiro, F.; Magnoux, P.; Guisnet, M. Catalytic oxidation of toluene over CuNaHY zeolites, Coke formation and removal. *Appl. Catal. B* **2001**, *33*, 149–164.

5. Tidahy, H.; Siffert, S.; Wyrwalski, F.; Lamonier, J.-F.; Aboukais, A. Catalytic activity of copper and palladium based catalysts for toluene total oxidation. *Catal. Today* **2007**, *119*, 317–320.

6. Carpentier, J.; Lamonier, J.-F.; Siffert, S.; Zhilinskaya, E.; Aboukais, A. Characterisation of Mg/Al hydrotalcite with interlayer palladium complex for catalytic oxidation of toluene. *Appl. Catal. A* **2002**, *234*, 91–101.

7. Torres, J.Q.; Royer, S.; Bellat, J.-P.; Giraudon, J.-M.; Lamonier, J.-F. Formaldehyde: Catalytic Oxidation as a Promising Soft Way of Elimination. *ChemSusChem* **2013**, *6*, 578–592.

8. Averlant, R.; Royer, S.; Giraudon, J.-M.; Bellat, J.-P.; Bezverkhyy, I.; Weber, G.; Lamonier, J.-F. Mesoporous silica confined MnO_2 nanoparticles as highly efficient catalyst for low temperature elimination of formaldehyde. *ChemCatChem* **2014**, *6*, 152–161.

9. Bai, L.; Wyrwalski, F.; Lamonier, J.-F.; Khodakov, A.Y.; Monflier, E.; Ponchel, A. Effects of β-cyclodextrin introduction to zirconia supported-cobalt oxide catalysts: From molecule-ion associations to complete oxidation of formaldehyde. *Appl. Catal. B* **2013**, *138–139*, 381–390.

10. Torres, J.Q.; Giraudon, J.-M.; Gervasini, A.; Dujardin, C.; Lancelot, C.; Trentesaux, M.; Lamonier, J.-F. Total Oxidation of Formaldehyde over MnO_x–CeO_2 Catalysts: The Effect of Acid Treatment. *ACS Catal.* **2015**, *5*, 2260–2269.

11. Zimowska, M.; Michalik-Zym, A.; Janik, R.; Machej, T.; Gurgul, J.; Socha, R.; Podobiński, J.; Serwicka, E. Catalytic combustion of toluene over mixed Cu–Mn oxides. *Catal. Today* **2007**, *119*, 321–326.

12. Lee, J.; Park, N.; Ryu, S.; Kim, K.; Lee, T. Self-oscillatory behavior in toluene oxidation on zeolite-NaX. *Appl. Catal. A Gen.* **2004**, *275*, 79–86.

13. Ribeiro, M.; Silva, J.; Brimaud, S.; Antunes, A.; Silva, E.; Fernandes, A.; Magnoux, P.; Murphy, D. Improvement of toluene catalytic combustion by addition of cesium in copper exchanged zeolites. *Appl. Catal. B Environ.* **2007**, *70*, 384–392.

14. Weber, G.; Bertrand, O.; Fromont, E.; Bourg, S.; Bouvier, F.; Bissinger, D.; Simmonot-Grange, M.H. TCE Adsorption on Hydrophobic Y and MFI Zeolites. *J. Chim. Phys.* **1996**, *93*, 1412–1425.

15. Fajula, F.; Plee, D. Application of Molecular Sieves in View of Cleaner Technology. *Stud. Surf. Sci. Catal.* **1994**, *85*, 633–651.

16. Benmaamar, Z.; Bengueddach, A. Correlation with different models for adsorption isotherms of m-xylene and toluene on zeolites. *J. Appl. Sci. Environ. Sanit.* **2007**, *2*, 43–56.

17. Jacobs, P.; Linarth, J.; Nijs, H.; Uytterhoeven, J. Redox Behaviour of Transition Metal Ions in Zeolites, Part 5: Method of Quantitative Determination of Bidisperse Distributions of Metal Particle Sizes in Zeolites. *J. Chem. Soc. Faraday Trans.* **1977**, *73*, 1745–1754.

18. Kieger, S.; Delahay, G.; Coq, B. Influence of co-cations in the selective catalytic reduction of NO by NH_3 over copper exchanged faujasite zeolites. *Appl. Catal. B Environ.* **2000**, *25*, 1–9.

19. Torre-Abreu, C.; Henriques, C.; Ribeiro, F.; Delahay, G.; Ribeiro, M. Selective catalytic reduction of NO on copper-exchanged zeolites: The role of the structure of the zeolite in the nature of copper-active sites. *Catal. Today* **1999**, *54*, 407–418.

20. Bulánek, R.; Wichterlová, B.; Sobalík, Z.; Tichý, J. Reducibility and oxidation activity of Cu ions in zeolites: Effect of Cu ion coordination and zeolite framework composition. *Appl. Catal. B Environ.* **2001**, *31*, 13–25.

21. Vijaikumar, S.; Subramanian, T.; Pitchumani, K. Zeolite Encapsulated Nanocrystalline CuO: A Redox Catalyst for the Oxidation of Secondary Alcohols. *J. Nanomater.* **2008**, *2008*, 1–7.

22. Abu-Zied, B. Cu^{2+}-acetate exchanged X zeolites: Preparation, characterization and N_2O decomposition activity. *Microporous Mesoporous Mater.* **2011**, *139*, 59–66.

23. Benaliouche, F.; Boucheffa, Y.; Ayrault, P.; Mignard, S.; Magnoux, P. NH_3-TPD and FTIR spectroscopy of pyridine adsorption studies for characterization of Ag- and Cu-exchanged X zeolites. *Microporous Mesoporous Mater.* **2008**, *111*, 80–88.

24. Dubinin, M.M.; Stoeckli, H.F. Homogeneous and heterogeneous micropore structures in carbonaceous adsorbents. *J. Colloid Interface Sci.* **1980**, *75*, 34–42.

25. Díaz, E.; Ordóñez, S.; Vega, A.; Coca, J. Catalytic combustion of hexane over transition metal modified zeolites NaX and CaA. *Appl. Catal. B Environ.* **2005**, *56*, 313–322.

26. Hammoudi, H.; Bendenia, S.; Marouf-Khelifa, K.; Marouf, R.; Schott, J.; Khelifa, A. Effect of the binary and ternary exchanges on crystallinity and textural properties of X zeolites. *Microporous Mesoporous Mater.* **2008**, *113*, 343–351.

27. Shpiro, E.; Grünert, W.; Joyner, R.; Baeva, G. Nature, distribution and reactivity of copper species in over-exchanged Cu-ZSM-5 catalysts: An XPS/XAES study. *Catal. Lett.* **1994**, *24*, 159–169.

28. Ning, P.; Qiu, J.; Wang, X.; Liu, W.; Chen, W. Metal loaded zeolite adsorbents for hydrogen cyanide removal. *J. Environ. Sci.* **2013**, *25*, 808–814.

29. Morales, J.; Sánchez, L.; Martín, F.; Ramos-Barrado, J.; Sánchez, M. Use of low-temperature nanostructured CuO thin films deposited by spray-pyrolysis in lithium cells. *Thin Solid Films* **2005**, *474*, 133–140.

30. Meda, L.; Ranghino, G.; Moretti, G.; Cerofolini, G. XPS detection of some redox phenomena in Cu-zeolites. *Surf. Interface Anal.* **2002**, *33*, 516–521.

31. Gruenert, W.; Sauerlandt, U.; Schloegl, R.; Karge, H. XPS investigations of lanthanum in faujasite-type zeolites. *J. Phys. Chem.* **1993**, *97*, 1413–1419.

32. Almeida, K.; Landers, R.; Cardoso, D. Properties of faujasite zeolites containing methyl-substituted ammonium cations. *J. Catal.* **2012**, *294*, 151–160.

33. Lillo-Ródenas, M.; Cazorla-Amorós, D.; Linares-Solano, A. Behaviour of activated carbons with different pore size distributions and surface oxygen groups for benzene and toluene adsorption at low concentrations. *Carbon* **2005**, *43*, 1758–1767.

34. Archipov, T.; Santra, S.; Ene, A.; Stoll, H.; Rauhut, G.; Roduner, E. Adsorption of Benzene to Copper in CuHY Zeolite. *J. Phys. Chem. C* **2009**, *113*, 4107–4116.

35. Zhang, W.; Qu, Z.; Li, X.; Wang, Y.; Ma, D.; Wu, J. Comparison of dynamic adsorption/desorption characteristics of toluene on different porous materials. *J. Environ. Sci.* **2012**, *24*, 520–528.

36. Boikov, E.; Vishnetskaya, M.; Emel'yanov, A.; Tomskii, I.; Shcherbakov, N. The Selective Catalytic Oxidation of Toluene. *Russ. J. Phys. Chem. A* **2008**, *82*, 2233–2237.

37. Kowenje, C.; Jones, B.; Doetschman, D.; Yang, S.; Kanyi, C. Effects of cation siting and spin-spin interactions on the electron paramagnetic resonance (EPR) of Cu^{2+} exchanged X Faujasite zeolite. *Chem. Phys.* **2006**, *330*, 401–409.

38. Duprat, F. Light-off curve of catalytic reaction and kinetics. *Chem. Eng. Sci.* **2002**, *57*, 901–911.

TiO$_2$-Impregnated Porous Silica Tube and Its Application for Compact Air- and Water-Purification Units

Tsuyoshi Ochiai, Shoko Tago, Mio Hayashi, Hiromasa Tawarayama,
Toshifumi Hosoya and Akira Fujishima

Abstract: A simple, convenient, reusable, and inexpensive air- and water-purification unit including a one-end sealed porous amorphous-silica (a-silica) tube coated with TiO$_2$ photocatalyst layers has been developed. The porous a-silica layers were formed through outside vapor deposition (OVD). TiO$_2$ photocatalyst layers were formed through impregnation and calcination onto a-silica layers. The resulting porous TiO$_2$-impregnated a-silica tubes were evaluated for air-purification capacity using an acetaldehyde gas decomposition test. The tube (8.5 mm e.d. × 150 mm) demonstrated a 93% removal rate for high concentrations (*ca.* 300 ppm) of acetaldehyde gas at a single-pass condition with a 250 mL/min flow rate under UV irradiation. The tube also demonstrated a water purification capacity at a rate 2.0 times higher than a-silica tube without TiO$_2$ impregnation. Therefore, the tubes have a great potential for developing compact and in-line VOC removal and water-purification units.

Reprinted from *Catalysts*. Cite as: Ochiai, T.; Tago, S.; Hayashi, M.; Tawarayama, H.; Hosoya, T.; Fujishima, A. TiO$_2$-Impregnated Porous Silica Tube and Its Application for Compact Air- and Water-Purification Units. *Catalysts* **2015**, *5*, 1498–1506.

1. Introduction

Photocatalytic environmental purification, particularly VOC removal, has received increased attention owing to its low cost and enduring stability. However, popularly used photocatalysts and photocatalytic filters are significantly limited in their application due to relatively low purification efficiency and difficulty in handling the powder. Thus, although extensive research has been conducted on photocatalytic air purification, the difficulty in creating a practical air purifier has rendered it ineffective for implementation in real-world industrial technology. We have reported various methods for the design and application of a TiO$_2$ photocatalyst to maximize its photocatalytic abilities [1–4]. Recently, we have succeeded in the simple fabrication of novel one-end sealed porous TiO$_2$-coated amorphous-silica (a-silica) tubes with large porosity using the outside vapor deposition (OVD) method [5]. The tube was evaluated through *Escherichia coli* removal and Qβ phage inactivation testing. The impregnation method was used to fill TiO$_2$ precursor deep into the pores of one-end sealed porous a-silica tubes. The porous a-silica tubes were assayed for their VOC removal ability through an acetaldehyde decomposition test. In addition, the water purification ability of these tubes was preliminarily evaluated through the methylene blue decolorization test. These tests revealed more efficient materials, with emphasis on their ability to remove VOC.

2. Results and Discussion

2.1. Characterization

The average bulk density and average porosity of the porous tubes were 0.84 g/cm^3 and 0.62, respectively. SEM images of the surface, secondary election images (SEIs) of the cross-section, and high-magnification SEIs of the cross-section of the porous TiO$_2$-impregnated a-silica tube are shown in Figure 1. Figure 1e–g shows high-magnification SEIs of cross-section of modified TiO$_2$ particles on the a-silica particles. White, gray, and black areas in Figure 1e–g represent TiO$_2$ particles, a-silica particles, and the resin intruding the pore, respectively. The estimated TiO$_2$ particle size is several tens of nanometers, which is smaller than the TiO$_2$ grain size in the TiO$_2$-coated a-silica tubes fabricated using the OVD method (several hundreds of nanometers) [5]. TiO$_2$ exists on the surface of a-silica skeleton even if it is located deep within a silica pore. However, in the deeper parts of the silica pores, the amount of observed TiO$_2$ declined. TiO$_2$ impregnation onto the porous silica tubes increase their pressure drops slightly but maintains a breathability sufficient to let the air or water pass through the tubes during purification or decomposition (Figure 2). The pore diameter of the tubes in this research can be estimated to 0.4 μm as the same as the previous research [5].

The Raman spectra of the TiO$_2$-impregnated a-silica tube and the TiO$_2$-coated a-silica tube by the OVD method are shown in Figure 3. The Raman spectrum of the TiO$_2$-coated a-silica tube by the OVD method is similar to the spectrum of the TiO$_2$ nanopowders with 60 wt. % of anatase content [5,6]. Repeating the heat process with a burner in the OVD method seemed to lead TiO$_2$ phase to rutile crystals. In contrast, the Raman bands of the TiO$_2$-impregnated a-silica tube at 142, 194, 396, 514, and 639 cm^{-1} are nearly identical to the spectrum of the anatase phase [7]. Thus, Raman spectroscopy indicates that the TiO$_2$ particles in the TiO$_2$-impregnated a-silica tube consisted of anatase crystals. Anatase crystals with exposed high-energy facets, including (001) and (010) facets, have attracted significant attention because of their high photocatalytic property [8,9]. The combination of TiO$_2$ particle size and crystal phase of the TiO$_2$-impregnated a-silica tube are more effective than the TiO$_2$-coated a-silica tube by the OVD method alone for photocatalytic capacity.

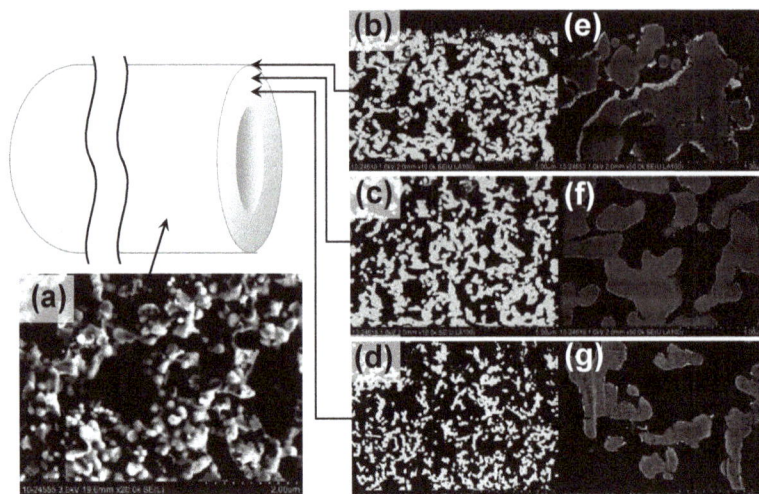

Figure 1. SEM images of the surface (**a**), secondary election images (SEIs) of the cross-section (**b–d**), and high-magnification SEIs of cross-section (**e–g**) of the TiO$_2$-impregnated a-silica tube. Cross-section images were obtained at 0 (**b**,**e**), 0.4 (**c**,**f**), 0.8 (**d**,**g**) mm from the surface.

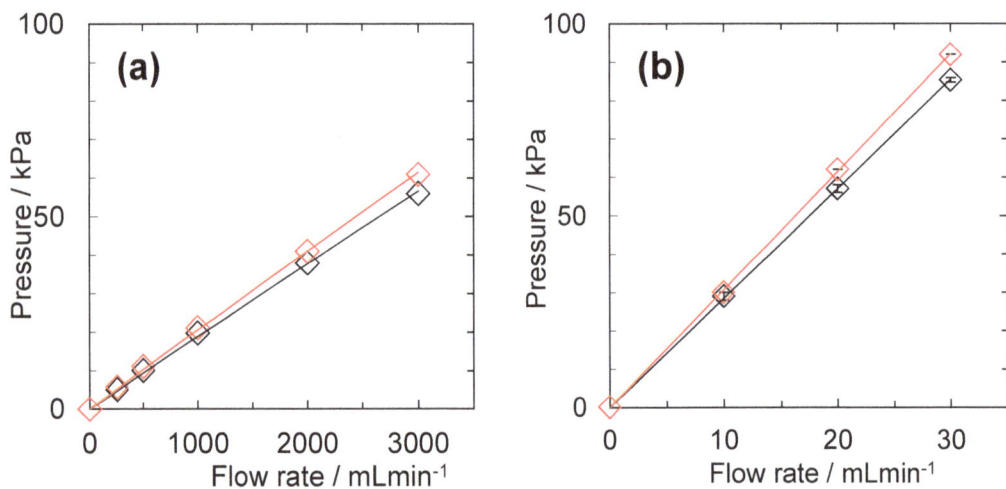

Figure 2. Pressure drops of the porous TiO$_2$-impregnated a-silica tube under the air (**a**) or water (**b**) flow. Black: a-silica tube without TiO$_2$ impregnation; red: TiO$_2$-impregnated a-silica tube.

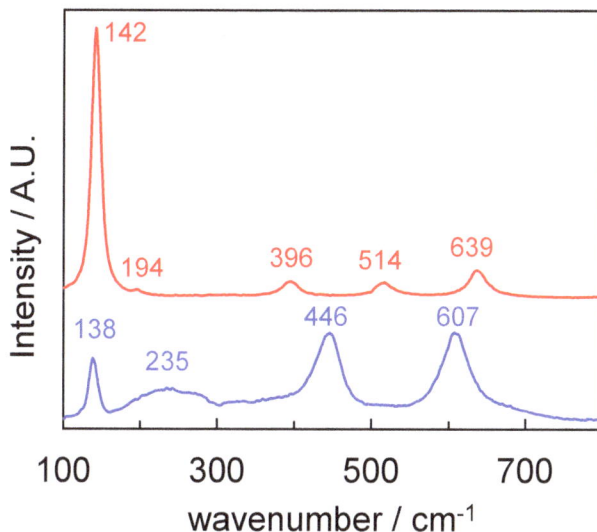

Figure 3. Raman spectra of the TiO$_2$-impregnated a-silica tube (red) and the TiO$_2$-coated a-silica tube by the outside vapor deposition (OVD) method (blue).

2.2. Results of Air- and Water-Purification Test

Figure 4a shows a typical data set of acetaldehyde removal (red) and CO$_2$ generation (green) by the TiO$_2$-impregnated a-silica tube with UV-C irradiation. The tube decomposed 100 ppm of acetaldehyde almost completely with a single-pass condition at a 250 mL/min flow rate. Under the same condition, the tube showed 93%, 78%, and 68% removal of 300, 700, and 1000 ppm of acetaldehyde, respectively (Figure 4b red). On the other hand, the a-silica tube covered with TiO$_2$/Ni-foam showed 89%, 46%, and 30% removal of 300, 700, and 1000 ppm of acetaldehyde, respectively (Figure 4b black). The TiO$_2$-coated a-silica tube by the OVD method could not remove high concentrations of acetaldehyde (Figure 4b blue). The significant difference between the TiO$_2$-impregnated a-silica tube and the TiO$_2$-coated a-silica tube by the OVD method may be caused by the particle size and crystal phase of the TiO$_2$. The high photocatalytic property of the anatase phase and smaller particle size of TiO$_2$ of the TiO$_2$-impregnated a-silica tube led to an effective decomposition of gaseous compounds [8,9]. However, the removal ratio of the tube was slightly decreased during long-term treatment (Figure 5). The data indicated that any type of deactivation process may occur. Now we are attempting to establish a re-activation method of the tube using a simple method such as washing or heating.

The methylene blue decomposition property of the TiO$_2$-impregnated a-silica tube exceeded that of the a-silica tube without TiO$_2$ impregnation during the experiments in which water passed through the tubes repeatedly (Figure 6). Both the decolorization behaviors occurred within the UV light and showed a similarity to the first order reaction equation. The reaction rate constant (k_1) of the TiO_2-impregnated a-silica tube (0.28, Figure 6 red) is 2.0 times higher than the k_1 of the a-silica tube without TiO_2 impregnation (0.14, Figure 6 white). These preliminarily evaluation indicate the potential for photocatalytic water purification ability of the tube [10].

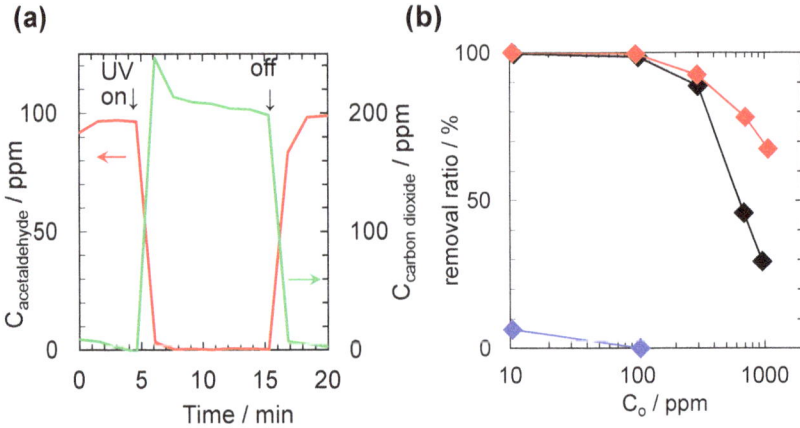

Figure 4. (**a**) Typical data set of acetaldehyde removal (red) and CO_2 generation (green) by the TiO_2-impregnated a-silica tube with UV-C irradiation. (**b**) Removal ratio *vs.* initial concentration of acetaldehyde by the TiO_2-impregnated a-silica tube (red), a-silica tube covered with TiO_2/Ni-foam (black), TiO_2-coated a-silica tube by the OVD method (blue).

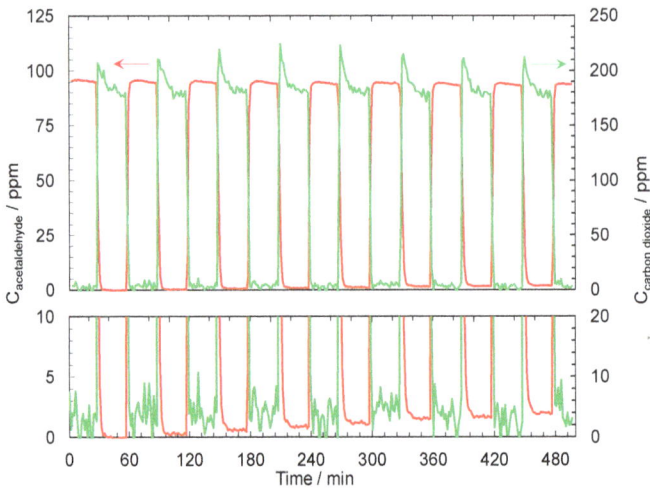

Figure 5. Data set of acetaldehyde removal (red) and CO_2 generation (green) by the TiO_2-impregnated a-silica tube with a 30/30-min on/off cycle of UV-C irradiation.

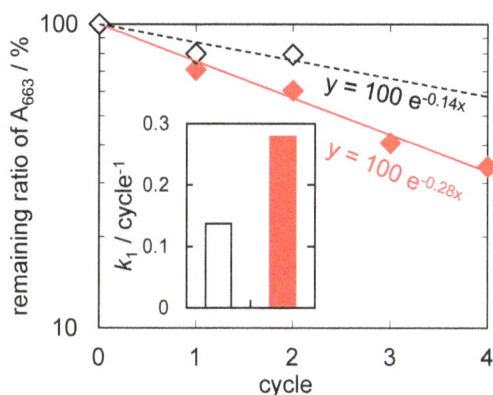

Figure 6. Plot showing the results of methylene blue decomposition test of porous TiO$_2$-impregnated a-silica tube (red) and a-silica tube without TiO$_2$ impregnation (white). Inset: Reaction rate constants (k_1) of the tubes.

3. Experimental Section

Figure 7 shows the method of fabricating the porous TiO$_2$-impregnated a-silica tube. A-silica tubes with an external diameter of 8.6 mm, a thickness of 1.3 mm, and a length of 300 mm were fabricated using the OVD method [5,11] (Figure 7a). One-end sealed porous tubes were obtained by pulling out the rod target from the soot body (Figure 7b). Then the tube was soaked in a 1 M titanium(IV) isopropoxide/ethanol solution, pulled out of the solution, vacuumed to dry, heated at 550 °C for 1 h, and dried again after soaking in milli-Q under the effect of ultrasonic treatment (Figure 7c). The porous structure of the tubes was observed using an FE-SEM (S-4800, Hitachi, Tokyo, Japan). Samples for cross-section observation were prepared by embedding the tubes in resin and then polishing them with a cross-section polisher (SM-09010, JEOL, Tokyo, Japan). For the structural characterization of the TiO$_2$ layer, Raman spectroscopy excited by 532 nm Nd:YAG laser (LabRAM HR-800, HORIBA JOVIN YVON, Longjumeau, France) was used. The pressure drops caused by the TiO$_2$ modification over the tubes were also measured.

Photographs of the TiO$_2$-impregnated porous a-silica tubes in air and water purification using decomposition tests of acetaldehyde and methylene blue are shown in Figure 8. For a continuous single-pass condition, a prescribed concentration of acetaldehyde gas was introduced into the TiO$_2$-impregnated porous a-silica tube at a flow rate of 250 mL/min and was exhausted after the reaction (Figure 8b). The TiO$_2$-impregnated porous a-silica tube was inserted into a quartz glass tube (27 mm i.d. × 30 cm length) and irradiated by a UV-C lamp. Acetaldehyde and CO$_2$ concentrations in the quartz glass tube were analyzed simultaneously and continuously by photo-acoustic infrared spectroscopy using an Innova AirTech

Instruments Multi-gas Monitor Type 1412 with suitable optical filters (Ballerup, Denmark). For comparison, the a-silica tube without TiO_2 impregnation, the TiO_2-coated a-silica tube by the OVD method [5], and the a-silica tube covered with conventional TiO_2-impregnated Ni-foam (TiO_2/Ni-foam) [12] were also evaluated. A made-to-measure helical UV-C lamp (Kyokko Denki Co., Ltd., Tokyo, Japan) was used as the UV light source. The UV intensity at 254 nm at the surface of the porous tube was measured using a UV-radiometer UVR-300 with a sensor head UD-250 (Topcon Corp., Tokyo, Japan).

Figure 7. Schematic of the method of fabricating porous TiO_2-impregnated a-silica tubes.

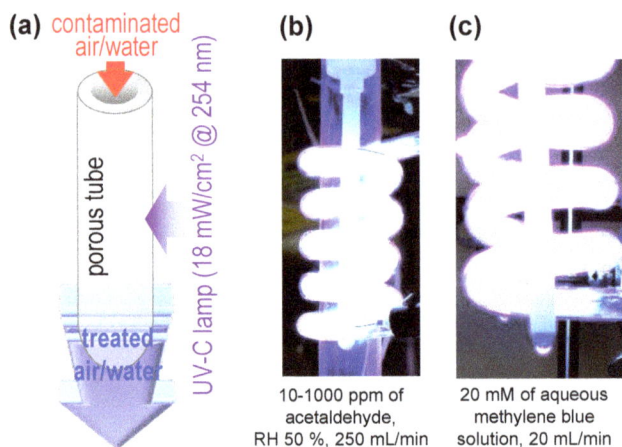

Figure 8. Schematic (**a**) and photographs of the air- (**b**) and water- (**c**) purification test for the TiO_2-impregnated a-silica tube.

The methylene blue decomposition test of the TiO_2-impregnated porous a-silica tube was carried out by passing 50 mL of 20 mM aqueous methylene blue solution through the tube at a flow rate of 20 mL/min with UV-C irradiation. The solution was then stored in a beaker (Figure 8c). The remaining ratio of methylene blue was calculated by a decreased absorbance at 663 nm using UV-visible spectrophotometer 2450 (Shimadzu Co., Kyoto, Japan). Then the stored and treated solution was passed through the tube again. Pseudo first order reaction rate constants (k_1) were calculated from the remaining ratio as a function of cycle number.

4. Conclusions

A convenient air and water purification unit that uses a TiO_2-impregnated porous a-silica tube was investigated. The tubes showed a possibility for air and water purification. In particular, VOC decomposition property was outstanding with a condition of high concentration acetaldehyde (78% at 700 ppm) and single-pass process. Moreover, a-silica glass can be welded to fused silica glass (Figure 9). Therefore, the tubes have a great potential for compact and in-line VOC removal and water-purification units.

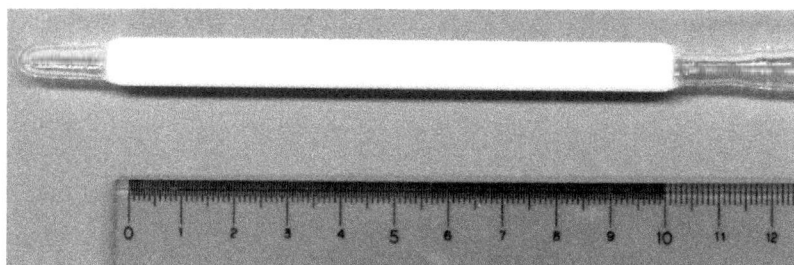

Figure 9. Porous a-silica tube welded to a fused silica glass tube.

Author Contributions: Tsuyoshi Ochiai, Hiromasa Tawarayama, and Toshifumi Hosoya participated in the study design and conducted the study. Data was collected and analyzed by Tsuyoshi Ochiai, Shoko Tago, Mio Hayashi, and Hiromasa Tawarayama. The manuscript was written by Tsuyoshi Ochiai and Shoko Tago. Akira Fujishima provided valuable discussions and advice on the manuscript.

Conflicts of Interest: The authors declare no conflict of interest.

References

1. Ochiai, T. Environmental and medical applications of TiO_2 photocatalysts and boron-doped diamond electrodes. *Electrochemistry* **2014**, *82*, 720–725.
2. Nakata, K.; Kagawa, T.; Sakai, M.; Liu, S.; Ochiai, T.; Sakai, H.; Murakami, T.; Abe, M.; Fujishima, A. Preparation and photocatalytic activity of robust Titania monoliths for water remediation. *ACS Appl. Mater. Interfaces* **2013**, *5*, 500–504.

3. Liu, B.; Nakata, K.; Sakai, M.; Saito, H.; Ochiai, T.; Murakami, T.; Takagi, K.; Fujishima, A. Hierarchical TiO_2 spherical nanostructures with tunable pore size, pore volume, and specific surface area: Facile preparation and high-photocatalytic performance. *Catal. Sci. Technol.* **2012**, *2*, 1933–1939.

4. Reddy, K.R.; Nakata, K.; Ochiai, T.; Murakami, T.; Tryk, D.A.; Fujishima, A. Facile fabrication and photocatalytic application of Ag nanoparticles-TiO_2 nanofiber composites. *J. Nanosci. Nanotechnol.* **2011**, *11*, 3692–3695.

5. Ochiai, T.; Tago, S.; Tawarayama, H.; Hosoya, T.; Ishiguro, H.; Fujishima, A. Fabrication of a porous TiO_2-coated silica glass tube and its application for a handy water purification unit. *Int. J. Photoenergy* **2014**.

6. Oh, S.-M.; Ishigaki, T. Preparation of pure rutile and anatase TiO_2 nanopowders using RF thermal plasma. *Thin Solid Films* **2004**, *457*, 186–191.

7. Balachandran, U.; Eror, N.G. Raman spectra of Titanium dioxide. *J. Solid State Chem.* **1982**, *42*, 276–282.

8. Liu, M.; Piao, L.; Zhao, L.; Ju, S.; Yan, Z.; He, T.; Zhou, C.; Wang, W. Anatase TiO_2 single crystals with exposed {001} and {110} facets: Facile synthesis and enhanced photocatalysis. *Chem. Commun.* **2010**, *46*, 1664–1666.

9. Zhu, J.; Wang, S.; Bian, Z.; Xie, S.; Cai, C.; Wang, J.; Yang, H.; Li, H. Solvothermally controllable synthesis of anatase TiO_2 nanocrystals with dominant {001} facets and enhanced photocatalytic activity. *CrystEngComm* **2010**, *12*, 2219–2224.

10. Ochiai, T.; Hoshi, T.; Slimen, H.; Nakata, K.; Murakami, T.; Tatejima, H.; Koide, Y.; Houas, A.; Horie, T.; Morito, Y.; *et al.* Fabrication of TiO_2 nanoparticles impregnated Titanium mesh filter and its application for environmental purification unit. *Catal. Sci. Technol.* **2011**, *1*, 1324–1327.

11. Petit, V.; le Rouge, A.; Béclin, F.; Hamzaoui, H.E.; Bigot, L. Experimental study of SiO_2 soot deposition using the outside vapor deposition method. *Aerosol Sci. Technol.* **2010**, *44*, 388–394.

12. Ochiai, T.; Fukuda, T.; Nakata, K.; Murakami, T.; Tryk, D.; Koide, Y.; Fujishima, A. Photocatalytic inactivation and removal of algae with TiO_2-coated materials. *J. Appl. Electrochem.* **2010**, *40*, 1737–1742.

MDPI AG

Klybeckstrasse 64

4057 Basel, Switzerland

Tel. +41 61 683 77 34

Fax +41 61 302 89 18

http://www.mdpi.com/

Catalysts Editorial Office

E-mail: catalysts@mdpi.com

http://www.mdpi.com/journal/catalysts